VALUE-BASED MANAGEMENT IN AN OPEN ECONOMY

Optimizing Strategy to Improve Business and Performance

VALUE-BASED MANAGEMENT IN AN OPEN ECONOMY

Optimizing Strategy to Improve Business and Performance

Prof. Post-Dr Walter Amedzro St-Hilaire

AAP | APPLE ACADEMIC PRESS

First edition published 2023

Apple Academic Press Inc.
1265 Goldenrod Circle, NE,
Palm Bay, FL 32905 USA

760 Laurentian Drive, Unit 19,
Burlington, ON L7N 0A4, CANADA

CRC Press
6000 Broken Sound Parkway NW,
Suite 300, Boca Raton, FL 33487-2742 USA

4 Park Square, Milton Park,
Abingdon, Oxon, OX14 4RN UK

Apple Academic Press exclusively co-publishes with CRC Press, an imprint of Taylor & Francis Group, LLC

Library and Archives Canada Cataloguing in Publication

Title: Value-based management in an open economy : optimizing strategy to improve business and performance / Prof. Post-Dr Walter Amedzro St-Hilaire.
Names: Amedzro St-Hilaire, Walter, author.
Description: First edition. | Includes bibliographical references and index.
Identifiers: Canadiana (print) 20220400237 | Canadiana (ebook) 20220400253 | ISBN 9781774912973 (hardcover) | ISBN 9781774912980 (softcover) | ISBN 9781003336198 (ebook)
Subjects: LCSH: Organizational effectiveness. | LCSH: Performance—Management. | LCSH: Strategic planning. | LCSH: Value. | LCSH: Industrial efficiency. | LCSH: Competition.
Classification: LCC HD58.9 .A44 2023 | DDC 658—dc23

Library of Congress Cataloging-in-Publication Data

CIP data on file with US Library of Congress

ISBN: 978-1-77491-297-3 (hbk)
ISBN: 978-1-77491-298-0 (pbk)
ISBN: 978-1-00333-619-8 (ebk)

Just because many are wrong, doesn't mean they are right.

—Prof. Amedzro St-Hilaire, Walter

About the Author

Prof. Post-Dr Walter Amedzro St-Hilaire
(Northwestern University, Evanston, Illinois, USA) (University of Ottawa, Ottawa, Canada) (PRISM-Pole SEE, Paris 1 Pantheon-Sorbonne University, Paris, France) (ExpertActions ExiGlobal Capital Group Co, Glasgow, UK) (Chair of Institutional Governance & Strategic Leadership Research, Gatineau, Canada)

Prof. Post-Dr Walter Amedzro St-Hilaire is the author of more than 20 books and around 40 scientific articles. His specialization areas include portfolio management (bank, telecom, health, energy, and agribusiness), project management, entrepreneurship policies, corporate and technology governance, business technology, strategic management, business economics, risk management, economic infrastructures, public administration, international development, and applied economics. He has taught at various universities: HEC-Montreal (Canada), University of Ottawa (Canada), Northwestern University (USA), and George Washington University (USA). He has served as Chief Resources Economist and Principal Officer of Procurement Operations for UEB United European Bank and as a Technical Advisor on Economic Strategies and Policy Development for the World Bank. He is also project economics and financial business expert for several institutions and international organizations.

Contents

Abbreviations

ABM	activity-based management
ATM	automated teller machine
BLOV	book value optimization leverage
BSC	balanced ScoreCard
CBAGDP	Central Bank Assets to GDP
CC	control of corruption
CEG	capital expenditure
CSR	corporate social responsibility
FE	fixed effects
GDP	gross domestic product
GE	government effectiveness
GICS	global industry classification standards
GMM	generalized method of moments
ID	informational differentiation
IDGDP	international debt issues to GDP
LCAG	Learned, Christensen, Andrews, and Guth
LQ	liquidity
LR	leverage ratio
MBVE	median book value
MCA	multiple correspondence analysis
MOL	market value optimization leverage
NIE	New Institutional Economics
NIT	neo-institutional theory
NPV	net present values
OLS	ordinary least squares
PBGDP	private bond market capitalization to GDP
PCA	principal component analysis
PF	profitability frontier
PS	political stability
RBV	resource-based view
RE	random effects
RL	rule of law
ROA	return on assets
ROI	return on investment

ROIC	return on invested capital
ROS	return on sales
RQ	regulatory quality
SBAs	strategic business areas
SBP	structure-behavior-performance
SG	services growth
SIND	structure independence
SMKGDP	stock market capitalization to GDP
SSIZE	structure size
TDR	total debt ratio
VA	voice and accountability
VRIN	value-creating, rare, inimitable, and non-substitutable
VRS	variable returns

Acknowledgment

The author thanks Northwestern University, ExpertActions ExiGlobal Capital Group, FordBridge University, the World Bank Group, and the Chair of Institutional Governance and Strategic Leadership Research for funding this research.

Preface

This book investigates the relationship between strategy optimization mechanisms and efficiency (performance and productivity) in the digital and open economy. The aim of this question on the interactions between strategies, performance, and market forms is to understand how, at the level of value-based management, productivity determines the optimized strategies and how the optimization process is modified according to the performance of the firm.

The book outlines the conditions under which this relationship is negative, neutral, or positive, in the banking sector. Applying economic and managerial principles, the book demonstrates that problems related to the competitive advantage, can be successfully analyzed like all other business problems, using traditional tools of economics, finance, and strategy (provided that the dynamics of industry-specific interactions are considered). The author argues that the ability of firms to improve their performance depends on the structure in which these organizations compete and their ability to optimize their value-based management.

This monograph is intended primarily for those who will be called upon to work as researchers or professionals in the fields of corporate governance and value-based management. The book identifies the technical practices and proposes mechanisms to encourage growth and improve performance. This way is part of renewing strategic approaches and value-based mechanisms in a digital and open economy.

Introduction

The question of the operations and optimization of imperfectly competitive markets and the behavior of firms operating (Church and Ware, 1999) has for a long time guided scientific research in value-based management and business economics. A field concerned solely with the supply side (Stigler, 1968), this field emerged following the questioning of the theories of perfect competition: at the end of the last century, the emergence of large modern groups was observed (Chandler, 1977), and the existing theories were unable to explain this fact sufficiently. The postulates of the theory of the atomicity of agents or the homogeneity of products were faced with differentiated products supplied by a small number of firms. This new situation gave them a contextual power that was not found in the theory of fair competition because they are price takers (Amedzro St-Hilaire, 2022).

In this respect, although the ultimate goal of this field of research is to determine when and why competitive inadequacy exists, the scientific orientation of the book is centered around three main areas of interest (Schmalensee, 1988): the determinants of firm behavior in the banking sector, international competition in an open economy and management practices. These three fields allow us to understand the mechanisms of creation, exercise, maintenance, and effects of market power in a comprehensive way. To achieve this, the traditional approach used is the structure-behavior-performance (SBP) paradigm. According to this paradigm, the structural characteristics of the market and the economy drive the behavior of firms and their strategic decision-making, which in turn influences their performance. However, the causality is not only one-sided. The way markets are organized (barriers, concentration, product differentiation, types of firms) affects the way firms behave and perform. But in turn, the behavior of firms influences the structure and organization of markets and their dynamism (Amedzro St-Hilaire, 2020). It is this retroactive link that the SCP model studies and that we mobilize to explain decision structuring in the banking sector.

How can we proceed with this analysis without making a synthesis of the main approaches as well as an elaborate quantitative and qualitative analysis of their contribution, limits, and empirical perspective? In an open market and digital context characterized by a flood of new products, services, and technologies that condemn companies to a race for competitiveness, organizations seem to demonstrate a permanent capacity to reinvent themselves in line with management challenges by devising innovative strategies that aim to seek out and obtain distinctive skills, a source of competitive advantage (guaranteeing their competitiveness and profitability over the long term). They, thus, define a global dynamic to be achieved and maintained: in order to improve this long-term dynamic, and to equip themselves with renewed strategic orientations, it becomes essential, in order to achieve set objectives, to re-evaluate the models on the complex and dynamic interactions between market design, strategic management and the behavior of the sectors which are ancient in industrial economics and strategic management (Mason, 1939, 1949; Bain, 1951, 1959).

In this respect, the search for evidence of the beneficial influence of optimization on business performance has been the subject of numerous studies since the establishment of the premises. Optimization is the strategic keystone for companies in their quest for openness, in the acquisition, allocation, and transformation of resources. Combined with rational analysis, optimization is a procedure established to produce consistent results, in which a process is broken down into successive steps. It is an elaborate, formal, and fragmented mechanism, a detailed plan that closely links the strategy formation process to a series of more operational steps.

Thus, strategy optimization would enable companies to improve their overall[1] performance, to cope with rapid environmental changes, to implement change, to solve problems related to human resources management, particularly resistance to change (Amedzro St-Hilaire, 2019). Theoretically, strategy optimization has been recognized by many management scientists working in the industry as a process for determining the main directions of a company, and providing companies with the means to evolve in their environment, while coping with change. But what does this mean in practice? Do existing processes actually result in rigorous optimization to efficiently achieve the set objectives? Does its adoption significantly change performance? Does it increase management efficiency? In other words, does the application of strategy optimization lead, in practice, to the success of companies and banks in an open economy context? This raises the question of the empirical evaluation of strategy and financial optimization: can the influence of the process on the performance of firms in the Canadian[2]

[1] Marketing, financial, and organizational.

[2] Some countries, such as Canada, are very open to foreign trade. Thus, in 2014, there was $EX = 462$ billion dollarsfor a Gross Domestic Product (GDP) of 1723 billion, i.e. $(462/1723) \times 100 = 26.81\%$. This proves that Canada is a country whose economy is very open to the outside world and in particular to the NAFTA countries and more widely to the industrialized countries. The same observation can be made for imports,

open economy be empirically proven or demonstrated beyond a reasonable doubt? If so, what explanations can be given for the divergence in assessment between theoretical and practical models?

In order to answer these questions of operational, economic, and managerial analysis of companies and strategies, the assessment of the interactions between the strategy optimization and the performance of companies is self-evident and therefore raises a double question: the first concerns the impact of the strategy optimization on the productivity of projects, while the second concerns the role of company performance in the productivity of optimization mechanisms. The aim of this question on the interactions between strategies, performance, and forms of operational management is to understand how, at the industrial level, the efficiency[3] of companies guides operational strategies and how they are modified as a function of the optimization process and the conduct of changes in an open economy.

The first approach attempts to assess the effects of strategy optimization on the qualitative and quantitative performance of companies. This question, at least with regard to the effects on the profitability of the company, has given rise to numerous debates, without having been fully resolved. It seems that part of the difficulty is that the answers differ depending on whether one is in the context of estimating the costs of the impact of the process on performance or whether one is in a framework assessing the impact of the characteristics of the process itself. Thus, at first sight, the conclusions drawn from studies on the usefulness of strategy optimization for the enterprise are rather mixed. However, it is less clear that at the level of the set of management practices with implications for performance, both micro and macro parameters of the firm may be affected by the optimization process. In particular, in a context of a zero-trade balance, it is impossible for the strategy optimization to have any effect other than a revitalizing effect on the firm's production and investment, although it may have a structural effect on the firm. On the other hand, when absolute export and import assumptions and interest rate parity are introduced, optimization simultaneously affects institutional and financial performance. Furthermore, it is in this context that the effects of decision-making structures on efficiency can be highlighted.

which in 2014 amounted to621 billion, or (521/1723) × 100 = 30.23% of GDP. We can see that this is even more important than exports. Canada is, therefore, a country open to the outside world. This openness is shown in 2014 by a deficit equal to EX- IM = 521–462 = –59 billion.

[3] Calvin (1994) emphasized upon the effectiveness and efficiency of the organizational unit. He was of the view that an efficient organization achieves its purpose with minimum waste of resources. He further said that effectiveness and efficiency are co-related and are both necessary for long-term success.

For example, it becomes possible to analyze the effects of the characteristics of strategy optimization on corporate investment.

In the second perspective, we are more interested in highlighting the role of internal and external factors of optimization in the determination and variation of strategic practices and decision tools. The question will mainly be asked from the point of view of the role of quantitative factors (profitability and costs) in the determination of optimal optimization mechanisms. But this question can be extended to the role of qualitative efficiency (staff input and stakeholder effectiveness) in the operational variables of the optimization and management process in its other strategic aspects in an open economy context. In particular, one can ask what are the implications of the heterogeneity of internal (productivity, profitability, industrial climate, staff involvement, etc.), and external (cost of materials, labor, capital, satisfaction of customers, shareholders, suppliers) optimization for the configuration of strategic best practices.

Under certain assumptions, it is then possible to show an interactive orientation. Under these conditions, the question of the role of performance (financial and non-financial) in the determination of optimal strategies seems even less well elucidated than that of highlighting the effects of strategy optimization on the multidimensional character of the profitability of firms. It seems, at least in part, that this difficulty can be solved if we take into account the close intertwining of these two questions. What seems paradoxical, from this point of view, is that theoretical approaches have never dissociated these two questions, whereas, on the contrary, most empirical studies consider only one of the two aspects.

This theoretical interlocking is immediately apparent. It is important to note that the true nature of the relationship has yet to be proven. The numerous empirical studies conducted tend to confirm one or other of the theoretical models. However, the bias introduced by the methodology limits the consideration of these different studies. Models based on an evaluation of optimization as an index of process efficiency seem particularly inappropriate for such an evaluation, given the consequences of applying strategy optimization. Although more recent models based on an assessment of the process characteristics themselves alleviate this deficiency somewhat, they are also limited by their approaches to obtaining a sample and responses that are sufficiently representative of reality. However, by incorporating the optimization system and contextual dimensions, these models offer a better representation of reality.

The problem is, therefore, to know to what extent decision-making choices, as well as the design, use, and development of new strategies, can modify or induce changes in the efficiency circuits of companies. It is therefore interesting to see how and in what form these elements are found in the management of financial[4] companies. Does this parameter change the models and the strategic process? Does it favor new action logics? Does it contribute to the development of knowledge and encourage the improvement of practices in the area of optimization management? What mechanisms are used to answer these questions? Do these mechanisms go beyond the usual tools? Do they make multi-criteria decision-making and knowledge development more complex? Overall, the aim here is to examine the current framework and its shortcomings, while evaluating, after integrating all the economic, operational, and financial considerations, the interest and the limits of the different options that can be envisaged for dealing with this problem.

In order to analyze this problem clearly, it becomes necessary to study an alternative model in which different hypotheses will be made explicit. Nevertheless, it seems quite intuitive that strategy optimization can, in such a case, force a certain performance choice. Moreover, every theoretical model offers its own answer to the two questions, without one being dissociated from the other. Thus, if we accept an integrated model based on a combination of organizational and market factors as an explanation of the reality of Canadian companies and banks, we are obliged, in interpreting the facts, to answer both questions simultaneously. If one can establish that the model is particularly well suited, one must be able to verify concomitantly at least the following three propositions: the effectiveness of decision-making practices, rational choice, and performance measurement. In principle, it would be sufficient for the facts to invalidate one of these three propositions to reject the theory. However, there are many theories, each of which produces a set of mutually consistent statements.

Our ambition is thus, firstly, to show, through the developments of the economic, managerial, and industrial theory, how the main models[5] have successively enriched the range of answers to the two questions, which are fundamental for the subject that concerns us. Then, in view of the need to go beyond empirical approaches that are too partial, an explanatory model integrating the economic (economic tools) and managerial (mana-

[4] Five economic agents exist: Households, financial enterprises, public administration, non-financial enterprises, and the outside world. For each economic agent, there is a type of financial economy: for households (personal and family finances), public administration (public and local finances), non-financial enterprises (industrial, agricultural, commercial, etc.), financial enterprises (banks and insurance companies), the rest of the world (external or international).

[5] Models based on a measure of optimization as an index of the effectiveness of the strategy optimization processand models based on a measure of process characteristics as an index of process effectiveness.

gerial practice) will be determined, which we will show makes it possible to interpret with coherence the reality of the Banking[6] and financial sector, with regard to the complex and dynamic interactions between the decision-making perspective and the nature of the optimization. Then, we will demonstrate that, with regard to this interaction, organizational, and conceptual factors frankly contribute to the effectiveness of the execution of the strategic plan for a better consideration of the company's performance, especially in an open economy where resources and flexibility are necessary even if they do not influence the effectiveness of the execution. This analytical work and the epistemological reflection that accompanies it was carried out through a field study, carried out in situ, during which we studied both the objective determinants of our problem (management instruments, economic constraints, production, and design processes, etc.), and the way in which the actors think about these situations, develop interpretations and act within the framework of these determinants and by making them evolve.

The aim of this book is to determine and analyze the interactions between strategy optimization and the performance of companies in the face of internal conditions and exogenous competitive shocks. The objectives are:

- Identify the factors favoring the role of performance in the strategy optimization;
- Explore the links between the dimensions of strategy optimization on business effectiveness;
- To identify the economic and managerial factors that influence the interconnection between strategy optimization and firm productivity in an open economy context;
- To understand how, at the organizational level, the efficiency of firms drives innovative strategies and how they change as a result of evaluating the optimization process and conducting change in an open economy.

---

[6] Banking, though an integral part of an open economy, is the most volatile business because the commodity it deals in changes value with variation in its circulation. Money value in the market, at times, is determined by its quantity, but the number of transactions carried out with money is of equal importance. Rarely, it happen that thequantity alone affects its purchasing power. The value of money is thus a function of both the supply and demandwhich together determines the trend of prices in the market. The supply of money results from the credit decisionof the banking industry as a whole which generally takes into account the market scenario depicting future economicactivities and the safety of banks' funds. No doubt, the scope of the market mostly depends upon the availability of finances, yet the market stability ultimately helps in credit expansion by promoting optimism and growth opportunities in the economy.

To this end, this book is structured around the following questions:

- Does the strategy optimization allow the objectives to be achieved efficiently?
- Does its adoption significantly change performance and increase management efficiency in an open economy context?
- What explanations can be given for the differences in appreciation between theoretical and practical models?
- How, at the industrial level, does the efficiency of firms drive operational strategies, and how do they change as a result of the optimization process in an open economy.
- What is the role of quantitative factors (profitability and costs) in determining optimization mechanisms?
- What role does qualitative effectiveness (staff input and stakeholder effectiveness) play in the operational variation of the optimization and management process in its other strategic aspects in an open economy context?
- What are the implications of the heterogeneity of internal (productivity, profitability, industrial climate, staff involvement, etc.), and external (cost of materials, labor? capital, satisfaction of customers, shareholders, suppliers) optimization for the configuration of strategic best practices?
- To what extent can decision-making choices, but also the design, use, and development of new strategies modify or induce changes in the efficiency circuits of companies?
- How and in what form do these elements find their way into the management of companies?

The following hypotheses were empirically tested:

- Concerning the relationship between optimization and competitive advantage: The variable "logistical disruption" was the subject of three hypotheses, all other things being equal:

  - **Hypothesis 1:** Logistical disruptions lower productivity.
  - **Hypothesis 2:** Logistical disruptions increase production costs.
  - **Hypothesis 3:** Logistical disruptions reduce profitability.

Regarding the dysfunction, three hypotheses are formulated, all other things being equal:

- ➢ **Hypothesis 4:** Dysfunction decreases productivity.
- ➢ **Hypothesis 5:** The malfunction increases production costs.
- ➢ **Hypothesis 6:** Dysfunction decreases profitability.

The assumptions for the strategy optimization are:

- ➢ **Hypothesis 7:** Optimizing strategies leads to a loss of competitiveness;
- ➢ **Hypothesis 8:** The effect of strategy optimization on competitive advantage depends on organizational and market factors;
- ➢ **Hypothesis 9:** The effect of strategy optimization on competitive advantage depends on the time factor (pre-planning strategic period, strategic plan year, and post-planning period).

- On the relationship between optimization and performance. Assumptions about internal and external institutional optimization systems:

- ➢ **Hypothesis 1:** A positive or negative relationship is expected between the optimization lever and productivity management.
- ➢ **Hypothesis 2:** A positive or negative relationship between income and dividend optimization is expected.
- ➢ **Hypothesis 3:** In institutional contexts with weak strategy optimization systems, greater productivity management would be expected to achieve certain objectives than in institutional contexts with relatively stronger governance systems.
- ➢ **Hypothesis 4:** Poor planning favors an alignment of interests and a low probability of productivity management.

Rigorous value-based management optimization, favors a risk of entrenchment and consequently the likelihood of opportunistic management of accounting results. This means that a U-shaped relationship between institutional optimization and productivity is expected.

- ➢ **Hypothesis 5:** The more rigorous the optimization, the less opportunism there will be through productivity management.
- ➢ **Hypothesis 6:** Optimized mechanisms control of less discretionary productivity should be observed following the adoption of accounting standards.

- **Hypothesis 7:** A negative association exists between legal enforcement, the effectiveness of the regulatory system, and the optimization of productivity.
- **Hypothesis 8:** Productivity optimization is lower if the banking system is not efficient.

- The assumptions regarding optimization mechanisms and management efficiency are:

  - **Hypothesis 9:** There is a significant positive relationship between firm effectiveness and the size of the optimized mechanisms structure in Banking and financial institutions.
  - **Hypothesis 10:** There is a significant positive relationship between corporate effectiveness and the independence of optimized mechanisms structures in Banking and financial institutions.
  - **Hypothesis 11:** There is a significant positive relationship between business effectiveness and the quality of management practices in Banking and financial institutions.
  - **Hypothesis 12:** There is a significant negative relationship between practice efficiency and decision duality in Banking and financial institutions.
  - **Hypothesis 13:** There is a positive relationship between business efficiency and optimized audit practices in Banking and financial institutions.
  - **Hypothesis 14:** There is a positive relationship between business efficiency and the frequency of audit optimization in Banking and financial institutions.
  - **Hypothesis 15:** There is a positive relationship between business efficiency and optimized compensation arrangements in Banking and financial institutions.
  - **Hypothesis 16:** There is a significant positive relationship between the effectiveness of optimized practices and equity in Banking and financial institutions.
  - **Hypothesis 17:** There is a significant positive relationship between the effectiveness of optimized practices and institutional input in Banking and financial institutions.

- **Hypothesis 18:** There is a significant negative relationship between the effectiveness of optimized practices and public financing in Banking and financial institutions.
- **Hypothesis 19:** There is a significant positive relationship between the effectiveness of optimized practices and foreign investment in Banking and financial institutions.
- **Hypothesis 20:** There is a positive relationship between management efficiency and optimized premiums in Banking and financial institutions.
- **Hypothesis 21:** There is a positive relationship between the effectiveness of optimized value-based management and the quality of external audits in banking and financial institutions.

- Regarding the role of performance in optimization: there were no explicitly stated assumptions.

That said, why this book? The first criterion of legitimacy comes from the fact that the interaction between the strategy optimization and the performance of companies does not seem to preoccupy economic management applied to banking risk[7] and constitutes an almost unexplored field.

---

[7] Banks as custodians of public money, on the one hand, play a significant role in determining the direction of the national economy and, on the other hand, are charged with the responsibility of safeguarding the interest of the stakeholders. As fund suppliers, they move the economy towards growth and stability. However, the organizational goal of the bank can be achieved only when money lent is recovered in time and without incurring additional expenses. Such an institution is, in fact, an ideal for the stakeholders, which ensures growth and promotes the interests of all the stakeholders. Therefore, the operational mechanism is so designed, and the procedure so carefully implemented those least chances of loss are left uncovered. So, every risk is properly analyzed, and preventive suggestions are made to avoid the occurrence of any event which may cause any loss to the institution. Since the banking business mainly revolves around money, therefore, its sensitivity and intricacies require extra vigilance to ensure the smooth flow of credit besides optimal utilization of other resources. For academic purpose, there are numerous risks,which needs attention but the risk involved in credit both pre and after disbursement has since attained paramount importance. The identification of risks helps in managing them through effective measures. But in some cases where the happening of events becomes beyond the jurisdiction of management, then alternative arrangements are made with the help of insurance companies. The most commonly known risks encountered by bankers, among others, are as follows:

1. Poor information risk. The pre-sanction/disbursement risks take their origin from the selection of borrowers. It is commonly observed that the infected portfolio of the bank comprises loan accounts owned by sponsors of high worthwith regular income flow. The default has never been due to the non-availability of funds with the borrowers, and rather it is always due to the weak will to pay back the bank's dues. Although will to pay is a psychological phenomenon, yet the past track record and market reputation of the prospective borrowers helps in determining the integrity and creditworthiness. In fact, an intensive and extensive inquiry into the borrowers before initiating a loan proposal saves the money from loss. State Bank, while introducing the KYC mechanism, has provided guidelines in obtaining information on the genuineness or otherwise of the borrowing applicant at the very initial

Regarding the banking sector in an open economy context, no theory really integrates their specific characteristics. The dearth of scientific knowledge on this subject may therefore express either a lack of interest or reflect neglect. The disinterest hypothesis can be associated with the idea that the interaction between strategy optimization and firm performance may simply be out of step with the current concerns of this scientific community. The hypothesis of neglect seems to us, in this respect, more likely and constitutes, in any case, an explanation more favorable to the legitimacy of our research. This interaction does not seem to be of interest to current empirical research, but could introduce serious variations around the mechanisms of strategic decision-making in an open economy and interest rate parity. An update of scientific knowledge is therefore needed on the relevance of the choice of correlation in the study of causality between these two notions because the theoretical models that have been put forward to explain the interaction between optimization and firm performance have not really integrated the specificity of firms in an open economy context.

This book, therefore, contributes enormously to the understanding of the interactions between the strategy optimization, the efficiency of companies, and the structure of markets, highlighting the role of performance in determining management practices, while questioning the scope of current theoretical models in the analysis of empirical evaluation cases, thus making it possible to take stock of this debate, while highlighting the current state of affairs and offering new avenues of reflection in industrial economics, competition management, the sociology of organizations, game theory, and operational banking management. Also, a request of another nature comes, in a more explicit way, to support the

---

stage. Although, in some cases, social pressure forced the defaulters to arrange repayment of the classified advances yet the common proverb "Nip the Evil in the Bud" will save the bank from a cumbersome recovery process or expensive litigation. It is, therefore, imperative that in the matter of lending decisions, the inherent risk in the loan proposals may be identified, and preventive measures are taken prior to taking a final decision of loan disbursement.

2. The income-generating capacity of the business entity for which the loan is being asked for needs to be considered because the incapability of the business renders the borrower incapable to liquidate the liability within a specified time period, which phenomena will ultimately put the credit at risk. It is because of these reasons that poor income-generatingcapacity leaves nothing to pay back the original loan amount what to talk about the interest and mark up and eventually adds to the liabilities of the defaulting borrower. In case of newly established business concerns, the financing manager has to take into consideration the market and the business trend of the products being produced because in case of non- disposal the stock will continue accumulating and will ultimately put the bank money at risk. The elasticity of demand for such products sends a signal of warning to the lending agency due to the availability of substitutes, or an increase in supply in the market may place the borrowing entity as a weaker competitor. Also, an infant business entity takes time toestablish its integrity and worth in the market.

legitimacy of our research: that of the stakeholders of the banking sector who were involved in this research with the aim of taking a step back and validating the scope of the management practices that emerged from their decision-making source. Our contribution, within the framework of the research mechanism set up with them, aims, among other things, to help them better appreciate the process-efficiency interactions. Stakeholders' demands are thus based on more or less practical concerns to which existing contributions do not yet seem to have provided a sufficient response. The objective, in order to remedy this, is to provide them with a global analysis framework enabling them to integrate practices stemming from heterogeneous logics.

The three axes mobilized situate the relevance of the book in relation to epistemological and methodological issues. Concerning the epistemological and theoretical issues, the concepts aimed at explaining the interaction between optimized strategic tools and the performance of firms are certainly attractive, but the phenomena to which they refer are difficult to access. The classical currents of business theory are full of useful insights for our purposes, but they are increasingly ill-suited to the realities of this relationship. They were developed at a time when the firm was viewed in a very reductionist way, with a concern for parsimony typical of the economic approach to the market. Today, companies are recognized as being much more complex and fuller of diverse actors, sometimes not very coherent with each other and displaying multiple logics that are more or less questionable, evolving towards a concept of global performance.

Theories developed in the simplified world of the past often do not apply well to today's concerns, hence the evolution of the concept of performance. This is the only real trend law in economics and management. When problems are defined differently, other theorizations are needed to explain them, especially in a particular context. In the name of realism, it is important that we provide ourselves with an analytical framework that integrates new contemporary considerations and visions. To achieve this, classical approaches provide the necessary foundations, which need to be complimented and enriched by empirical data. There are also methodological issues underlying this research. The methodologies used to obtain the fragments of discourse sought are very feasible in an open economy context. They lend themselves perfectly to the type of fieldwork we have adopted. They do not require overly complex means or devices and are, therefore, relatively simple to negotiate with the actors. In short, all these reasons reinforce our desire to question the interaction between strategy optimization and the performance of companies in a context of economic openness.

It is important also to emphasize at this point that our case (the banking sector) is somewhat atypical of financial institutions, which are characterized by great diversity and a strong risk[8] perspective. For a while, we thought of integrating this diversity into our research. The challenge would have been to highlight the characteristics of the optimization of Canadian companies through institutions chosen on the basis of their differences. However, this ambition did not seem to us to be in phase with the spirit of a methodological mix. Taking on a sample that was too varied would have implied multiple variables determining significantly different universes. This is certainly an advantage, but the comparative analysis of such disparate elements in our case would probably not have yielded very relevant knowledge insofar as the scope of the differences separating them would most certainly have annihilated the explanations given to our problem. In addition to this constraint inherent to the approach followed, the banking and financial sector[9] in open

---

[8] Risk is a concept widely used in matters associated with assets and liability management. The term denotes a phenomenon wherein danger of loss in the value of assets is apprehended, or there is an apprehension that a substantial addition to the liability may arise due to the happening of a presumed event. It is a potential danger of loss associated with the happening of a particular event. It is a contingent loss which may or may not arise. According to Khan, the risk is a probable loss in income or assets. He further elaborated his views and said that it is unexpected loss which comes in the definition of risk, and the expected loss is not included. Nevertheless, it needs to be guarded against. The term is normally used for the losses which arise during the course of business due to imprudent management. To mitigate the repercussion or eliminate the chances of occurrences, shrewd managers use various instruments and techniques to minimize the quantum of loss, although danger cannot be eliminated altogether. The practices and use of such techniques which minimize the chances of loss in the value of an asset or additional impact of liability are, in fact, the management of risk. In banking, both the assets and liabilities aspects are equally important to be taken care of. The banking business largely depends upon the volume of deposits secured, and any such event which may result in the loss of deposits, may it be with drawl by the customers due to declining trust or quality of service, may attract the initial attention of the managers. Under such circumstances, the managers are always anxious not only to retain the existing portfolio but also to make all-out efforts to increase the deposits. This is because the ever-increasing deposits give strength to the management to increase and enlarge its investing and leading post folio and thereby generate maximum revenue to add to its profitability. Profit is no doubt the major goal of bank, yet it can be increased only by disbursing quality finances and keenly caring of its investment portfolio. The curiosity of the manager in selecting customers with integrity and worth is an artful job and demands extra skills to predict and identify the risks associated with a particular post folio and prevent the occurrence of the contingent event, which may possibly cause a loss to the institution.

[9] The Canadian financial services sector is among the strongest in the world. The sector includes banks, loan, and trust companies, insurance companies, credit unions, securities dealers, loan, and leasing companies, pension fund managers, mutual fund companies, and independent insurance agents and brokers. The World Economic Forum has named Canada's banking system the soundest in the world seven years in a row. In 2012, the Toronto Stock Exchange and TSX Venture Exchange ranked first among international exchanges in terms of new listings for the fourth consecutive year, according to data from the World Federation of Exchanges. Canada's financial services sector is underpinned by strong regulatory frameworks, and the entire financial sector is subject to an annual review by the Bank of Canada. The

economy, unlike its counterparts, is characterized by multiple specificities such as tools, practices, stakeholders, and the logic of action. These variables determine, in particular, at the level of decisional optimization, a very singular reality.

The book is therefore divided into two main complementary parts: one theoretical and one more practical. It begins with an overview of the literature that may help us in our discussion. Thus, the first part constitutes an explanation of the main models describing the nature of the relationship between strategy optimization and the performance of companies. Each of these models provides an interpretation and characterization of efficiency that makes it possible to anticipate and define the practices of a given company in a given context in terms of efficiency. After a focus on a key concept essential to the understanding of industrial strategies (efficiency), we return to the foundations of value-based management and their application to empirical studies essential to the understanding of the behavior of companies and markets. These parameters are discussed, and the main lines of their evolution are traced through the different currents of thought that have contributed to a better conceptual representation. This section outlines the evolution of business theory on the interaction between strategy optimization and industrial efficiency. The relationships (as well as the factors underlying these relationships), as perceived by the different models, will be reviewed.

These chapters will provide us with the essential basis for understanding the interactions related to our problem. In particular, the aspects emerging in theoretical discourses and more or less important nowadays in consideration of the conceptual framework of the notions related to strategy optimization and industrial performance, in particular, the study of the impact of strategy optimization on project productivity with a review of the models dealing with the relationship between strategy optimization and competitive advantage, and then we will present the essential models dealing with the role of performance in the development of management strategies. It also reviews the empirical perspectives available to address and conduct research on the issue of the interaction between management practices and industrial efficiency. This section, while tracing the contributions of these different empirical perspectives, demonstrates that they are too incomplete (too

---

operating costs associated with professional services in Canada are very favorable. Indeed, Canadian-based international financial services firms enjoy operating costs that are typically 17.8% lower than similar firms. Canadian-based international financial services firms enjoy operatingcosts that are typically 17.8% lower than those of similar firms in the US.

partial) for the purposes of our research. Thus, it seeks to go beyond these partial empirical approaches to develop an adequate model. Among other things, it details the main paradigmatic and methodological foundations that guided us throughout our research, as well as the main concerns that led us to adopt these foundations. It describes with relevance and rigor the course of the research, and the approach applied, the problems encountered, as well as the ethical considerations that we took into account.

The second part summarizes the main findings at the end of the research and skillfully analyzes the results in the light of theoretical knowledge. This part exceptionally focuses on the essential points and on the avenues of reflection to be taken into account in order to study the complexity of the problem between the strategy optimization and the performance of companies in the face of competition from foreign[10] banks in an economy. In particular, it proposes a redefinition of the concepts, at many levels, of the elements underlying this plural and multi-conceptual interaction.

[10] Another component of the banking sector is the foreign banking sector, which includes subsidiaries and representative offices of many of the world's largest banks. Foreign banks own almost 10% of the assets of the Canadian banking sector. While there is one foreign bank subsidiary with a large retail branch network (Hong Kong Bank of Canada), the traditional foreign bank is small and focuses on wholesale banking and high-end commercial lending. Since they have few shareholders, all foreign banks are included in Schedule II of the Bank Act. Foreign banks were first allowed toestablish branches in Canada in 1980. Since then, several important restrictions have been removed [6] as part of trade liberalization, first with the United States and Mexico and, more recently, with World Trade Organization signatories. The establishment of subsidiaries is seen as a necessity, and the entry of foreign banks into Canada is seen as a routine phenomenon for the establishment of strong, well-managed, and recognized institutions. However, the number of foreign banks operating inCanada has declined in recent years. This is primarily due to difficulties in competing with the Big Six banks, which have well-established business relationships and networks, but it also reflects a general international downturn in banking in the late1980s and early 1990s.

CHAPTER 1

The Concepts of Optimization and Efficiency in Digitalized Economy

The relationship between strategy optimization and firms performance is essential to the understanding of corporate strategy, to the analysis of their performance and to the characterization of their management system. The very nature of the problems analyzed in the context of the optimization of firms in an open economy requires an approach that is both macro and yet includes a sufficient level of disaggregation. Macro, industrial organization, and value-based management are by construction: how can we analyze the different aspects related to the structural mechanism of the analytical reasoning of optimization without considering each company or institution as a global entity?

Micro-organizational theory and optimization management are so by necessity: how to highlight the optimized sources of the efficiency of firms and institutions in association with descriptive reasoning without distinguishing at least endogenous and exogenous factors? How to dissociate what is structured (operational) from what is analytical without distinguishing at least the processual factors of resource transformation? The formalization of these models, largely developed, has immediately made it possible to place all the problems in a context compatible with the concepts of process and rational strategy, long absent from the macro-traditional models of industrial strategy. Under these conditions, the study of the interactions between strategy optimization and the performance of firms is greatly facilitated in an open economy context. Whether it is the formal model, the adaptive analysis proportion models, or the conceptual mechanism model (measurement of process characteristics), the implications of optimization on the efficiency of firms and institutions or, on the contrary, the role of the profitability factor

Value-Based Management in an Open Economy: Optimizing Strategy to Improve Business and Performance. Prof. Post-Dr Walter Amedzro St-Hilaire (Author)

in the determination of the best management practices can immediately be highlighted. However, until recently, only process situations could be developed in the framework of these models. It is only after the work based on performance measurement that financial and marketing profitability situations could be modeled. The globalization and the digitalization of the economy, illustrated by the rise of transnational companies, marks a new stage in the history and evolution of capitalism. Services, goods, and the corresponding factors of production (capital, labor, knowledge, etc.), have become more mobile, making it possible to envisage exchanges on a planetary scale and thus to envisage unhoped-for business and growth opportunities. Also, globalization and internationalization have become strategic management tools likely to increase the overall performance of banking institutions, but their real impact on financial performance is still somewhat obscure. In this very dynamic context, what can be retained from the conceptualization of the notions of strategy optimization and efficiency that are essential to identify the nature of their interaction?

The optimization is an effective tool in terms of its contribution to project productivity, or the final outcomes that the plan is initially intended to achieve. These outcomes are usually initially set by the strategic system as a range of social objectives. In their research, many studies have focused on the difference between formal (or explicit) optimization and implicit planning. Formal optimization is an explicit and continuous industrial process with several components (including goal setting, strategy generation and evaluation). An effective optimization system must be able to link long-term strategic objectives with medium-term objectives and operational optimization. Therefore, controllers collect data, make forecasts, model, and construct alternative future scenarios. It is these activities that will enable companies to outperform those that are not engaged in a strategy optimization process. However, this view is not universally shared.

Optimization can only objectively be carried out in the short term, due to budgetary constraints, the inability to predict the future, and the lack of objectivity in planning, which is usually biased by the vision and desire of the designers and the hierarchy. With environmental constraints, the needs of the company are constantly changing. They cannot be defined after an assessment of the company's strengths and weaknesses, but rather gradually. Optimization, based on the needs of the company, must therefore respond to these needs, and must take into account their gradual evolution. The real role of performance is to develop and articulate the consequences of a pre-existing strategy: optimization does not create strategy. He further argues that optimization is an adaptive process, which evolves gradually and emerges from the different directions the company takes in response to its context. It is this emergent process that will lead the company to success. There is no need for explicit optimization, which would hinder the company's growth. The manager should focus more on the evolution and optimization of the company's capabilities and efficiency, rather than on optimization, which is likely to be limiting. On the other hand, it can be believed that room should be left for emergent strategies because they are part of the explicit optimization, just as unforeseen expenses are included in a pre-set budget.

Optimization is a concept that is increasingly used in the corporate world as part of decision making. Although most models of thought recognize this concept as a means of reorienting the evolution of a company, not everyone agrees on the importance to be attached to it, on the means of conceiving and then implementing it, or on the usefulness of optimization in a decision-making process. The first argument is based on a critique of the design model, whose principles governing the implementation of a strategy optimization. Thus, in the conception of this model, companies must make a diagnosis of their situation by taking into account their strengths and weaknesses on the one hand, and on the other hand by considering the contextual expectations and constraints. Once this diagnosis has been made, several strategies likely to respond to the situation are created and then evaluated. These strategies should be simple (accessible, i.e., understandable to all human resources) and explicit (a necessary condition for a positive or negative evaluation). If the strategy is not contested after the evaluation, it will be implemented. The prescriptive model, according to which strategy optimization is a conceptual process of optimization and analysis, and the descriptive model, which deals with and describes a particular aspect of the optimizing process. All governing schools are the same. Although it recognizes a certain evolution of the different optimizing models, they still do not differ significantly. They limit themselves in their possibilities of adaptation. He argues against the explicit formulation of strategies.

The first major divergence between the two models of thought lies in the formulation, explicit or otherwise, of a strategy optimization. The first model advocates the explicit formulation of strategies optimization, and makes it a necessary element for the success of companies. While, admits that the principles of the designing are no longer entirely appropriate to the present, it maintains that context is the dominant force dictating the aggressiveness of the strategy optimization to be adopted to lead the company to success, since each company has its own context characterized by particular uncertainty and unpredictability (different environmental challenges require different strategic solutions and, as a consequence, different planning approaches. Explicit formulation must take place in a context of uncertainty, because this uncertainty puts the future of the company at the mercy of the hazards of the future. This explicit formulation makes it possible to take these hazards into account, and would prevent the firm from deviating from its objectives in an unfortunate manner. It is impossible to formulate an explicit and plausible optimizing strategy in an unpredictable context, without an experimental process being set in motion. Explicit formulation is only feasible in a stable context, under the control of the company, but that in a predictable environment, explicit formulation is not necessary. But, in the open and digital economy it is impossible to measure the unpredictability of an environment in order to adapt an explicit strategy to it. Moreover, a company cannot be sure of its strengths and weaknesses until it has tested them experimentally. A rational approach cannot lead to business success. One cannot identify in advance whether opportunities not anticipated by the explicit strategy are beneficial opportunities or harmful deviations. An explicit formulation makes strategies rigid, constitutes a brake, a limitation for the visions of the decision-makers and closes the door to unanticipated opportunities.

Those strategies should not be explicit but rather emerge as they are developed, depending on the conditions, experiences, and mistakes experienced by the company. They should be the result of a learning process and non-predetermined actions. It is important to recognize the experience in optimization ('successful practitioners of strategy typically use a strategic value-based mechanism which revises the strategy in the light of experience), and the importance of having both emergent and, in contrast, deliberate optimization strategies. The purpose of deliberate strategies (which guarantee the willingness to act rationally) is to enable firms to raise the emergent strategies to be adopted by their firm, and the emergent strategies adopted by competitors. But it is the emergent strategies that will lead the firm to success. In the banking sector, there should be room for emergent strategies because they are part of explicit optimization, just as unexpected expenses are included in a pre-determined budget. For the latter, the manager should focus more on the evolution and optimization of the company's capacities and efficiency, rather than on optimization, which is likely to prove limiting. Strategy optimization is a creative process. It therefore requires formal techniques and analysis, which companies do not have in an open economy. Indeed, optimization is more a matter of rigorous coordination of future events than of a conceptual system intended to guide the growth of a company in a coherent manner.

Optimization cannot be achieved in the short term because of budgetary constraints, the inability to predict the future and because mechanisms are rarely objective, but are more biased by vision, desire by their designers, and hierarchy. In the digital era, a strategy that is constantly adapting puts the company at a disadvantage. The adaptation time, the waiting time during the realization of the experiments is a hindrance for the company in a market where competitors have already optimized their advance. The reason is that when they arrive on a market with a new product/service, such firms find the market pre-empted by more foresightful competitors, who had optimized their strategic moves is advance. Given the rapid contextual changes, learning from mistakes would be enormously costly. A company that develops an explicit strategy for contextual uncertainty and unpredictability is more aligned with it and more likely to succeed than a company that adapts as it goes along. Optimization facilitates adaptation in the face of rapid change and reduces costs. Finally, it is important to denounce the danger of using strategy optimization. This optimization process must be carried out by those directly in touch with the context, not by strategists who are ignorant of the realities on the ground. There should be no separation between strategy development and implementation. The real role of performance, is to develop and articulate the consequences of a pre-existing strategy: optimization does not create strategy. Finally, explicitness is very likely to create misunderstandings.

Understanding the complexity of corporate and institutional decision-making and value-based management requires an understanding of how key concepts developed over time and their evolution account for industrial efficiency: one of the most important concepts in institutional economics and corporate management is that of economic efficiency. Economic efficiency is a hypothesis of strategies providing a set of canons for judging actions. It is often assessed in terms of the goal pursued, the means used and the results of the actions. Over time, economic theories of business have proposed different types of efficiency: limited, contextual, procedural, substantive, etc. For our purposes, classical economic efficiency will be our starting point. It will be criticized by the so-called postmodern, institutionalist, and contemporary critical approaches. The aim is to show that there are different ways of conceiving efficiency, and that theories based on maximizing efficiency alone are at best incomplete for our purposes. The postulate of plural efficiency is not unanimous, however, and we shall see throughout the effects of this controversy on the theoretical positions available. The efficiency of economic actors is at the heart of industrial theory and has supported its expansion. It is used to explain and account for the strategies of economic agents. In classical theory, the economic agent is an efficient being, seeking to maximize his profit, his well-being, his objectives, or the utility he derives from a product, a service, an activity, etc. Efficiency comprises two models: the strategic model where economic actors seek to maximize a set of values assumed as a vector, and the neoclassical model where the value to be maximized is unique. The efficient decision-maker is a maximizer: he makes choices that maximize his satisfaction. This principle of efficiency implies that governance is driven by a purely selfish desire to make the most of the situations they encounter. In other words, economic agents are equipped with functions/objectives that allow them to measure the profit they derive from a situation, taking into account the constraints imposed by that situation and the resources at their disposal. In doing so, they are able to determine the profit they would make in a given situation, and adopt the strategies that will allow them to maximize their profit, and achieve optimal utility. This assumes that all actions are generated by conscious optimization and deliberation. The unconscious has no place in the generation of actions. The economic agent is considered to be capable of performing all the calculations necessary to make optimal choices. He is capable of apprehending, on a case-by-case basis, all the variables that can inform him about the situation, those that can influence the situation and the implications of each choice.

It then analyzes all these variables and derives the optimal choice. Under this assumption, knowledge of the context is sufficient to know what an agent's choice will be. Taking into account only their own interest, economic agents are autonomous decision-making units, whose decisions are defined independently of any macro habits and constraints. The open economy consists of individuals acting efficiently. The idea of efficiency already appears in Adam Smith's reference to the strategy of the efficient actor to justify free trade. According to him, the maxim of every actor is never to make at home, which would cost him less to buy than to make. The efficiency of actors is related to the efficiency of the system. The interaction between efficient agents and the mechanisms of pure and perfect competition leads to the optimal use of resources, to the maximization of the utility functions of each given the situation, and to equilibrium. Similarly, his analysis of the invisible hand and the division of labor is based on the maximizing strategy of the actors.

Although the concept of efficiency has been developed with regard to individual actor strategies in some cases, the dramatic increase in open economy operations has refocused interest at the level of firms and institutions. Actors increasingly rely on firms and institutions to provide the services they need in all aspects of their lives. Furthermore, proponents of value-based management argue that companies and institutions are stepping stones for the development and freedom of actors. Companies and institutions are the main actors in socio-economic activities. They are gradually replacing the community and even the family. Most of life takes place in enterprises and the strategic decision-making issues raised by their accelerated growth are at the basis of institutional economic theories and models. Applying the principle of efficiency at the level of firms and institutions, classical theories place well-being at the forefront: all actors' and firms' strategies should be directed to promote the health of the firm and institutions. Actors specialize according to the needs of the firm and institutions. The rise of capitalism, which advocates the virtues and value of work (open economy), and considers that the search for profit is dictated by a strategy of efficiency and systematic among actors, favors the push of modernism. Science acquires immense power. It is efficient and a source of knowledge and truth. Modernism is characterized by a firm belief that any institutional problem can be formally solved by state intervention and the application of socio-scientific principles, with the choice of economic actors arising from logical consequences that can be effectively anticipated and explained.

Other characteristics of this modern movement are the specialization and delimitation of work tasks, a clear hierarchy and the centralization of management, administration, and decision-making at the top of the hierarchy. Modernism is thus based on efficiency and the use of technical knowledge by a bureaucratic administration. At the same time, by analogy with Darwin's theory of evolution, the classical movement considered that competitiveness between actors and between companies (institutions) determined which actors were best able to survive in the different contextual situations faced by all companies and institutions, and which actors were best able to form the ruling elite. Faced with a constraining context, only the best adapted survive. Since the hierarchical elite is specifically 'selected' according to social Darwinism, absolute trust is placed in them. Its decisions are primarily concerned with welfare and it sets the management code. Values such as obedience, submission, loyalty, and fidelity to the hierarchical elite become more important for success and professional achievement than personal competence, creativity, expression, and skills. The management elite has the responsibility to reconcile the interests of each stakeholder and to integrate them in a more or less contractual way in the decision-making process. The modern manager as leader strives to craft and communicate a coherent vision. He or she is viewed as psychologically complete by him or herself – autonomous – and as bestowing gifts, wisdom, and leadership on the followers (Calton and Kurland, 1996). It manages the relationships and interactions between stakeholders in order to serve a unitary and transcendent socio-economic goal: the maximization of profit and the well-being of the company and its institutions.

Simon's conception of efficiency constitutes a major break with many schools of thought in economics and management. Simon's work on efficiency consists of a reflection on efficiency based on the observation of facts. This leads Simon to turn away from the classical, substantive efficiency, which assumes that the company or institution has all the information necessary to make the optimal decision in all circumstances. Simon begins by noting the existence of several forms of efficiency: A decision can be said to be 'objectively'… if it actually presents the correct behavior that will maximize given values in a given situation. It is 'subjectively'… if it maximizes the chances of achieving a desired end according to one's actual knowledge of the subject. It is 'consciously'… insofar as the adaptation of means to ends is a conscious process. It is "intentionally"… insofar as the enterprise will have deliberately made this adaptation. It is… "from the company's point of view" if it serves the company's objectives. Finally, it is 'personally'… if it obeys the designs of the individual' (Simon, 1947).

In an efficient choice theory, in order to make an optimal choice, the agent must know the set of possible choices, predict the set of possible outcomes, and associate a payoff with each or at least rank each outcome relative to the others. Simon argues that it is impossible to have perfect efficiency. Simon argues that actors and institutions are unable to process all the information they face. Every organism that produces millions of bits of new information every second, but the bottleneck of the perceptual apparatus certainly does not admit more than 1,000 bits per second and probably less (Simon, 1959). Moreover, it is not always possible to know the possible gains in advance, as information is imperfect. Indeed, earnings are not always comparable due to the lack of a standard to measure them. Moreover, the market, even if it reduces the knowledge needs of the actors through its price mechanism, does not make it possible to obtain all the information necessary for the decision and the agents can be confronted with contradictory choices. Markets are not optimal mechanisms, and in particular, they do not allow externalities to be dealt with, which require other regulatory mechanisms: Markets can only be used in conjunction with other methods of social control and decision-making (Simon, 1983). This will result in gaps between action and goal achievement.

For example, the information needed very often goes beyond simple price information. The quality of the product or service, the set of substitutable products or services available, the conditions of production, the good faith of retailers, etc., are all information capable of influencing consumer choice. First, actors may have (and most of the time will have) incomplete or erroneous information about the situation and potential changes in the situation over time. Second, even if information is complete, an actor may be unable (and usually will be unable) to calculate all the consequences of the action. Third, actors do not usually have only one goal, and there may be incompatibilities between goals, with the achievement of one of them interfering with the achievement of the others. Fourth, an actor may fail to achieve a goal because of ignorance of the means to act (Simon, 1991b).

According to classical theory, the efficiency of actors helps to make decisions in situations of imperfect information. It presupposes an immense computational capacity that ultimately reduces market uncertainty. Agents are able to calculate the costs of searching for information. In Simon's view, it is necessary to recognize the limitation of the computational capacity of individuals and the inability of agents to create perfectly efficient and foolproof rules: It will be recalled that, for neoclassical economics, one way of dealing with uncertainty is to say: Oh yes, there is uncertainty and people will seek to reduce it to the point where the marginal value they can expect from an additional search for information will be equal to the marginal cost of that information search. Far from simplifying the computational problem, far from taking into account bounded rationality, neoclassicals have simply exploded the computational problem. Decision-makers will now not only need to know the probability distributions of alternatives and their consequences, but will also need to know the marginal productivity of information seeking (Simon, 1984a). Furthermore, actors often make errors in judgment and do not always achieve their goals: Of course, as Freud (and many laboratory experiments) told us, people can deceive themselves. Real goals can be different from what they are supposed to be. To say that there are goals to actions means that there is a connection between actions and goals (values, utility function). Actions increase the possibility that some of these goals will be achieved. However, even in what we may call rational behavior, there may be real gaps between action and goal achievement (Simon, 1991b).

Faced with a multitude of stimuli, the perception of gains can vary according to the actors. An agent's perception of his or her context will help determine the nature of his or her choice and the way in which it is made: If we accept that both the knowledge and computational power of the decision-maker are severely limited, then we must distinguish between the real world and the actors' perception of it and reason about this. Our theory must include not only the reasoning process but also the process that generates the actor's subjective representation of the decision problem (Simon, 1997). An actor's perception of the world (and the choices and decision process he adopts) is highly dependent on the context in which that individual evolves. The organizational environment in which an individual makes a decision determines the consequences she will expect, those she will not expect; the choices she will consider and those she will ignore (March and Simon, 1958). To this context, he adds the emotional context of the agent. Simon integrates the world of emotions into his analysis of efficiency and reminds us that agents are not only efficient machines. He shows that emotion can also guide an agent's choice. Indeed, emotion can draw an agent's attention to one contextual aspect rather than another, and thus alter the agent's perception of it.

A stimulus, whether internal or external, draws attention to certain aspects of the situation at the expense of others that might guide the choice in a different direction (Simon, 1947). Efficiency has its limits. And it is with this limited efficiency that actors will make their decisions, depending on the goals they are aiming for and the subjective analysis of the context related to this decision. A decision is only made in consideration of a part of the available contextual data. Effectiveness belongs to one's experience: A real-life decision consists of some goals or values, some facts about the environment, and some inferences from the values and facts. The goals and values may be simple or complex, consistent, or contradictory; the facts may be real or assumed, based on observations or reports by others; the inferences may be valid or false (Simon, 1959). The recognition of the limits of efficiency requires a reconsideration of the agents' strategy when faced with a choice. Rather than looking for the optimal choice, agents initially look for a subset of solutions from the set of possible solutions that seems satisfactory to them. They select a few solutions without analyzing the whole range of possibilities and make their choice based on a level of aspiration. The latter is a function of the difficulty of finding alternative solutions: if it is easy to find satisfactory alternative solutions, the aspiration level rises, if it is difficult to find satisfactory alternative solutions, the aspiration level falls. Human nature implies a diversity of strategies in a given business context. It is impossible to characterize the strategies of the actors precisely. Yet, in Simon's view, it is possible to make a scientific description of the way agents make decisions.

First of all, an important aspect remains: despite the stated limitations, human beings are still efficient since they can explain in general the choices they make: in a broad definition of efficiency, practically all behavior of agents is efficient. Agents have reasons for doing what they do, and, if questioned, they can give their opinion on what these reasons are (Simon, 1991b). Simon then puts forward a new concept, that of procedural efficiency. This concept of procedural efficiency implies the existence of procedures, norms, rules, and a system of values aimed at facilitating the choice of economic agents in a context of uncertainty and imperfect information and at compensating for the limits of efficiency. These procedures are considered efficient and scientifically analyzable. We are concerned with how humans can be efficient in a world where they are mostly unable to predict the relevant future exactly: they can only adopt an efficient choice procedure, including a procedure for forecasting, or at least adapting, to the future (Simon, 1976).

A new emphasis is placed on cognitive theory. The latter highlights the fact that information alone is not enough to understand a phenomenon. There is a set of perceptions and thoughts that must surround the information to give it meaning. The cognitive capacities of the agents subsequently enable them to activate a choice procedure according to the goals and values they have set themselves, the stimuli they have perceived and retained from the environment and the analysis they have made of them. For a given stimulus, an automatic action may correspond, but for more complex or unusual decisions, a deliberative procedure is set up. Nevertheless, the effectiveness of actors cannot be reduced to established procedures. The latter, being put in place by rationally limited agents, may contain flaws. Experience allows the improvement and strengthening of existing procedures. They are capable of innovation. Simon's work stands out considerably from existing currents and marks the beginnings of a new form of thinking on economic efficiency. Such an approach to efficiency requires taking into account the limits of agents' capacities to understand economic phenomena. It brings greater realism to the analysis of economic actors and calls for a re-evaluation of the economic theory of institutions and firms on the one hand, and of the economic theory of market functioning on the other.

Postmodernism emerges as a result of, among other things, the inability of managers to prevent corporate and institutional failures, the crisis of capitalism, the problems caused by resource scarcity, the change in mentality that accepts less and less the legitimacy of the modern goal of profit maximization, corporate efficiency, and the weaknesses of optimization theory in explaining the choices of actors and the increasingly recurrent problems.

Postmodernism characterizes a way of life or a style of thinking and expression that reflects the movement or context that follows classical modernism and introduces a break with the latter. It challenges the very foundations of modern knowledge, in science, theology, history, and management. The term postmodern is applied to emerging forms of enterprise that are flexible, adaptable, interactive, networked, and move away from hierarchical forms of mass production. Postmodernism embraces various forms of commodification and looks at business and management in a new way. It changes the way of seeing the world, fragmenting it, leaving it open to interpretation and giving way to the sensitivity, creativity, and symbolism of the actors. Innovation is at the center of this culture. On this basis, the theories and foundations of classical theory are revised and approached in a new light for the purpose of reconstruction. This is particularly the case for management practices.

Postmodernism criticizes, among other things, the management of stakeholders, notably the centralization of the manager's power in reconciling stakeholder interests and in decision-making, the low decision-making power allocated to stakeholders in this deeply hierarchical structure and resulting in their low contribution to the processes of managing institutions and the social contract. Postmodernism demonstrates the flaws in this management: the failure to take into account the unequal power of different stakeholders when confronted with each other and the assumption that each stakeholder acts efficiently and is willing to make some sacrifices for the good of the companies and institutions. The latter implicitly assumes that stakeholders put the interests of the company ahead of their own and do not engage in opportunism. In contrast, with postmodernism, a more participatory management is advocated. It is a question of encouraging dialog and conversation between stakeholders by decentering the managerial vision that imposed a privileged unitary industrial goal.

In doing so, communication offers opportunities for everyone to express themselves to the best of their abilities. It provides a way to interact and learn from others, to build trust and to discover how the relationship can enrich each party. It is about building together a relationship that works for everyone, and determining together what goals are worth pursuing. This vision thus emphasizes the interdependence of the stakeholders and the company. In this perspective, the manager is no longer the unilateral decision-maker, but has the responsibility to facilitate dialog and to promote and maintain the bonds of trust between stakeholders, thus reducing opportunistic strategies. He/she is the guarantor of a context where all stakeholders cooperate to create together a vision, a common goal, a set of norms and determine the direction the company will follow. Postmodernism thus undertakes a dissolution of the classical industrial substance based on the perfect efficiency of economic agents and a destitution of the myth of value-based management. It promotes the emergence of new forms of industrial structures. Transnational institutions, a new international division of labor, automation, and new information technologies, industrial networks, and the contracting out of work are increasingly envisaged. Within companies, reengineering, and restructuring are dissolving the hierarchy and bureaucracy that characterize modern companies.

The institutionalist approach specifically attacks the definition of classical efficiency to highlight its weaknesses and its inadequacy in explaining the strategies of actors. Based on the work of Commons and Veblen, it criticizes the mechanistic, static, and uncreative vision that orthodox theories offer of the efficient economic agent, marginally adjusting to circumstances in a closed system. It takes into account the personal skills of actors and their capacity to adapt, which it situates in an institutional framework. In doing so, it introduces the role of institutions in the generation of strategies. Institutions are management structures based on rules, norms, values, and systems of beliefs and symbolism. More specifically, institutionalists suggest the importance of concepts and norms in the thought processes preceding actions. It is worth noting that there is considerable variety within contemporary institutionalism. There are several kinds of institutionalism that share a common skepticism towards reductionist approaches, and a common view that institutional processes and arrangements are important.

By way of illustration, the new institutional economics (NIE) takes up the hypotheses of microeconomic theory to which it adds new considerations such as transaction costs. It considers the transaction as the primary unit of analysis. Neo-institutionalism has a sociological essence and, despite its institutional roots, diverges substantially from the institutionalist tradition. Considering the work of several theorists with affinities to neo-institutionalism, it can be concluded that this work has contributed considerably to the advancement of action theory. However, the emphasis has been on certain institutional constituents at the expense of others, and these theories have neither retained and integrated aspects of traditional institutionalism nor provided a sufficiently powerful analytical framework to explain the correspondences between people and their places in the division of labor. Positive institutional theory is concerned with how political institutions influence political decisions, thereby eliminating the instability inherent in the majority voting system. However, this theory is based on a fixed and formal structure of the political process.

Regime theory, on the other hand, applies institutional principles to international relations. Moreover, there are variations within each of these institutional approaches. The reason why contemporary institutional theory is so varied is that it is connected to and constantly enriched by the contribution of recent theoretical developments such as new culturalism, cognitive theory, etc. Institutional theory explains industrial strategy as governed by an adoption of the values and practices of the institutions governing life. Institutionalism considers that institutions, by acting on the way actors see things, modulate their habits and ways of thinking. Institutionalism also emphasizes the importance of routine in cognitive processes, the limits of efficiency in taking into account the entire context and the role of institutions and routines in selecting and understanding the world. As actors evolve in a constantly changing context, their conduct overtime is influenced by institutional relationships and institutional change. Faced with the complexity inherent in many of the situations in which actors find themselves, they, with their limited knowledge and intellectual capacities, take into account only a tiny part of this complexity, which they reproduce in the form of a simplified model, from which they will make their choices. But the modeling and data integration processes of each individual are highly influenced by their previous personal experiences, identities, and skills (Vanberg, 1993), and are often imperfect. In these complex situations, rules facilitate the decision-making process by limiting the list of circumstances and variables to be taken into account in order to make an optimal choice. It is a matter of acting on the basis of selective knowledge. Compliance with the rules becomes the best option for maximizing choices for imperfect agents.

These norms and regulations are not the result of subjective, self-reflection, but rather come from institutions and interactions. They are the result of a selection process that takes place at the institutional level. Conformity to rules describes choices as responses to situations perceived as examples of a larger set of similar and more or less recurrent situations that the actor has already had to deal with. Conformity to norms is a rule. Economic agents seek institutional arrangements that allow them to minimize their costs and justify their legitimacy. This approach considers that situations are not dealt with independently of each other but categorized and grouped according to their similarities into groups of situations where habit, routine, and norms direct the agents' choices. Over time, industrial, and corporate strategy becomes institutionalized. Particularly in the case of companies and institutions, this process of institutionalization leads to an increasingly pronounced homogenization of companies, without increasing their profitability. Institutions passively conform to the institutional demands and circumstances of the external context. However, the rules may be imperfect and in turn lead to imperfect choices. A general model of trial-and-error learning where agents learn from their experiences over time and then form better rules is advocated by the institutionalist approach. The latter introduces adaptive efficiency, which takes into account the process of development and evolution of individuals based on their experience, and emphasizes the importance of the institutional framework in shaping this development.

Economic inequalities pervade the rest of social life (Richard, 1975). Critical theories belong to the radical humanist paradigm and recognize the importance of subjectivity and emotions in the production of agents' strategies. They argue that the analysis of companies cannot be done independently of the socio-historical context in which companies and institutions are embedded. They question the established order and advocate emancipation. Critical theories are based on a critique of the production society as alienating and as a source of suffering and frustration. Therefore, actors, in search of their well-being, are led to resist these alienating conditions arising from the structure of the context. Efficiency in this sense is a force of resistance and a means to achieve emancipation. It is limited by the established order and by the historical conditions of production. Critical theories seek to replace the classical analytical framework with a theoretical framework that takes into account the evolution of actors, i.e., a more dynamic and open framework. This framework should also emphasize moral or ethical issues in transformation processes. A greater emphasis on and inclusion of economically disadvantaged or marginalized groups completes this new framework. Their approach proposes to define a more coherent project of companies and institutions based on the so-called radical critique of the pluralist-liberal and human resource management approaches, and incorporating new research avenues. Inspired by Marx's thought, these theories are based on class struggle and the governance of the means of production, but go beyond pluralist concerns with topics as diverse as structural inequalities and labor market segmentation, the role of the state and ideology in industrial organizations, workers' consciousness and conflict, labor as an institutional problem. They also differ by recognizing the heterogeneity and internal evolution of each class.

While the basic premise of working-class unity is constituted from a common relationship of class interests opposed to capital and the bourgeoisie, critical theories interpret changes in the composition and structure of the working class and their implications for the enterprise and workers' collective action. Thus, political economy theory seeks to redefine its field of research to include conflict, its causes, its consequences. It is a question of taking into account the structure of classes, the dynamics of accumulation and crisis, the relations of force. The theories of the debate on the labor process take into account the modalities and consequences of the transformation of the labor power bought by the capitalist and maintain that the main function of management is to control and organize the labor process in order to extract the maximum surplus-value. The feminist approach is based on gender inequalities in the labor market, denounces the perverse effects of these inequalities and proposes solutions to restore the balance. It denounces the dominance of male stereotypes in the corporate and institutional world; stereotypes that lead to the adoption of bureaucratic, hierarchical, and competitive practices that combine to produce industrial and decision-making processes that fail to take into account many contextual issues. Following Marx's thinking, critical theories recognize the existence of social classes structured and separated by inequalities: "Those who work in a variety of manual occupations, in clerical position, as technicians, or in minor supervisory grades: men and women who make an obvious contribution to production which is not adequately reflected in their pay and conditions. And those whose property allows them to live from the labor of others... pay themselves salaries which far exceed any contribution they make to the production process (Richard, 1975).

Between these two classes there is a radical conflict of interest which underlies the problems of context. Class relations become power relations, where the possession of capital, means of production or other means of pressure confers economic power and advantage on its owner. The power structure in the markets must be understood as the result of the more or less complex interactions between: the uneven development of capitalist growth, the consequent variations in employers' strategies towards workers and the patterns of enterprise and worker resistance. The relations between agents are thus constantly modulated by the means of pressure they can use to increase their power. The labor market is more than just a place governed by a narrow economic process of supply and demand, it also includes relations of power and control. The quote does not fit smoothly into the text "In every workplace exists an invisible frontier of control, reducing some of the formal powers of the employer: a frontier which is defined and redefined in a continuous process of pressure and counter pressure, conflict, and accommodation, overt, and tacit struggle (Richard, 1975).

If admits, on the one hand that optimization is a procedure established to produce coherent results, a procedure in which it is a question of breaking down a process into successive stages, and on the other hand that optimization is associated with a rational analysis, it is clear that for some models, are part of a rather conservative analysis of strategy optimization, whereas others models stand out more as an environmental progressive. Also, since efficiency is at the heart of the theory of business economics and strategic management, the evolution of this concept has supported the emergence of new currents of thought, endeavoring to integrate into the reflections an increasingly realistic representation of economic actors when confronted with contextual issues. From the pure and simple search for profit to the consideration of various realities, the questions raised by these emerging currents bear witness to the complexity, variety, and inconsistencies of companies and institutions. They show that an issue requiring corporate attention involves a power struggle over the allocation of resources, and that the issue goes far beyond simple profit maximization. In order to cope, institutions are forced to deal with the constraints by using multi-faceted efficiency. Indeed, to be realistic, we need to consider the stakeholders of the firm in the totality of their strategies, some of which do not easily respond to the maximization of a stable objective function. The industrial strategy of the actors can be explained less by exclusive reference to a maximizing calculation than by the good reasons given by the actors. Agents' choices can be based just as much on calculations as on values, habits, or rules of strategy considered legitimate, and often despite their cost. Several models have been proposed to try to account for the way in which companies and institutions deal with their context.

KEYWORDS

- **Darwin's theory**
- **digital economy**
- **global entity**
- **globalization**
- **micro-organizational theory**
- **optimizing strategy**

CHAPTER 2

The Main Models of Value-Based Management and Their Application to Empirical Studies

Value-based management is the study of the operation and optimization of imperfectly competitive markets and the behavior of firms operating in them (Church and Ware, 1999; Amedzro St-Hilaire, 2022). According to Amedzro St-Hilaire (2019), it is the field of business economics concerned solely with the supply side. This field emerged as a result of the questioning of theories of perfect competition because, for some reason, there was not enough competition. Indeed, at the end of the last century, the emergence of large modern firms was observed (Chandler, 1977; Amedzro St-Hilaire, 2018) and the existing theories were not able to explain this sufficiently. The postulates of the theory of pure and perfect competition such as the atomicity of agents or the homogeneity of products were faced with differentiated products supplied by a small number of large firms. This new situation gave them a market power that was not found in the theory of perfect competition because they are price takers.

Although the ultimate goal of value-based management is to determine when and why there is insufficient competition, it focuses on three main areas of interest (Schmalensee, 1988): the determinants of firm behavior, imperfect competition, and public policy towards firms. These three fields ultimately allow for an understanding of the mechanisms of creation, exercise, maintenance, and effects of market power. The traditional approach used in industrial economics is the SBP paradigm. According to this paradigm, the structural characteristics of the market drive the behavior of firms which in turn influences their performance. However, the causality is not only one-sided. The way markets are organized (barriers, concentration, product

Value-Based Management in an Open Economy: Optimizing Strategy to Improve Business and Performance. Prof. Post-Dr Walter Amedzro St-Hilaire (Author)

differentiation, types of firms) affects the way firms behave and perform. But in turn, the behavior of firms influences the structure and organization of markets and their performance. It is this feedback link that the SCP model studies.

The aim of this analysis is to provide a synthesis of the main theories of industrial economics as well as a very elaborate quantitative and qualitative analysis of their contribution, limitations, and empirical perspective. The theories of interest are based on the theory of oligopoly, product differentiation, strategic behavior, and their corollaries. More specifically, we will seek to determine the empirical implications of the analysis of these theories in the dynamics of the industrial economy. In reality, the organization of markets is such that we generally find ourselves in an intermediate situation between pure competition and the total absence of competition. Section II presents the theory of oligopoly, which is the basic theory for understanding this intermediate structure. Sections IV and V will deal with product differentiation and its consequence, monopolistic competition. Section III will deal with the market power resulting from the new market structure in which firms operate. The following sessions will deal respectively with barriers to entry, strategic behavior, and empirical studies on market performance. The last part will conclude our study.

Oligopoly theory deals with the market structure between monopoly and pure and perfect competition (Church and Ware, 1999; Martin, 2001). In a monopoly situation, a single firm is responsible for supplying a multitude of consumers. As a result, it defines the quantity produced and sets the selling price. In pure and perfect competition, the market is represented by the atomicity of agents (suppliers and demanders). Firms are then recipients of a price that is determined at the intersection of the supply and demand curves. In an oligopolistic market, few suppliers ensure the supply of goods and services on the market. Nowadays, the theory of oligopoly, which is based on either price competition or quantity competition, is the theoretical foundation on which industrial economics is based. It is therefore appropriate to study it. To do so, we will first study static oligopoly models and then dynamic models. Static models have the particularity of introducing the concepts of interdependence of earnings and interaction of earnings during a single period (Church and Ware, 1999). Static theories of oligopoly show how the tension between cooperation and competition is resolved in favor of competition. Non-cooperative game theory has been used to understand such models. There are two models: the Cournot model and the Bertrand model.

The Cournot model deals with a particular oligopoly comprising only two firms (Church and Ware, 1999; Martin, 2001). It is based on the assumption that the homogeneous quantities produced are the only strategic variables that the firms use and that conjectural variations are zero (each firm knows what its rival produces, decides on the quantity that will maximize its own profit and expects that the other producer will not react when it has fixed its quantity). Moreover, the two firms have exactly the same cost functions. Cournot's model leads to an equilibrium point that we would call Cournot-Nash because at this equilibrium point, each firm provides its optimal reaction function according to the quantity produced by the other producer. The quantities produced are equal given the equality of the cost functions. Their sum is higher than in the monopoly but lower than in the competition. But unlike the competitive situation, the duopoly allows firms to exercise some market power. Each firm in the duopoly has market power and their equilibrium price exceeds the marginal cost of each firm (Church and Ware, 1999; Martin, 2001). However, this market power is limited by the elasticity of demand. Firms with the lowest marginal costs have a larger market share and the most efficient ones are larger. The more competitors there are, the lower the market share and market power of each firm. By reducing its market share, the increase in the number of firms increases the elasticity of residual demand of a given firm. This indicates the importance of barriers to entry on the market power of firms. The higher the barriers, the lower the number of competitors and the greater the market power of the firms in the market.

Cournot's model is one of the first to explain market organization using competition by factors other than prices (Church and Ware, 1999; Martin, 2001). The market structure based on a quantity strategy leads firms to behave uncooperatively because it prefers competition. Since the price and quantity obtained are between the monopoly situation and that of pure and perfect competition, the Cournot duopoly is to some extent socially desirable in the absence of cooperation between the firms present. This model makes firms, in terms of quantities produced and prices paid by the consumer, much more efficient than monopoly but less efficient than pure and perfect competition. However, the static nature of the model does not allow for the effects of optimization in turn on the behavior of firms and the structure of the market in the following period. The Cournot model assumes symmetry of cost functions (Church and Ware, 1999; Martin, 2001). It would be difficult, however, given technology and experience, for even two firms producing the same good to have exactly the same cost function. In this case, only firms with lower production costs would have an incentive to stay in the market because their profits would be higher and they would have more market power. In the Cournot duopoly, there is a positive structural relationship between market power and market share (Church and Ware, 1999; Martin, 2001). Overall, the quantity of the market will increase and so will the profit of firm 1. If the costs of the second firm 2 are too high for it to produce, then all monopoly output will be provided by firm 1. However, if N firms have the same cost function, in equilibrium they will all produce the same quantity of goods. They will all have the same market share.

In this case, when the number of firms increases, market power decreases. In the end, we would end up with a situation of competition in which the price equals the marginal cost. There is a decrease in quantities for each firm, a total increase in market quantities, a decrease in the market price, a decrease in the profit of each firm because of the decrease in price and quantities that they suffer. This shows the importance of barriers to entry in this model so that it does not eventually resemble a situation of perfect competition. What happens to firms that decide to enter the market? A firm that wants to enter the market anticipates the post-entry competition and its profits. The entry equilibrium is the point at which the entry of a new firm into the market results in negative profits for the firm. At this equilibrium, all firms present maximize their profits in the sense of the Nash equilibrium, but they earn zero profits (Church and Ware, 1999; Martin, 2001). As the number of firms on the market increases, the price falls, and the exercise of market power decreases. However, to maintain market power, Cournot's model would require the presence of economies of scale or fixed entry costs to limit the entry of new entrants. Furthermore, if we relax the assumption on conjectural variations, we end up with the Stackelberg model where choice is sequential (Church and Ware, 1999; Martin, 2001; Amedzro St-Hilaire, 2011).

In this model, one of the companies 'the leader' knows the reaction function of the other, which persists in Cournot-like behavior 'the follower.' The follower who persists in Cournot-like behavior persists in proposing solutions that do not improve its profit. The leader, on the other hand, has a higher profit than the follower. When the follower incurs fixed costs that are not borne by the leader can make the follower leave the market by making his profit fall towards zero. This strategy of limiting quantities can lead to maintaining a supply position indefinitely. However, this is not a monopoly position because if the price rises, it would attract new entrants. But if the fixed costs are small relative to the market, the quantity limitation on entry is close to the quantity of competition. Thus, the size of the fixed costs relative to the size of the market will guide the quantity leader in his quantity constraint strategy. If the costs are large compared to the size of the market, the leader would gain more profit by discouraging entry, otherwise he would prefer to allow entry and enjoy the profits of the oligopoly. The Cournot model also works by homogenizing products (Church and Ware, 1999; Martin, 2001). However, it is truer to encounter differentiated products than perfectly substitutable products. In this case, the price cut by firm 1 does not cause firm 2 to take all of its output because it behaves like a monopoly in its segment and this affects the performance of the market. Another limitation of Cournot's model is Bertrand's model which pointed out that in reality, firms will tend to compete on price and not on quantity (Church and Ware, 1999; Martin, 2001).

The major problem with Cournot's model is that it is not Pareto optimal and this poses a problem for its viability. The Cournot model as presented is not optimal for the firms involved because there are quantities of output that increase their profits. Since the demand function is strictly decreasing, if both firms simultaneously decrease their quantities, they benefit from a higher profit since the price increases (Church and Ware, 1999; Martin, 2001). This new situation is preferable to the previous one since their revenue increases. This reasoning, which is different from the previous situation, shifts the game from a non-cooperative to a cooperative aspect. In a collusive situation, the marginal cost is constant and the profit that the firms obtain is higher than in a duopoly situation. This profit corresponds to the monopoly profit. However, since this equilibrium is not a Nash equilibrium, it is not sustainable because each firm would be tempted to increase its quantity unilaterally to enjoy a higher profit than the second firm. And vice versa. From a simple situation, we can end up with a situation that involves a 'prisoner's dilemma' type of reasoning. This shows the oversimplicity with which Cournot presented the market structure. His model with its rigid assumptions did not take into account the rationality of firms and their emergent strategies. Finally, Cournot's model is subject to the 'business stealing effect' because the equilibrium price of each firm falls as new firms enter the market. This can lead to more firms than in the socially desirable situation (Church and Ware, 1999; Amedzro St-Hilaire, 2011).

In the Bertrand model, strategy is based on prices (Church and Ware, 1999; Martin, 2001). Products are homogeneous, firms have the same cost function and there are no capacity constraints. Nash equilibrium is achieved when the price equals the marginal cost. This is Bertrand's paradox because this situation is equivalent to the competitive situation. To find the equilibrium, we derive the optimal response functions. The Nash equilibrium price will simultaneously satisfy both prices of the optimal response functions. Bertrand's model, while explaining the organization of the market under the assumption that competition is through prices, introduces another type of strategic behavior that firms use. Since each firm can lower its price to capture the entire customer base, each firm will lower its price until it reaches the equilibrium point where the price equals the marginal cost. This model introduces price predation which works up to a certain level since firms have the same cost functions and therefore cannot crowd each other out (Church and Ware, 1999; Martin, 2001). As much as it is an alternative to Cournot's model, it helps to understand market organization through non-cooperative games.

When products are differentiated, firms realize that they cannot undercut their rivals and capture the whole market. The severity of price competition is reduced and each firm exercises market power in equilibrium. The elasticity of demand depends on the willingness of consumers to substitute. The less differentiable the products, the greater the willingness to substitute and the greater the price elasticity of demand. In a collision situation, the price that would prevail in the monopoly situation is higher than in the Bertrand model. Because each firm does not internalize the effects of their price decline into the profit of their rivals, they will be tempted to move from the collusion situation to their best response function which will unilaterally increase their profit. Since each firm can lower the price to capture all the customers, each firm will lower its price until it reaches the equilibrium bridge where the price equals the marginal cost. Another limitation of the Bertrand model is the assumption of unlimited capacity (Church and Ware, 1999; Martin, 2001). Edgeworth introduced the phenomenon of limited capacity into the Bertrand model because in reality, firms cannot produce more than their capacity. Indeed, it is often very costly to increase production beyond the capacity imposed by the inputs. The last limitation concerns the presence of economies of scale (Church and Ware, 1999).

According to Kreps and Scheinkman (1983, cited in Church and Ware, 1999), the two models are stages in a two-stage game. First, the equilibrium implies that each firm invests in capacity equal to its Cournot quantity. In the second period, the Nash equilibrium in prices given capacity, firms compete through prices. The K and S model places us in a situation where each model is appropriate. The Cournot model is appropriate when firms have capacity constraints and capacity investment is slow. The Bertrand model is appropriate in the situation where returns to scale are constant and firms are not capacity constrained. However, the use of static models is inappropriate. Strategic behavior can be defined as the investment in resources to limit the rival's choices (Martin, 2001; Amedzro St-Hilaire, 2018). Oligopoly firms use it because their profit depends on the behavior of other firms in the market and because they are aware of this interdependence. This strategic behavior can be seen in the costly investments that a firm may make to deter the entry of a new firm into the market. These strategies may include quantity expansion (quantity limits), predatory pricing, brand proliferation, advertising, technology selection, R&D, etc. In order to deter the entry of new rivals, the incumbent firm may use excess capacity by imposing higher costs on rivals.

Masson and Shaanan (1986, cited in Martin, 2001) estimated a system of three equations explaining excess capacity, marginal price-cost, and market share. They found that excess capacity has a positive effect on the marginal price-cost, implying that excess capacity discourages entry and increases market power. Gilbert and Lieberman (1987, cited in Martin, 2001) find that investment behavior depends on firm size as measured by market share. Small firms tend to invest when other firms invest (Paraskevopoulos and Pitelis' wagon effect (1989, cited in Martin, 2001)) whereas large firms invest less when others invest because their investment is mainly aimed at maintaining their market share. Thus, deterrence is only in the short term. Lieberman (1987b, cited in Martin, 2001) also finds that in the chemical industry, firms invest less when a firm already present on the market builds a new plant, but accelerate their investments when a new firm decides to enter the market. These results show that capacity investments can be used to influence the investment decisions of rivals. They can deter or delay investment by rivals and increase the market power of the firm that uses them.

Collusion refers to the situation in which firms coordinate their actions (Church and Ware, 1999). In order to ensure the success of this endeavor, they must reach an agreement or enforce an agreement in case some are tempted to break it. Enforcement requires that firms are able to detect and punish those who deviate. Cournot's model does not maximize the profit of the industry. From the firms' point of view, the Cournot equilibrium is inefficient because it is inside the profitability frontier (PF). If a firm does not earn at least as much as in a non-cooperative equilibrium, it is unlikely to decide to enter into a collusive relationship. Efficient agreements correspond to the point on the PPF that is Pareto optimal, i.e., one firm's profit could not be increased without reducing the profit of at least one other firm. The impact of collusion on the reduction of market power (as competition between the parties to the agreement is eliminated or reduced) depends on (Church and Ware, 1999):

- The elasticity of demand (as in Bertrand's model). However, the inelasticity of demand leads the collusion to increase its market power;
- The larger the size of the firms in the collusion, the more market power the collusion will create. On the other hand, if the firms outside the agreement are more numerous than those inside, the elasticity of demand of the cartel will be reduced;
- If there are no barriers to entry, efforts to increase market power will be futile.

An agreement can be tacit or explicit (Church and Ware, 1999). Since explicit agreements are mostly prohibited by law, there is a proliferation of subterranean or tacit agreements. Tacit agreements occur when firms coordinate their behavior by anticipating that of their rivals. The Nash equilibrium in the dynamic game results in greater coordination and higher industry profit than that prevailing in the static game. Several factors can prevent or complicate the achievement of an agreement. These include legal prohibitions, which depend on the severity of penalties, the ability to persuade and the resources available to the taxing agencies. However, if the profit from collusion exceeds the cost of the penalties, collusion will take place. There are conditions other than laws that may or may not favor the achievement of explicit agreements (Church and Ware, 1999). These are: structural costs such as cost asymmetry, product heterogeneity, innovation (possible changes in product characteristics, production costs, and demand), incomplete information, uncertainty, asymmetry in preferences, the social structure of the industry through conventions in which prices are set, seller concentration, taxation. When it is difficult to achieve efficient agreements, second-tier agreements can be reached where the revenue from collusion is Pareto higher than that from competition.

Stigler (1968b); and Chamberlin (1933) have criticized static oligopoly models. According to them, the essence of an oligopolistic interaction is that it is repeated (Church and Ware, 1999). For Chamberlin this implies that firms must recognize and act on their interdependence and maximize their joint profits. For Stigler, collusion must be enforced because firms tend to want to cheat by reducing their prices or increasing quantities. Generally, the ability to monitor an agreement depends on detection, the speed with which the punishment takes effect and the strength of the punishment. This section looks at the causes of market dominance and the exercise of power by certain firms. To maximize profits, the monopoly produces where its marginal revenue equals its marginal cost. At the monopoly price Pm demand equals the quantity QM that maximizes profit. This quantity is much lower than the socially desirable quantity, which leads to a loss of efficiency. But a market in which profits are positive is attractive (Church and Ware, 1999). If potential competitors have access to the same technology as the firm already in the market and there are no barriers to entry, in the long run they will remove the firm's market power and lower the price to the point where it equals its marginal cost. At this point, profits are zero. In the absence of economies of scale, market power is only possible if there are barriers to entry that prevent competition (Church and Ware, 1999; Amedzro St-Hilaire, 2016).

Barriers to entry can be the result of either corporate strategy or public policy (Church and Ware, 1999). The aim of the strategy is to retain market power while maintaining positive profits. Profitable barriers to entry depend on the interaction between structural barriers and firm behavior. These profitable barriers to entry may be natural or the result of strategic investment. The government creates barriers to entry when it grants exclusive production rights to a firm and uses its legal power of coercion to prevent the entry of other firms (Church and Ware, 1999). This type of monopoly is found in local services (electricity, gas, water, television) as well as in telecommunications. The government grants these exclusive franchises for several reasons: natural monopoly in order to minimize costs, source of revenue in order to share in profits, redistribution of profits and to protect intellectual property rights.

Barriers to entry may be the consequence of structural characteristics that make potential future firms anticipate negative post-entry profits (Church and Ware, 1999). Market entry depends on both the structural characteristics of the market and the nature of post-entry competition. The nature of the competition depends on the ability of the firm already present to be credible in implementing, for example, a price cut. However, the credibility of this threat depends on the structural conditions of the industry, which are: economies of scale, sunk costs of the entrant such as fixed costs, cost advantages that allow the incumbent firm to benefit from lower costs than the entrant, sunk costs for consumers and product differentiation (Church and Ware, 1999). Furthermore, since the profitability of entering a market depends on the nature of the post-entry competition and the behavior of the incumbent firm, it is possible that the incumbent firm will behave ex ante in such a way as to raise barriers to entry and reduce the profitability of entering the market. To do this, it will use one of three strategies: aggressive post-entry behavior by making sunk investments or by choosing to peg its marginal costs to its experience; by raising its rivals' costs or by reducing their revenue by reducing demand for their product (increasing the switching cost of consumers to their product) (Church and Ware, 1999). In theory, a monopoly holds all the market shares, but in reality, we are dealing with quasi-monopolies. Those firms that hold almost all the market shares have enough market power to influence the price. Two factors explain the presence of dominant firms (Church and Ware, 1999):

- Dominant firms are more efficient than their rivals and therefore have a considerable cost advantage. They can benefit from economies of scale; and
- They have a superior product in quality.

These dominant firms are generally faced with a group of small firms that we will call marginal. Their aim is to reduce the market power of the dominant firm and not to eliminate it (Church and Ware, 1999). The persistence of their presence on the market will have the effect of reducing the price that maximizes the profit of the dominant firm. Indeed, the difference between market demand and the quantity produced by the latter is the quantity supplied by the marginal competition or residual demand of the dominant firm. Thus, when it increases this price, its quantities sold fall and it suffers a loss equal to the margin of this increase. This drop is understandable because, on the one hand, an increase in price makes it more profitable for marginal firms to increase their production, thus reducing the dominant firm's residual demand, and on the other hand, the quantity demanded falls as the price increases.

The market power of the dominant firm is determined by three factors (Church and Ware, 1999):

- The elasticity of demand especially in the presence of substitutable goods;
- The supply elasticity of marginal competition because the greater the response of marginal firms, the lower the market power of the firm. But this elasticity depends on the marginal cost. The less the marginal cost increases for an additional quantity of output, the greater the elasticity of their supply; and
- The more efficient the dominant firm is in terms of low marginal cost, the greater its market power.

Although the number of marginal firms is assumed to be constant, the possibility of having positive profits may attract new firms, thus reducing the power of the dominant firm in both homogeneous and heterogeneous markets if costs are symmetric. However, if it retains a cost advantage, it will be able to maintain a significant market share. Market power can also be influenced by durable goods (Church and Ware, 1999). The presence of such goods has two consequences:

- It causes the monopoly to compete against itself in the future since the good is reusable more than once. Its reuse leads to the emergence of a second market which determines its future market power and the quantities the monopoly has to produce; and
- At the level of consumers, their attitude to buying the property today will depend on their anticipation of its price in the future. If they expect the price to fall in the future, they will postpone their purchase.

In the Coase conjecture the good is durable forever (Church and Ware, 1999). In the first period, the monopoly sells it at the monopoly price. In the second period and thereafter, this price falls, so that it can use up all its stock, until it reaches the competition price. At all times, the monopoly makes a profit by practicing intertemporal price discrimination. In the first period it offers its product only to consumers willing to pay a high price. Thereafter, it gradually lowers the price until it reaches the competitive price to exhaust its stock. On the consumer's side, if he anticipates that the monopoly will reduce the price of the good from period t+1, he will have an incentive to postpone his purchase in the future. The consumer's strategy is to benefit from the surplus resulting from the price reduction. If the consumer implements this strategy, he is able to force the monopoly to reduce the price in the first period and prevent it from engaging in intertemporal price discrimination. If the waiting time between periods is very short, the monopoly's demand becomes perfectly elastic. The consequence is that a durable monopoly in the Cournot conjuncture has no monopoly power because it is forced to apply the competitive price if the time between price adjustments is small. To counteract the consumer's strategy, the firm can credibly convince the consumer in the first period that it will not lower the price even if it ultimately applies intertemporal price discrimination. It can do this by leasing the durable good or investing in its reputation, entering into contracts, limiting capacity, discontinuing production, reaching new customers, creating obsolescence by introducing new versions of the product.

The Pacman model is the opposite case to Coase where a finite number of buyers face a sufficiently patient monopoly (Church and Ware, 1999). The consequence is maximum market power and perfect intertemporal price discrimination. Von der Fehr and Kuhn (1996, cited in Church and Ware, 1999) have shown that depending on the number of consumers (finite or infinite) and the behavior of the firm, one or the other of the models can be applied. When the good, instead of being durable, is recyclable, the market power of the monopoly becomes constrained (Martin, 1982). In the first period the monopoly supplies the good. But in the second period, the recycling sector that recovers the good and sells it to consumers competes against the monopoly's primary production. If the primary product is not recoverable or its recovery requires a reduction in quantity, then the monopoly retains its market power because the recycling sector will not be able to stay in the market sustainably. Otherwise, they will have to compete. Thus, the more efficient the recovery, the stronger the constraint on market power. Finally, the measure of monopoly inefficiency is the deadweight loss (small quantity produced). Rent-seeking is the assumption that an additional social cost arises from market power because of the efforts made by firms to acquire and maintain a monopoly position (Tullock, 1967; Posner, 1975). Under these conditions, the inefficiency of monopoly is the deadweight loss plus this rent. One advantage of market power is the presence of economies of scale. Another advantage is the investment in R&D that makes possible the production of new products and technologies essential for economic growth and a substantial increase in the quality of life (Church and Ware, 1999).

Normally, a dominant firm with low costs and no capacity constraints should strategically drive its marginal competitors (FM) at constant average cost out of the market. However, many dynamic models conclude that it will constrain the price positively so that a number of marginal firms remain in the market. In the static model, they would all exit. In the theory (Gaskins and others) marginal firms have adaptive expectations. But empirically, Berck and Perloff (1988) find that when firms have rational expectations given the same assumptions as in Gaskins' model, the dynamic model resembles the static model. The firm in its interest must reveal its intentions. This reduces the probability of acting as a price predator (price below marginal cost). In the literature on dynamic firm models, the rate of entry of FMs depends on the current price. In the Gaskins model, the dominant firm with low costs and no capacity constraints initially sets a high price which it gradually lowers to the limit price so that FMs remain and persist forever. Here while the firm has rational expectations, the FMs are myopic and only look at instant profits.

Yet several studies have considered the rationality of FMs (Flaherty, 1980; Judd and Petersen, 1986; Karp, 1987). Market power can also be studied at the level of other links in the product supply chain. Compared to the firm that produces a good, the latter represents a countervailing power. One of the most important manifestations of countervailing power, according to Galbraith, is the presence of large and powerful retail organizations. Indeed, it has been observed that the power of retailers grows through their purchasing power. According to Galbraith, by exercising this power, they are able to reduce the price they pay to suppliers and pass it on to consumers. This type of counter-power is socially desirable.

Stigler (1954); and Hunter (1958), criticized this model because of the existence of contracts between retailers and suppliers based on a two-part tariff. Indeed, both can benefit from joint profits by using this type of contract. In this case, the redistribution of profit between the two has no impact on the price paid by the consumer. For Von Ungern-Sternberg (1996); and Dobson and Waterson (1997), concentration at the retail level does not necessarily lead to lower prices for consumers. On the contrary, under certain conditions, it can lead to an increase in the price paid by consumers. However, their model does not take into account the polarization of shop size because all retailers are symmetric. Moreover, these models are characterized by a symmetrical reduction in the number of firms, whereas in reality, new firms enter the market and because of their marketing power and technological superiority, manage to reduce their costs, which allows them to achieve a position of dominance and purchasing power vis-à-vis suppliers. When the number of retailers is reduced, they gain both countervailing power vis-à-vis suppliers and monopoly power vis-à-vis consumers.

According to Berck and Perloff (1988), in their study of retailer countervailing power, the power possessed by the dominant retailer lowers the price paid by the consumer but not on the same grounds as Galbraith. The lowering of the price is not a deliberate act of the dominant retailer. In fact, an increase in the power of the retailer reduces the joint profit sharing of the two-party tariff contract described by Stigler. In order to recoup its loss, the supplier increases its sales to marginal retailers by lowering the wholesale price. This decrease in their costs shifts their supply curve to the right, thus reducing the retail price. Therefore, the fall in the retail price is not the result of the dominant retailer's will but the result of the suppliers' will to make up for their loss of profits caused by the increase in purchasing power. Contrary to Galbraith's model, countervailing power is not always socially desirable depending on the total surplus. Therefore, competition and the limitation of this power is important to protect consumer welfare. To do this, the presence of marginal retailers is essential to ensure that countervailing power lowers the retail price. The presence of a large number of DMs is in fact intended to improve the total surplus. In this model, goods are no longer defined as homogeneous, but each firm chooses the attributes and characteristics of the goods it produces and sells. By doing so, they determine the number and variety of products available on the market. The goods are then differentiated by the consumers although they remain similar in the eyes of the consumer who can substitute them because their functions are interchangeable. In fact, these differentiated goods are imperfect price substitutes (Tirole, 1988; Church and Ware, 1999). Differentiation is a multidimensional concept because it concerns location, proliferation, quality, advertising, etc. (Tirole, 1988; Amedzro St-Hilaire, 2022). However, it can be formulated within two main groups: horizontal differentiation and vertical differentiation (Tirole, 1988; Church and Ware, 1999). In the former, consumers have heterogeneous preferences with regard to the different attributes they prefer. There is no consensus among consumers about the best product or brand because their tastes are asymmetric and prices are identical. Products are vertically differentiated if consumers unanimously agree on the preference for one type of product or brand. This situation corresponds to the state in which consumers agree on the quality index. Thus, if all products have the same price, consumers will buy all the same brands. Products can be differentiated both horizontally and vertically. One approach used to specify consumer preferences when products are horizontally differentiated and which will be presented in this section is the goods approach or monopolistic competition.

Horizontal differentiation occurs when the consumer must travel to acquire the product (Tirole, 1988). Two firms can produce the same homogeneous good but they will be different in the eyes of the consumer if the cost of transport to acquire one or the other is very high. If this cost increases, the customers living in the firm's area, by falling back on the firm, grant it a monopoly power that allows it to increase its prices. But if the cost of transport is zero, there is no differentiated product and we end up with the Bertrand model. In this case, competition can be spatial as presented by the Salop circular city (1979) and linear competition (Tirole, 1988). These models show how location can be used by a firm to differentiate its product from its competition by moving closer to its customer. In these models, an increase in fixed costs leads to a decrease in marginal profits and therefore in firms. On the other hand, an increase in consumers combined with constant or low fixed costs leads to an increase in firms. By focusing on the number of firms entering the market, a firm can issue several brands instead of one and fill the space allocated. The aim behind such behavior is to fill as many niche markets as possible to reduce competition. This leads to a monopolistic market structure with several brands (Tirole, 1988).

Also, one observation of these models is that firms will tend to locate where the demand is, in a framework that allows them to benefit from the same externalities (example: supermarket) (Tirole, 1988). When competition is not based on price, firms have no incentive to differentiate products. However, they will try to have the best position to sell their quantity (Tirole, 1988). Thus, in equilibrium, they will all be at the same point. Furthermore, differentiation can be informational and lead a consumer to remain loyal to a particular brand that they have already tried if the cost of trying a third brand is too high. Advertising is one of the main dimensions of non-price competition. By providing information about the product and its location, it reduces the consumer's search costs, reduces the differentiation associated with lack of information and promotes competition.

Benham (1972) notes that the cost of eyewear in US states where advertising is prohibited is significantly higher. The opposite view is that advertising is designed to deceive the consumer by creating a false differentiation (Galbraith, 1967). It thus reduces competition and increases barriers to entry. Bain (1956) points out that informational differentiation (ID) can be a barrier to entry because consumers will tend to remain loyal to pioneer brands. Bagwell (1985) shows that even a firm with low product quality can deter the entry of a firm with higher quality by using ID. In a Cournot or Bertrand oligopoly, product differentiation has the advantage of increasing the degree of stability of collusion because it reduces the additional profit that would be gained from a defection (Martin, 2001). Indeed, product differentiation isolates markets and reduces the extent to which a firm can capture the customers of its rivals. When products are differentiated, market power is fundamentally firm-specific and depends on the degree of differentiation, the assumptions of rivals, the price elasticity of demand and market share (Martin, 2001). However, there are several limits to differentiation which are fixed costs and location (Tirole, 1988). Another limitation of the model is simultaneous entry and brand uniqueness. However, serial entry and brand proliferation is a reality that is not taken into account in the linear and circular competition model (Tirole, 1988).

Monopolistic competition exists when its assumptions are met (Church and Ware, 1999): there is a large set of all possible differentiated products over which consumer preferences are defined (taste variety) and consumer preferences over this possible set of brands are symmetric (close substitutes). Asymmetric preferences assume that the elasticity of substitution is constant and equal between products. Also, two conditions must be satisfied for the free entry equilibrium (Church and Ware, 1999): profit maximization and the free entry condition (the profit of a potential entrant must be negative and that of a firm in the industry positive). If all firms keep the same price but only one increases its price, this firm could see its sales fall without becoming zero because consumers will substitute them (Church and Ware, 1999). Each firm is a monopoly in its segment and since each product is in competition with the others, we are facing monopolistic competition. Given rivals, each firm maximizes its profit when its marginal revenue equals its marginal cost (Church and Ware, 1999). As firms grow, each firm experiences a small decline in demand. The number of firms in equilibrium depends on the size of the economies of scale and the elasticity of substitution (Church and Ware, 1999). When the elasticity of substitution increases, products are less differentiable, which reduces the market power of firms and equilibrium prices. The reduction in the marginal price-cost ratio leads to a reduction in the profitability of entering the market. An increase in the scale of economies of scale reduces the number of firms and therefore the variety of products, and requires firms to increase their prices and market shares to break even. The monopolistic competitive equilibrium is criticized for being insufficient or for having an excess of product variety compared to the socially desirable situation (Church and Ware, 1999). There are two effects that act on these characteristics but which operate in opposite directions: business stealing, which provides firms with an excessive incentive to enter the market, and non-appropriability of the total surplus, which implies that firms, when they introduce a new product, are unable to appropriate all the surplus because it is captured in part by consumers. This second effect contributes to the lack of incentive to enter the market and reduces variety.

Predatory pricing is the strategy that a dominant firm uses by reducing the price below the average cost of its rivals even if it means incurring short-term losses to drive them out of the market (Martin, 2001). According to Bain (1949) the price limit is the strategy that firms implement by deliberately foregoing a high price for fear of making their market too attractive (Martin, 2001). The deterrence strategy is based on the fact that if the entrant believes that the dominant firm will actually carry out its predatory pricing threat, it will prefer to stay out of the market. For the dominant firm, the best strategy is then to announce that it will react aggressively to entry. However, the threat could not work in the latter period because the firm would not be able to recover its lost gains. It is forced to cooperate in this market (Martin, 2001). Kreps and Wilson solved the chain shop paradox by finding that under uncertainty about the conduct of the incumbent firm, if the incumbent firm has always fought for its interests, it will continue to do so even in the last period (Martin, 2001). Milgrom and Roberts also looked at imperfect information in the chain shop paradox and found that the potential entrant will stay out of the market because it is not quite sure what kind of firm is in the market (strong or weak). If there is a small chance that it will fight then the entrant will defy entry (Martin, 2001). In Benoit's model, predation occurs because the firm believes that it is capable of crowding out its opponents (Martin, 2001). In these models, predation occurs because the dominant firm invests in its reputation to discourage entry by future entrants. In addition, the financial constraint makes reputation-based predation work because capital markets are not perfect (Martin, 2001). However, the possibility of merger and acquisition instead of predation or price limitation is an interesting alternative for companies. Although this alternative is costly, compared to predation or price limitation, it is neither time consuming nor risky.

Hotteling (1929) studies the stability of competition. In traditional oligopoly models, it is argued that when a firm raises the price of its product relative to its competitors, it allows all of its customers to be lost to them because the goods are homogeneous. However, given the spatial or product differentiation, when a firm increases its price, it will gradually lose its customer base and not instantly as found in the theoretical literature. Indeed, a certain number of consumers will remain loyal to it because of a certain number of reasons not necessarily related to the product. Hotteling (1929) points out that once the quantities sold are considered as a continuous function of the difference in prices, the hypothesis of instability that prevails in the model of Cournot, Amoroso, and Edgeworth. He thus supports the theory of horizontal differentiation by proving that transport costs or differentiation in the product itself (taste, shape, color, etc.), contribute to the stability of competition and therefore to monopolistic competition. Spence (1976) studied the effects of fixed costs and monopolistic competition on the selection of products and product characteristics in a set of interconnected markets. Fixed costs contribute to imperfect competition in markets and are a source of non-price competition. Furthermore, they restrict the number and variety of products that are feasible or desirable to offer. Thus, the products that are designed will be those that generate sufficient revenues to cover both fixed and variable costs. He studies the effect of monopolistic competition on complementary products. He finds that monopolistic competition tends to reduce the supply of complementary products. The reason is that when firms in such a market maintain quantities and raise the price of their products above marginal cost, the demand for complementary goods decreases.

The models we will study in this section are part of the theory of uncontested markets. Markets are said to be perfectly contestable when there are absolutely no costs to entering or leaving the market. On the other hand, they are imperfectly contestable when they are determined by the magnitude of sunk costs incurred by an entrant (Church and Ware, 1999). It is on the basis of this 'anomaly' that firms make strategic investments to prevent a competitor from entering their market. To do this, they can either use excess capacity or overinvest to reduce their short-run marginal production costs so as to produce the quantity limit if entry were to occur. The first models to address this issue were those of Spence and Dixit, which sought to determine whether or not firms would strategically use capacity investment to prevent entry. For Spence, firms in the second period are price takers, whereas in Dixit's model, they are in a Cournot game (Church and Ware, 1999). In the Dixit model, the firm commits itself in the first period to excessive investments so that these constitute a barrier for potential entrants. There are several types of barriers: natural barriers and structural barriers. The latter include economies of scale, product differentiation, and cost advantage (Church and Ware, 1999). According to Demsetz (1982, cited in Church and Ware, 1999) these exist because of information costs and information asymmetry.

Barriers to entry exist when established firms can exercise market power that adversely affects the post-entry profit of potential entrants. Thus, in order to protect both their market power and their economic profit, firms are often encouraged to engage in strategies aimed at crowding out their competitors (Church and Ware, 1999). Since entry depends on the entrant's expectation of post-entry profit, the incumbent firm may engage in investments even before it enters the market to discourage it. This aggressive strategy described in Dixit's model leads firms to overinvest in capacity relative to the monopoly level. The costs of capacity in the second period are sunk and high. Also, the marginal cost of the firm is lower than that of the entrant. The cost advantage makes the firm credible vis-à-vis the entrant provided that its marginal revenue is not lower than its marginal cost excluding capacity costs (Church and Ware, 1999).

According to Bain, barriers to entry are socially undesirable. By limiting entry and protecting the economic profits of firms, they prevent prices from falling to the long-run average cost. However, Von Weizsäcker introduces the notion of incremental cost and defines a barrier in this framework only if it leads to a decline in welfare (Church and Ware, 1999). Spence (1977) presents barriers to entry as a combination of structural and technological factors on the one hand and obstacles on the other put together by a firm. He finds that entry will be prevented if the firms in the industry have sufficient capacity to make entry unprofitable. However, this capacity need not be fully utilized in the absence of potential entry. The consequences are higher costs, higher prices, and lower quantity levels. Furthermore, he finds that when goods are homogeneous, firms will have a preference for capacity whereas in a market of differentiated products, they will have a preference for marketing strategies and advertising. However, for this strategy to be credible, the investment must in any case be sunk.

Dixit (1980), in studying the role of investment, finds that its role is to alter the initial conditions of the game in the second period to the advantage of the already established firm. Spence (1979) shows that in this game, capacity will be acquired slowly by the firms. The difference in the acquisition of firms in this respect will determine how the industry evolves, including whether the second firm could enter and what kind of equilibrium would result. Dixit's study in the power of the firm to change the initial conditions in its favor faces certain constraints such as capital markets. Klemperer (1987) studies a crowding-out model based on consumer switching costs. His reasoning is similar to Dixit's model but uses the phenomenon of switching consumer costs to explain the price limit. An established firm will charge lower prices in order to persuade firms to pay for its products and thus prevent entry. It may also overinvest if, in the second period, it can capture all consumers. In Dixit's model, increasing second-period quantity beyond first-period capacity does not reduce cost savings, but in the cost switching model increasing second-period quantity beyond first-period capacity does.

NEIO is different in that it focuses on the estimation of market power within a market (Church and Ware, 1999). The degree of market power is estimated simultaneously with the marginal cost used in oligopoly models. Structural estimation involves the use of theory to specify the relationship between demand and supply from which estimates of demand elasticity, marginal cost and firm behavior are derived. NEIO uses the results of comparative statistics of different market structures to identify the behavior of firms and their degree of market power. Non-parametric and reduced form approaches use comparative statistics to distinguish firm behavior and market power but do not require the estimation of a structural model (Church and Ware, 1999). These approaches are particularly necessary when there is a specification problem in structural models or when the data needed to estimate them is not available (Church and Ware, 1999). However, they are inherently limited in that they often result in a determination of what market structure or the degree of monopoly is not and do not suggest what it is. The limitations of the SCP approach, in the face of the complexity of firm behavior, to explain market performance has led to the NEIO approach coming increasingly to the fore. While according to Schmalensee (1989) the SCP approach only gives a description of what the market looks like and not how it works (Martin, 2001), the NEIO is more effective in establishing the existence of market power in individual markets. Although NEIO does not focus on determining the sources of market power, it does appear to depend on barriers to entry and understanding their determinants and how they are influenced by the behavior of firms. This leads to a consideration of the long-term strategies and competition between firms.

This chapter began by analyzing the basic theory of industrial economics. However, despite the strength of Cournot and Bertrand's models in explaining firm behavior and market structure, their limitations have led economists in this field to push their thinking further. Thus, new models based on product differentiation, investment strategies, strategic substitutes and complements, entry barriers, and market power have made oligopoly theory more realistic in its explanation of imperfect competition, firm behavior, and firm optimization.

KEYWORDS

- **Cournot model**
- **informational differentiation**
- **monopolistic**
- **oligopoly theory**
- **profitability frontier**
- **structure-behavior-performance**

CHAPTER 3

Strategy Optimization, Project Productivity, and Business Performance in an Open Economy

The change in the open economy and the need to cope with new competitive requirements have forced companies and institutions to evolve continuously (Amedzro St-Hilaire, 2014). The globalization of markets, the growing importance of the knowledge economy, and the integration of new technologies present companies and institutions with new challenges. Among these challenges, the internationalization of competition requires a global presence in the market and a dynamization of imports and exports. This confrontation pushes actors to reorganize, in particular by improving the management of strategies in order to be more efficient in the market. This has led to a growing interest in and evolution of competitive strategy models: the orientation in strategic decisions is mainly that of the company with its competitors. Companies and institutions in an open economy have to deploy strategies and resources to meet new challenges posed by a new competitive context, to differentiate themselves through competitive advantage, and thus to achieve a higher level of performance than the competition.

The effect of strategy optimization on the performance of firms in an open economy is thus exercised through institutional restructuring under the new conditions of international competition. The question of the optimization of firms and institutions remains, in this respect, an important subject in strategic research of the last decades and leads to several questions, in particular, how do firms and institutions define their strategies, how do they behave, and how are they managed in a globalized context? These questions contribute greatly to the theoretical problem of the impact of strategy optimization on institutional performance. This chapter consists of two complementary

Value-Based Management in an Open Economy: Optimizing Strategy to Improve Business and Performance. Prof. Post-Dr Walter Amedzro St-Hilaire (Author)

elements. The first section focuses on models of strategy optimization and performance in an open economy, in particular, it presents the conceptual framework of the notions of strategy optimization and performance in the first two subsections. The third subsection is devoted to the study of the impact of strategy optimization on the performance of firms and institutions. In the second section, a strategic analysis of firms and institutions in an international competitive context is carried out with regard to the models dealing with the relationship between strategy optimization and the competitive advantage of firms, then the different models evoking competitive strategies and their recent development are identified while focusing on the impact of competitive advantage on the performance of firms and institutions.

In the value-based management literature, strategy optimization is broadly defined as a logical and continuous process involving a number of sequential steps that enable companies to achieve their objectives (such as: defining the mission and long-term objectives, contextual analysis, generating, and evaluating alternative strategies, implementation, and finally monitoring the results). Models of strategy optimization in a holistic way show that companies and institutions that use them achieve their objectives, gain market advantages, and perform better (Amedzro St-Hilaire, 2014). In the context of an open economy (economies oriented towards international markets and exposed to international competition) even more so, companies and institutions are under pressure to develop competitive strategies. In this respect, what are the economic and managerial models that show that the company has an advantage in internationalizing to be competitive? This section presents the conceptual framework of the notions of strategy optimization and productivity of projects and institutions in an open economy context. It should be noted at the outset that the study of the impact of strategy optimization on project productivity is controversial in the management literature.

This chapter defines the notions of strategy and strategy optimization process, then presents the classical strategy optimization models and finally gives an overview of the schools of strategic thought. The generic concept of strategy has attracted the interest of several management researchers (Chandler, 1962; Johnson, Scholes, Whittington, and Frery, 2005; Porter, 1991; Strategor, 2005). Chandler (1962) states that strategy consists of the determination of the long-term goals and objectives of an enterprise, the adoption of the means of action and the allocation of resources necessary to achieve these objectives. According to Strategor (2005), developing a company's strategy means choosing the areas of activity in which the company intends to be present and allocating resources in such a way that it can maintain and develop itself there. In its original sense, strategy deals with the survival of companies and institutions and the microeconomic behavior of market players. It consists of making a lasting commitment to a direction, i.e., choosing the demands that the company wants to satisfy and the offers that it will make to this end.

This choice involves the managers and is based on the analysis of the company's potential in relation to the economic context. For some authors, corporate strategy began with the S.W.O.T. model (Strengths and Weaknesses of the Organization in the light of Opportunities and Threats in its environment) in 1965. This model is characterized by a dichotomous approach that contrasts the analysis of the external context of the company (a number of opportunities and threats) with the internal analysis of the company qualified through its strengths and weaknesses (Siegel, 2008). This approach to strategy is similar to that of the 'founding fathers' of the Harvard School and was developed by Learned, Christensen, Andrews, and Guth (LCAG). According to these authors, goals are set by the owners and implemented by the managers after examining the internal and external situation. The policy thus consists of formulating general objectives, identifying the most important problems, choosing the best solution, and implementing it. Some authors, such as Porter (1991), believe that strategy should integrate all the functions of the company, providing a set of objectives and policies to prevent them from working in centrifugal directions. For Johnson et al. (2005), strategy is the long-term orientation of the activities of companies and institutions: it is a strong commitment to their futures. It is about gaining competitive advantage by reconfiguring the resources and skills of firms and institutions in a changing context to meet the needs of the market and the expectations of different stakeholders (owners, employees, financiers, etc.).

In this perspective, we cannot talk about strategy without associating risk with it, hence the notion of risk strategy. For Amedzro St-Hilaire (2022), the risk strategy is the capacity of the entrepreneurial system to mobilize fragile and destabilizing events, within the framework of three possible scenarios: (a) setting up favorable conditions for competitiveness; (b) maintaining its activity within acceptable limits; and (c) redeploying itself to activities of another nature. In the first scenario, it is a question of an optimization approach. In the second scenario, the risk strategy functions as a resource to activate homeostasis processes. As for the third scenario, the risk mobilizes its strategic resources in the capacity to approach bifurcations and take a different direction. This risk strategy is in time: before, during, after the risk and functions as a preventive or curative means. In a situation of optimization, it cannot be reduced to the linearization of actions. It brings the emergence of strategic resources, it conveys a sense of optimization, and it allows leaders to achieve extraordinary things by using their skills in difficult situations. The risks related to entrepreneurship and the optimization of decisions over time are closely linked in an intertwining of interactions.

What about the strategy optimization? Strategy optimization is the dominant logic during the 1960s: a rapidly expanding context associated with more diversified competition. It is a process by which strategy should be formally developed and operationalized. Originally, strategy optimization took a so-called 'long term' form based on a three-to-five-year business forecast. It was carried out on the company's own market, using techniques that were, in practice, equivalent to simple reasoned extrapolations of past trends. Basically, the term strategy optimization referred to new management methods in companies, intended to replace traditional methods, developed in a context that paid little attention to the acceleration of change, the opening of markets and the rise of uncertainty. The best-known initial model is the LCAG model (1965): it defines strategy optimization as necessary for the strategic positioning of companies and institutions, i.e., the choice of activities considered most interesting for them. This general approach will be enriched and clarified by many authors, who defines strategy optimization as the process that sets the main orientations allowing companies and institutions to improve, modify or consolidate their competitive positions. The process of strategy optimization involves the realization of:

- a diagnosis of the competitive position of companies in its various activities;
- the drafting of a strategic plan that specifies the place the company wants to occupy in each segment;
- the development of operational plans that programed and coordinate the actions to be taken to achieve the strategy;
- the use of a budgeting system to implement and monitor short-term actions and of traditional financial tools (profit and loss accounts and balance sheets).

The aim is to help managers project themselves into the future with the help of a 'strategic plan.' This involves identifying the strengths and weaknesses of companies and institutions in order to analyze their context and assess the opportunities and threats that may arise. Mintzberg (1979) describes strategic situations as being characterized by novelty, complexity, openness, by the fact that companies and institutions know little about the situation and the route to a solution, what the solution might be and how to evaluate it. These are situations where decisions are made in ambiguity or nothing is given or easily determined. Those who develop strategies effectively have a more intuitive than rational approach. He defines strategy optimization as a mediating force between the firm and its environment: a force expressed as integrated patterns, sequences of decisions made or taken to fit the context. Porter (1980) believes that an objective analysis of the external and internal context favors the creation of appropriate structures for companies and improves decision-making. Thus, the implementation of a strategy optimization process facilitates efficient allocation of resources, competitive advantage, and innovation. Porter (1985) notes that although the concept of strategy optimization was widely criticized in the 1970s and 1980s, it remains useful and needs to be improved or reformulated.

Wendy (1997) explains the strategy optimization as a process of developing and maintaining coherence between the objectives, opportunities, and resources of companies and institutions. He further believes that it is an approach to doing business that leads to satisfactory profits and growth. He further explains that the strategy optimization process consists of three components that help companies to make their missions and visions more concrete. These are strategic analysis, strategic choice, and strategy implementation. Strategic analysis includes the parameters of management in terms of visions, missions, and objectives. Strategic choice refers to the creation, measurement, and selection of the most appropriate strategy. Implementation consists of the establishment of relevant policies and the formulation of theoretical foundations for the realization of the goals and objectives of companies and institutions. Strategies optimization thus allows to look into the future and to develop plans to act effectively in a competitive environment. It can help companies and institutions avoid costly mistakes in order to survive in highly competitive environments (Aram and Cowan, 1990). The traditional classical architecture of essentially normative optimization systems is based on:

- an internal and external diagnosis of companies and institutions in terms of strengths and weaknesses;
- the establishment of long-term operational budgets necessary to achieve the various objectives agreed;
- strict control of related activities.

The LCAG Model created in 1965 by the Harvard Business School is still an important reference point as it is the first model to help in strategic formulation. The LCAG model offers a logical reasoning in five phases:

1. **External Evaluation:** This involves not only identifying threats and opportunities in the context of companies and institutions, but also identifying key success factors.
2. **Internal Assessment:** This consists of identifying the strengths and weaknesses of the companies in relation to the competition (and overtime) and identifying the distinctive competencies in relation to the competition.
3. Creation and evaluation of all possibilities for action (strategies).
4. Clarification of contextual and managerial values.
5. Choice of strategies optimization according to resources and implementation of strategies.

The LCAG model allows for an internal and external strategic diagnosis. Companies must first compare their strengths and weaknesses with the opportunities and threats in their own context:

1. **Business (Internal) Analysis:** To identify its strengths and weaknesses, the company should carry out an internal analysis. This includes an examination of the company's resources, activities, and performance. Strengths are those factors that make the company more successful than its competitors. Weaknesses are areas where the company may be struggling in comparison to the competition.
2. **Contextual (External) Analysis:** The external analysis aims to detect opportunities and threats in the environment. It focuses on the general context (demographic, economic, institutional, natural, technological, and cultural environment) and the competitive universe. Opportunities correspond to favorable trends which open up new development prospects from which the company could benefit. Threats refer to problems posed by a transformation of the environment which, in the absence of an appropriate strategic response, may deteriorate the company's position.
3. **Aim:** To react strategically: the company must use the internal and external analysis carried out with the help of the SWOT matrix to make strategic decisions that will enable it to counter threats and seize certain opportunities. It must determine the strategic areas of activity that it plans to maintain, develop, or abandon. Then, on the

basis of this selection, the company will have to choose how to carry out the selected activities.

This model, although simple in its presentation, represents a rich and complex process of strategic analysis. It proposes a step-by-step method that should lead to considered conclusions. The sharp line between external and internal analysis has been challenged repeatedly by more contemporary authors. Indeed, some research has shown that strategists do not really distinguish between internal and external analysis in formulating their strategic choices. Mintzberg, Ahlstrand, and Lampel (2009) criticize the LCAG model for denying some fundamental aspects of strategy, including emergent strategy, the influence of the existing structure on strategy, and the involvement of actors other than the CEO in strategy development. For these authors, conscious thought plays a central role in strategy development, it necessarily precedes action and, therefore, the firm must separate design work from execution work. The presentation of this sequence follows a reflection by the author on the nature of strategic decisions. He distinguishes three main types of decisions: strategic, administrative, and operational. The purpose of strategic decision or strategic planning is to ensure that the firm's products and markets are well chosen. This emphasizes the relationship between the firm and the context. Administrative decision's structure the company's resources for optimal success and ensure the development of these resources: financing, equipment, personnel, raw materials. Finally, operational decisions aim to implement the operation under optimal conditions of capital profitability. These different dimensions of the strategic decision are found in model, which proposes a "funnel" approach to explain the process of strategy in companies and institutions, with the following stages:

- Formulation of objectives: Strategic decisions determine strategic objectives. They are centralized, non-repetitive, and taken by a small number of actors, usually at the highest level of the hierarchy.
- Identification of the strategic problem: this stage makes it possible to assess the company's capacity to achieve the strategic objectives previously set.
- Identification of a set of possible solutions: these are the guidelines that the company can take to achieve the objectives set. This involves relating the resources to the strategic problem that has been identified in order to assess their suitability and possibly consider a reorganization

or development of current resources, or even the acquisition of new resources.

- Assessment of the solutions considered: they are then assessed in terms of feasibility, both in terms of financial aspects and in terms of time (implementation time).
- The choice of a solution: it must be chosen from among those evaluated and the choice must be justified by the analysis carried out.
- The implementation of the chosen solution: the decided strategy will be implemented within the company, which will be translated into a set of operational decisions.
- Mintzberg et al. (2009) have classified strategic management approaches into 10 schools, including.
- The normative schools (the design school, the optimization school, the positioning school) which "insist more on the way strategies should be designed than on the way they are actually constituted."
- The descriptive schools (the cognitive school, the learning school, the power school, the cultural school, the contextual school) which allow "understanding how strategy development takes place."
- The configuration school, which 'conceives of this process as a transformation that incorporates much of the normative literature and practice of strategic change.

The design school is the development of strategy as a design process. It is based on the effectiveness of managers. This school bases strategy development on the notion of strategic diagnosis. Thus, the elaboration of a strategy consists of finding the best possible match between internal strengths and weaknesses and external threats and opportunities (well-known concept of SWOT analysis). By cross-referencing these data and in the light of the values retained, scenarios are defined and then strategies are chosen. Finally, we move on to implementation. One of the criticisms of this school is that conscious thought plays a leading role and separates the strategy development phase from the implementation phase. As a result, it denies some fundamental aspects of strategy development such as incremental development, emergent strategy, the influence of the existing structure on strategy and full stakeholder participation. The optimization school is the development of strategy formation as a formal process. It essentially takes up the hypotheses of the design school by breaking down its development into distinct stages and by giving it a more rigorous formalization in the form of checklists. The 'strategic planning' team thus replaces the manager, who is only marginally involved. Here, strategy is considered as an objective process, organized, and optimized in its smallest details (Marchesnay, 2004). This school of thought gives a clear meaning to strategy and allows not only for a good allocation of the company's resources, but also for an ex-post control of the implementation on the basis of what was optimized.

The positioning model is the development of the strategy as an analytical process. By analyzing the industry, the sector and thus the competition, the strategy will determine a positioning, choosing a place where the potential for development is highest. This school emphasizes the interactions with the environment on the study prior to action. The context is understood essentially in economic, market, and competition terms. Defining a strategy is above all positioning oneself in a competitive context. Porter (1980) gave impetus to this school in the wake of other work carried out on the theme of strategic positioning, both in the academic world and in that of consulting firms (notably by the Boston Consulting Group). For this school, strategy is reduced to a certain number of generic positions, chosen through a formal analysis of situations.

What is the "open economy"? An open economy is an economy that interacts with the rest of the world through the purchase and sale of goods, services, and financial assets. It is an economy oriented towards international markets, exposed to international competition, and interacting with other economies in the world. According to Strategor (2005), four factors are dominant in international trade: free trade and public policies, market, and demand developments, cost factors and the internationalization of competition. Deblock (2013) distinguishes four main forms of internationalization of competition:

1. **International Trade:** It has evolved considerably, according to Maddison's (2006) estimates, international trade represented about 8% of world GDP in 1913; today it represents one third. International trade, consisting of imports and exports, is an engine of growth for developed economies and growth in turn promotes international openness.
2. **Direct Investment:** It occurs when companies open subsidiaries abroad or buy companies abroad. It is a way for companies to get closer to distant markets, to exploit natural resources locally, to have a presence in financial markets or to produce goods at lower costs that they will then re-export.

3. **Outsourcing:** This is the process by which a company entrusts an external service provider with the management of an area or function that it previously carried out internally.
4. **Financial Openness:** Financial liberalization has led to strong investment growth and allowed more countries to access financial markets, but it has also led to much instability and risk-taking to the point of imprudence and breakdown.

In an open economy, the exchange rate regime is fixed and there is complete freedom of capital movements. In an open economy, residents can buy either domestic or foreign securities, which are assumed to be perfectly identical and therefore substitutable. Since foreign and domestic securities are identical, as long as there are no obstacles to the exchange of securities, the arbitrages of agents ensure that foreign and domestic securities will be exchanged at the same price and therefore provide the same return. Hence the concept of interest rate parity, which is the equalization of returns between domestic and foreign securities under the assumption of perfect capital mobility. The theory of interest rate parity establishes a link between the foreign exchange market and the international money markets (Taylor, 1987). Krugman (1979) introduced monopolistic competition to international trade theory. Trade was based on consumer demand for variety, which could not be satisfied in autarky. Monopolistic competition models assume that within a given industry, a large number of firms are engaged in production. Each firm offers a specific 'variety' of the same good. The presence of increasing returns ensures that each producer has a monopoly on the production of its own variety. For their part, consumers perceive the differences between these imperfectly substitutable varieties and seek to diversify their consumption. Trade openness allows consumers to have access to all varieties produced abroad. This demand for foreign varieties gives rise to simultaneous import and export flows within each product category (Krugman, 1979).

At the heart of this new theory of trade is a representative firm with a monopoly on the variety it produces, but subject to competition from other varieties in the absence of barriers to entry. The opening of trade has placed firms in a situation of increased international competition. The generalized opening up of markets and the catching up by emerging countries has led to the emergence of new competitors, including at the research and technological innovation stage. Definitions of globalization link this phenomenon very closely to that of international trade. Indeed, globalization can be defined as the process by which the interdependence between the markets and production of different countries increases as a result of trade in goods and services and financial and technological flows. This is not a new phenomenon but a continuation of a long-standing trend (Thompson, 1999). In the same vein, Ayoub (1998) defines market openness as the spread of the free movement of goods, services, capital, people, and ideas between all countries regardless of the political borders that separate them.

The growth of international trade has been made possible by the gradual removal of various barriers to trade between the world's major countries. Firms, which are essential players in this international trade (it is firms that trade with each other), have had to adapt to this global field of action by developing specific competitive strategies. The opening of borders to trade pushes national companies to develop competitive strategies (Markusen, 1981). Thus, the opening up of contemporary economies is associated with an increase in the competitive pressure experienced by companies and institutions. However, strong competition can have a negative effect on the company, in the sense that activities such as exports or innovation require high sunk investments. These can temporarily jeopardize the viability of the company, so that in a situation of strong competition, a company may prefer not to invest. Exposure to competition means that countries not only agree to open their borders to foreign companies, but also give them the same rights, obligations, and security as domestic companies. It also requires economies to be competitive (Deblock, 2013).

The objectives of companies are both economic and non-economic. The economic objectives are manifold and include growth in turnover, production volume, market share, new markets, and profit. Non-economic objectives include research and development, employment, remuneration, etc. According to Carroll (1979), the company should first fulfill its economic (making a profit) and legal (complying with the law) responsibilities, then ensure ethical behavior (being loyal and respectful in business life) and finally strive for philanthropic actions (being a good corporate citizen serving the community). Corporate performance is a fundamental element of strategic management research. It is not considered in the same way by all company managers. It is a multidimensional concept that can take several forms depending on the objectives assigned to the company (Bourguignon, 2000). Thus, Otley (1982) believes that performance is an ambiguous term that does not have a single definition. Burlaud (2000) attributes three meanings to it: 'a success,' i.e., performance does not exist in itself, it is a function of the representations of success, which vary according to the companies and the players; 'a result,' i.e., an ex-post evaluation of the results obtained; 'an action,' i.e., the process which leads to success. Hall (1980) distinguishes two major ways of conceiving performance: the goal approach and the resource approach. According to Bourguignon (1995), performance is the achievement of organizational objectives, whatever the nature and variety of these objectives. This achievement can be understood in the strict sense or in the broad sense of the process. Thus, a successful company is one that achieves its objectives.

Performance is a multidimensional concept. Management studies generally propose two approaches to measuring performance: an objective approach and a subjective approach. The objective approach concerns the use of financial indicators to assess performance, while the subjective approach concerns the non-financial indicators of the company. Increasingly, studies are using both approaches to assess company performance for consistency (Dixon, Nanni, and Vollmann, 1990; Ittner and Larcker, 1999; Banker, Potter, and Srinivasan, 2000). According to Kaplan and Norton (1999), traditional financial indicators measure the past performance of the company, while non-financial indicators determine the future performance of the company. Thus, the coexistence of financial and non-financial indicators reflects the different dimensions of optimization. Financial performance is assessed on the basis of indicators derived from stock markets or produced by accounting systems. The concept of performance is reduced to essentially financial and economic considerations aimed at identifying the company's net wealth creation. In the neoclassical view, which considers the organization as an instrument for maximizing profit (Friedman, 1984), the performance envisaged is typically financial and must satisfy the profitability requirements established by the shareholders. Financial performance can be measured by profitability, productivity, and several financial indicators, namely working capital, cash flow, shareholder value creation, return on investment (ROI), return on assets (ROA), return on sales (ROS), and return on equity (ROE) (Hopkins and Hopkins, 1997; Kald and Nilson, 2000).

But these financial indicators have been heavily criticized for failing to capture the effects of the intangible asset on wealth creation in a timely manner. They also do not provide sufficient information for management to direct and control the intangible asset. In addition, they are historical, offer little indication of future performance, do not take into account the intangible elements of a company's value, and are not linked to the strategy pursued by management (Kennerley and Neely, 2003; Mbengue and Ouakouak, 2012). When it has economic considerations, this performance can be measured by competitiveness. Competitiveness is the ability of a firm at a given point in time to outperform its competitors. Competitiveness is therefore a potentiality that is characterized by an advantage over competitors in its market. Increasingly, studies are attaching significant importance to performance indicators that are not derived from performance or economic statements and that are associated with non-financial performance (Fernandez, 2000). Non-financial information is essential for many of the board's responsibilities, particularly those related to optimizing strategies and monitoring the effectiveness of the implementation of strategic plans (Hart, 1992). Furthermore, since accounting data is not always available, non-financial indicators must be used to assess performance (Gauzente, 2000). Non-financial indicators reflect the investment in intangible assets and may be much stronger than accounting information in predicting financial optimization. They should therefore be used to complement financial indicators (Wallman, 1995). Non-financial measures include quality, customer satisfaction, shareholder satisfaction, employee satisfaction, efficiency in production or performance, market position, flexibility, on-time delivery, sales distribution, innovation, environmental friendliness, human capital (Johnson and Kaplan, 1987; Kald and Nilson, 2000). These measures point to other performance indicators. These indicators are organizational, social, and political:

1. **Organizational Performance Indicators:** Organizational performance is the ability of a company to act according to a wide range of optimality criteria in order to achieve a result. It also refers to the result and the actions that made it possible to achieve it (Bourguignon, 1995). A successful company must be both effective and efficient. Effectiveness means achieving the objectives set by the company (De La Villarmois, 2001). These objectives are part of the defined strategy. Efficiency adds the notion of the means used. The most efficient company will be the one that uses the least amount of means to obtain a result.
2. **Social Performance Indicators:** The social approach derives from the contributions of the human relations school, which emphasizes the human dimensions of the organization. It is the effective implementation of the social mission of an institution in accordance with social values. The term thus refers to the implementation of corporate social responsibility (CSR). Quinn and Rohrbaugh (1981) indicate that this approach integrates the activities necessary to sustain the organization.

3. **Political Performance Indicators:** The political approach is relativistic and considers that each individual may have his or her own criteria for measuring the productivity of projects (Morin, Savoie, and Baudin, 1994).

The productivity of projects has always been at the heart of the preoccupations of managers, scientists, researchers, shareholders, etc., each of them taking an interest in it from their own angle. This probably explains why performance has always been a controversial subject (Nazik FADIL, ATER). As Bourguignon (1995) notes, the concept of performance has been widely used for centuries without a unanimous definition. It is indeed a vague and polysemous concept (Bourguignon, 1995). Etymologically, the word performance comes from the Latin "perfomare," and was borrowed from the English "to perform" in the 15th century and meant to accomplish. Performance therefore meant accomplishment. Subsequently, the notion of performance was inserted into the language through horse racing in relation to the results of a horse, which led to a conception of optimization in the sense of an exploit, an exceptional achievement. The Petit Robert defines this concept as the numerical results obtained by a racehorse, an athlete, at each of his public exhibitions. In the field of management, the concept of performance has for several decades been linked only to the notions of profit or profitability of the company. However, this vision was rather simplistic and reductive of the term performance; as stated by MARMUSE (1997), "performance" does not exist. It is a contingent and multidimensional notion, but one that is necessary to evaluate any decision taken. Thus, many authors have tried to define the concept of performance in the company. For Albanes (1978), performance is the reason for management positions, it implies efficiency and effectiveness. Miles (1986) defines performance as the ability of the organization to achieve a minimum satisfaction of the expectations of its strategic customers. An equally interesting definition is given by Chandler (1992), who sees performance as a combination of functional and strategic effectiveness. Functional effectiveness is about improving the products, purchasing, production processes, marketing function and human relations within the company. Strategic efficiency consists of getting ahead of competitors by positioning oneself on a growing market or withdrawing from a declining market. Finally, for Philippe Lorino (1998), performance in a company is everything that contributes to achieving strategic objectives and therefore performance in a company is everything that contributes to improving the value/cost ratio.

From these different definitions, two important elements seem to emerge to characterize performance, namely effectiveness and efficiency, which leads to the general agreement with Luthaus (2003) that performance is a combination of the effectiveness and efficiency of the company. Effectiveness refers to what produces the expected effect, i.e., the relationship between the results obtained and the objectives set, whereas efficiency is the capacity to produce the maximum results with the minimum effort expended, in this case the relationship is no longer between expectations and results but rather between results and the resources implemented to obtain them. In the same vein, the Albane management manual, which has long been a reference in North America, states: we often use the words efficiency and effectiveness when talking about optimization. Efficiency means doing well and doing without loss, no matter what it does. It is the 'more, better, faster, cheaper' side of optimization. Effectiveness goes further than efficiency and considers the effect of work on people, the relevance of goals, the long-term results, and the norms and values implicit in the work of goals (Albane, 1978). Thus, performance, taken as a combination of effectiveness and efficiency, is the ability of a company to achieve its results in relation to the objectives it has set itself, but to achieve these results with a minimum of cost (Bouquin, 2001).

All these attempts at a more or less clear definition show that the notion of performance must be approached with a certain amount of caution and according to the objectives of the research. Moreover, the difficulty in understanding the concept of performance lies not only in its rather complex definition but also and above all in the polymorphous or multidimensional character that performance in companies can have. Indeed, performance taken as a whole is a contingent notion, and cannot therefore be limited to its accounting or financial aspects alone, as was the case a few years ago. The current dynamic environment in which increasingly complex organizations are evolving, where material, financial, and human factors interact, leads us to consider the notion of performance in a more global manner by integrating several dimensions to performance, namely: the economic, financial, social, human, managerial, technological, commercial, societal, strategic, and organizational dimensions. This last dimension, which translates organizational performance, is of particular interest to us in the context of this work, which aims to present the impact of the use of Knowledge Management practices on the organizational performance of the company. To this end, the following section will focus on organizational performance and mainly on the modes of evaluation of the said performance.

The concept of institutional performance, although central and sometimes unavoidable, remains one of the most elusive in organizational theory. Indeed, as the functioning of an organization is not described by a single conception, various conceptions of institutional optimization have emerged (Morgan, 1989), making its understanding even more complex. This has led to a range of indicators and models for assessing institutional performance. After an attempt to clarify what institutional performance is (I), this section will also analyze, in general terms, how institutional performance can be assessed (II). Institutional performance has been at the heart of many debates. Inspired by the studies of Taylor and Fayol, this approach to optimization dominated the first half-century. It fed into the work of the so-called OST school, introduced in Europe after the First World War. Subsequently, at the beginning of the 1980s, this institutional approach to optimization was reinforced by the resource-based theory (Barney, 1986, 1991, 1995), according to which the differences in performance between companies in the same sector of activity stem from the exploitation of internal resources rather than from an adaptation of the organization to the market and the environment. Thus, the analysis of the literature on institutional performance, fed by these different movements, has made it possible to identify a certain number of definitions delimiting the field of application of the institutional approach to optimization.

Indeed, most authors tend to consider that the performance of the company perceived in its institutional dimension, results from the value of its organization. According to Bouquin, this is the capacity of an organization to determine and implement good strategies within the framework of the goals it pursues (Bouquin, 1997). In other words, institutional performance concerns the way in which the company is organized to achieve its objectives and the way in which it manages to achieve them. It is therefore clear that institutional performance should not be confused with project productivity or organizational performance, which refers rather to the overall performance presented in the first section of this chapter. Institutional performance should be seen as the set of measures that relate directly to the institutional structure and not to its possible social or economic consequences (Kalika, 1998). From these different definitions, it appears that institutional performance is particularly interested in the arrangement of the functional organs of the company, the distribution of personnel and consequently the allocation of resources. Thus, Kalika (1998) puts forward four factors of institutional effectiveness, namely: the respect of the formal structure, the relations between the components of the organization, the quality of the information flow and the flexibility of the structure. However, since the objective of this work is to see the impact of knowledge management practices on institutional performance, it is necessary, after this overview of the different definitions, to consider the methods of evaluating the said performance.

Measuring performance, whatever the dimension, is a perilous and delicate exercise. H. Bouquin defines performance measurement as the ex-post evaluation of the results obtained (Bouquin, 2004). Bescos et al. use the terms "monitoring" or "results measurement" to designate the process of budget monitoring and variance analysis, and "performance measurement" or "performance evaluation" for the three processes of value-based management, i.e., the setting of objectives, the results measurement system, and the sanctions/reward system (Bescos, 1994). Thus, performance cannot be described simply in terms of a comparative measurement between the value of inputs and outputs. The possible list of elements to be taken into account is thus without precise limits. However, from the analysis of the literature, there seems to be agreement between the different authors on the methodology or measurement approach that applies to all dimensions of optimization. Thus, according to Morin, Guindon, and Bouranne, institutional performance, like all other variants of overall performance, must be based on a measurement model comprising three levels of abstraction (Morin, Guindon, and Boulianne, 1996; Spriggs, 1994). The first level is to establish a general description of performance and productivity in line with the context of the research in order to determine precisely the dimensions of the concept to be measured. In other words, the aim is to identify the components that best capture institutional optimization. The second level is the transition from the definition of the performance domain and the parameters for understanding it to the measurement problems. The measurement criteria must then be chosen from among the many categories of measures available (quantitative or qualitative, objective, or perceptual, gross, or standardized) and the means of collection. Once these first two steps have been completed, the various indicators or items making up the measurement scales must be generated.

For P. Lorino, a performance indicator is defined as information that should help an individual or, more generally, a group of actors to steer the course of an action towards the achievement of an objective, or that should enable them to evaluate the result (Lorino, 2001). Thus, these indicators can be directly observable indicators as well as specifically created items. Moreover, indicators are not necessarily numbers. They can take any informational form that fulfills one or other of the two functions mentioned in the definition (conducting the action, evaluating the results): qualitative judgment, binary yes/no sign, graph, etc. Finally, it should be noted that these indicators should be cognitively or ergonomically effective, which means that they should be easy to read, understand, and interpret by the agent for whom they are intended. Adopting this methodology, some authors have proposed a framework for measuring institutional performance. This work will be limited to the presentation of the dimensions developed by Estelle M. Morin, in particular because the other authors propose similar measurement frameworks, but subtracting one or more dimensions depending on the case. Morin et al. (1994) also propose an instrumentalization of their concepts which aims to give performance measurement a certain credibility. The difficulties of operationalization are numerous, the concept being multidimensional, hierarchical, antinomic, and operative. Yet richer performance indicators can be developed. So, let us agree with Pierre Voyer that when a performance measurement system is implemented, it must be borne in mind that it can only be interpreted rigorously and correctly in the context in which it was carried out and which served to define it. Lenz also states that performance is a construct (LENZ, 1981). However, it is important to remember that this construct crumbles when it comes to choosing the indicators that will enable it to be measured. To this end, performance, and in this case institutional performance, must be approached in terms of the objectives targeted by the researcher.

The performance criteria and the assessment made of them can also vary for the same stakeholder, depending on its expectations, the level of competition and more generally the context of the company. Given that we are working in a context of international competition, we will study the competitiveness criterion more closely in order to evaluate the performance of a company in an open economy. With the growth of international trade, there is increasing talk of the globalization of trade in capital, products, and services. In this new competitive context, competitiveness is becoming an unavoidable imperative. Companies must react or risk being outclassed by more efficient companies. In an open economy, competitiveness can be the ability of a company to compete both on the national and international markets. It therefore refers to trade competition. Martinet (1984) defines competitiveness as "the ability to sustain competition: the competitive firm possesses a set of capabilities that allow it, as the case may be, to enter, maintain or develop in a competitive field constituted by the set of forces traversing its environment and likely to oppose its objectives, its projects, and its operations. Porter (1982) developed a more comprehensive and integrated analysis of competition and competitiveness. Competitiveness may be associated with other criteria such as profitability, productivity, market share, exporting, innovation, product quality (McFetridge, 1995). Porter (1982) emphasizes that innovation is the key to the competitiveness of firms because it determines their ability to maintain sustainable competitive advantages in changing markets. Innovation enables firms to strengthen their competitive position in markets. Indeed, innovation allows companies to increase their productivity, improve the quality of their products or services and develop key competences. According to Christensen, Anthony, and Roth (2004), innovation is anything that creates or improves resources, processes, or value within the company. It can provide a competitive advantage to the organization and an opportunity to enhance its competitive position in the market.

Internationalization corresponds to the action of going international, i.e., generalizing the action of the company to all or some nations. However, this term remains confused and many definitions exist. Internationalization, a real strategic process, is one of the main ways in which companies develop and a major fact in the dynamics of the world economy (Lemaire and Petit, 2003). There are several economic and managerial theories that describe and explain the internationalization of companies. Some models explain the internationalization process through factors internal to the company. This is the case first of all with Hymer's (1976) monopolistic advantage. The latter explains how a firm which sets up abroad and incurs the costs of relocating its production can remain competitive with local firms which produce on their own market and have a better knowledge of the environment, in fact, this firm which sets up abroad exploits specific advantages of various kinds: technological, capital resources, economy of scale. The OLI paradigm (Ownership, Location, Internalization) of Dunning (1988) proposes an explanation of the choices of international deployment of companies. This choice is thus a function of the presence or absence of specific advantages of the company, specific advantages of the countries, and advantages of internalization.

The Uppsala model or the stage model of internationalization developed by Johanson and Vahlne (1977) explains the internationalization of the firm based on two concepts. These are psychological distance, which takes into account all the cultural and linguistic differences that influence decision-making in international transactions, and the learning process, which explains that experience of foreign markets is acquired progressively in a sequential process. Knickerbolker's (1973) theory of oligopolistic behavior states that when in a sector with an oligopolistic structure, a so-called leader of the oligopoly establishes itself abroad with the aim of modifying the structure of the market to its advantage, by reaction the other companies in the sector imitate the leader by also investing abroad. Strategor (2005) mentions five main reasons why companies go international: 1) the conquest of new markets; 2) the reduction of costs and the improvement of the company's competitiveness; 3) the opportunity to exploit abroad the competences at the origin of a competitive advantage developed on the national market; 4) the distribution of risk in countries with different economic cycles; 5) the need to be present on the large international markets because the competitors are there Another example is Porter's (1990) nation advantage, which shows that a company can gain competitive advantages by going international. Porter (1990) has identified the determinants that allow each nation or company to claim a competitive advantage. These determinants make it possible to identify the conditions that a nation or a company must meet in order to be more competitive on the international level. Considering Porter's work on competitive advantage, Strategor (2005) has suggested four strategies corresponding to four types of firms.

1. **The Global Strategy:** It pursues economies of scale abroad in search of the best locations for cheaper manufacturing. It seeks efficiency on a global scale, while maintaining local specificity.
2. **Multinational Strategy:** This is a firm that manufactures all or part of a product abroad through a subsidiary. A firm becomes multinational when it sets up a production unit abroad. It is characterized by a 'differentiation of activities from one country to another, close adaptation of products to local specifications' (Stratégor, 2005).
3. **The International Strategy:** It exploits the know-how, technologies, human, and financial resources of the parent company which it disseminates and adapts to new foreign markets. It is valid if cost competition is low and if there is a customer base for a fairly standard product.
4. **Transnational Strategy:** This type of firm exploits a configuration of capital and activities based on a distribution of roles and responsibilities within the company. It optimizes the two global objectives: global synergy and cost reduction and adapts them to local specificities.

The concepts of strategy and strategy optimization are debated in the management literature. The review of the managerial literature on strategy optimization carried out in the first section presents the benefits that a company applying strategy optimization can have. However, empirical studies on the relationship between strategy optimization and performance have been debated for four decades in the management literature. Early studies on the impact of strategy optimization on performance showed that strategy optimization improves project productivity (Herold, 1972; Thune and House, 1970). These studies showed that financial performance was better in planning firms than in non-planning firms. Following the work of Thune and House (1970), several studies on the same topic have published empirical results on the relationship between strategy optimization and project productivity (Glaister, Dincer, Tatoglu, Demirbag, and Zaim, 2008; Falshaw, Glaister, and Tatoglu, 2006). The results of these studies are varied, ambiguous, and contradictory. Some studies find that strategy optimization is of great interest for project productivity (Greenley, 1994; Capon, Farley, and Hulbert, 1994; Hopkins and Hopkins, 1997), others find that strategy optimization has no interest for project productivity (Fulmer and Rue, 1974; Bresser and Bishop, 1983; Whitehead and Gup, 1985). Other studies show that there is no relationship between strategy optimization and performance (Bresser and Bishop, 1983; Whitehead and Gup, 1985). In the following, we will present the different results obtained in the studies.

The managerial literature on strategy optimization implies that there is a positive relationship between strategy optimization and firm performance (Greenley, 1994; Amedzro St-Hilaire, 2022). There are two opposing schools of thought, the normative school, and the descriptive school. A company that implements and practices long-term strategy optimization better achieves its objectives (Stonehouse and Pemberton, 2002). Some studies from the normative school suggest that firms should use high-level optimization (Gavetti, Levinthal, and Rivkin, 2005). The descriptive model expresses the need for flexibility as it is difficult for an organization to plan for the future. The strategy optimization is effective in a non-turbulent environment since in this case it is easy to organize and allows the anticipation of future operations under certain circumstances. Some years later, studies have shown that a hostile and turbulent environment is a threat to project productivity (Haigang, 2001; Tsai, MacMillan, and Low, 1991). A hostile business environment is characterized by fierce competition, lack of market opportunities, unfavorable business climate (Govin and Slevin, 1989; Khandalla, 1977). This often results in low profits and a high failure rate (Hall, 1980). Greenley (1986) has identified the potential benefits to firms of performing strategy optimization. Strategy optimization is then seen as a managerial process given the optimization to be achieved. Managers of companies that develop strategy optimization believe that it contributes to efficiency by allowing them to have better control and by giving them a sense of confidence under the assumption that strategy optimization improves project productivity (Falshaw et al., 2006).

Other studies on strategy suggest that strategy optimization is only valuable if managers apply it with a certain intensity, thus the focus on strategy optimization leads to financial performance (Steiner, 1979). Hopkins and Hopkins (1997), in their study, defined the intensity of strategy optimization as a relative emphasis placed on each dimension of the strategy optimization process. Strategy optimization researchers have established three dimensions for the conceptualization of strategy optimization: formalization, comprehensiveness, and strategic control. These three dimensions have emerged with great frequency in the literature as characteristics of strategy optimization (Powell, 1992; Papke-Shields, Malhotra, and Grover, 2002). It has been shown that if managers apply the strategy optimization process with equal intensity in each dimension of strategy optimization, this leads to a positive impact of strategy optimization on project productivity (Dimma, 1985). Supporting this idea, Hopkins (1987) found that financial performance tends to be high in firms where differences in the intensity of each component of strategy optimization are negligible. In the same vein, Capon et al. (1994) showed that the higher and more sophisticated the strategy optimization process, the better a firm performs. On the other hand, other studies have shown the relationship between strategy optimization and firm performance to be negative in the sense that strategy optimization, by its very characteristics, could create rigidities and encourage excessive bureaucracy (Bresser and Bishop, 1983). The study by Whitehead and Gup (1985) established that planning firms perform worse on some measures than non-planning firms. Amstrong (1982) considered 12 studies showing positive, negative or no relationship of strategy optimization on project productivity and concluded that these studies support the importance of strategy optimization on performance. However, he raised serious research issues arising from the fact that few studies reveal how and when to apply strategy optimization (Falshaw et al., 2006).

Greenley (1994) made a classification by considering 29 relevant empirical studies on the relationship between strategy optimization and performance, and he classified these studies into three groups. The first group consists of nine studies where researchers found no relationship between strategy optimization and firm performance. In the second group, 12 studies supported the relationship between strategy optimization and firm performance, while in the third group, nine studies concluded that optimizing firms performed better than non-optimizing firms. After reviewing these results, Greenley concludes that these studies are evidence of a relationship between strategy optimization and firm performance. He believes that methodological weaknesses may be at the root of the results obtained by the first group of studies (Falshaw et al., 2006). Other studies have shown that there is no relationship between strategy optimization and project productivity. Thune and House (1970) studied 36 companies in six industry groups. The results showed that the evaluation of optimization before and after the strategy optimization process gave similar benefits on project productivity (Boyd, 1991). There is a stream of studies on strategy optimization that takes into account the time taken by the firm to perform strategy optimization. These studies have shown that the time taken has no impact on project productivity.

In the study by Fulmer and Rue (1974), researchers compared financial performance in service industries over a three-year period, 50% of the firms studied had used the strategy optimization programed two years prior to the study. Given the lack of a positive relationship between strategy optimization and financial performance, the researchers concluded that firms have no incentive to make optimization efforts. Gup and Whitehead (1989) also tested the effect of the time taken to implement strategy optimization on bank performance, they found statistically no positive relationship between the time taken by banks to optimize strategies and their financial performance (Hopkins and Hopkins, 1997). Other studies, notably those of Pearce, Freeman, and Robinson (1987); Shrader, Taylor, and Dalton (1984); Boyd (1991) show that the effect of strategy optimization on project productivity is very weak or non-existent. In view of the controversial empirical results, the question of the relationship between strategy optimization and performance remains unresolved and problematic. In the value-based management literature, empirical studies on the relationship between strategy optimization and performance yield controversial and sometimes contradictory results. Researchers have found that these contradictions may stem from methodological and conceptual errors.

With regard to methodological errors (Mbengue and Ouakouak, 2012), these include incomplete and unreliable operationalization's for both strategy optimization and performance (Powell, 1992), heterogeneous, and small sample sizes (Peel and Bridge, 1998), or inappropriate statistical procedures (Powell, 1992). Most previous research has described companies according to the degree of their strategy optimization process. Thus, planning firms are those that have programed a high level of strategy optimization while non-planning firms have not. However, the presence of an elaborate program does not necessarily mean that the strategy optimization process is effective in the company (Falshaw et al., 2006). Burns and Stalker (1961) distinguish two main types of structure corresponding to two different types of environments. The mechanical structure based on work specialization, formalized procedures and centralized decision-making is adapted to a stable environment because its evolution is predictable. The organic structure with decentralized decision-making and a flexible hierarchy is adapted to an unstable environment because the company must be able to adapt its products quickly to the evolution of the market and respond to new needs arising from unpredictable changes. A study by Khandwalla (1977) showed that high-performing firms adopt an organic structure in a hostile environment, while a stable environment is more appropriate for firms with a mechanical structure.

Traditionally and historically, financial performance has been used as the only measure of performance despite the relevance of other non-financial indicators (Falshaw et al., 2006). However, some studies have used a subjective or objective approach to measuring performance (Greenley and Foxall, 1997). Objective performance refers to the financial reports of the company (Wall et al., 2004) while subjective performance refers to the whole company. According to Falshaw et al. (2006), the objective measure of optimization is not appropriate for this type of research. Furthermore, studies on the subject show that authors do not use the same variables to measure performance, yet some variables are more significant than others in the correlation between strategy optimization and performance (Shrader et al., 1984; Falshaw et al., 2006). The lack of uniformity in the measurement of performance would lead to a lack of visibility in the relationship and contribute to mixed results (Pearce et al., 1987). According to Pearce et al. (1987), the size of institutions is the most important characteristic of an organization, and greatly influences the relationship between strategy optimization and performance. Size is important and must be taken into account when designing a strategy optimization program. The optimization process is therefore different for large and small organizations. Much of the research on optimization in SMEs points to the existence of strategy optimization in SMEs but describes it as incomplete, unstructured, irregular, sporadic, and rather reactive and informal (Sexton and Dable, 1976; Shuman, 1975) and that these organizations do not engage in a structured and formal process of strategy optimization as prescribed by the normative models of the traditional strategy perspective (Robinson and Pearce, 1984). Powell (1992) showed a strong correlation between strategy optimization and performance in large firms rather than in small firms. However, research by Miller and Cardinal (1994) showed that institutional size is not a significant variable in the relationship between strategy optimization and performance.

Studies have shown that the business environment can be turbulent and strongly influence the relationship between strategy optimization and performance (Pearce et al., 1987). However, these studies have produced contradictory results, with one group of authors believing that strategy optimization is more likely to have a positive impact on performance in a stable environment where future conditions are easier to anticipate and the information needed to make strategic decisions is more readily available (Fredrickson and Mitchell, 1984). Other authors argue that the correlation between strategy optimization and performance may be stronger or more useful in a turbulent environment than in a stable environment (Miller and Friesen, 1983; Eisenhardt, 1989; Miller and Cardinal, 1994; Boyd, 1991). Between the two groups, there is another group of authors who have argued that rational strategy optimization is associated with high performance whether the environment is turbulent or stable (Andersen, 2000).

As with other businesses, competition in the banking sector means that banks are increasingly offering new and better services to their customers. Bank managers focus on their internal and external context, their missions and visions, and the evaluation of their strategies (Hector, 1991). According to Hopkins and Hopkins (1997), these activities correspond to the components of the strategy optimization process and the fact that bank managers are actively involved in these activities shows their (conscious or unconscious) consideration of strategy optimization and its impact on project productivity. Some authors who have studied strategy optimization in the banking sector have studied it on the basis of the phases of the classical LCAG model approach, using the term "dimensions." Notably, Ramanujam and Venkatraman (1987); Awino, Muturia, and Oeba (2013) in their studies, established seven dimensions of strategy optimization that can ensure the effectiveness of strategy optimization and lead to improved business optimization. Others have studied it by taking into account the intensity with which bank managers engage in strategy optimization (Hopkins and Hopkins, 1997), these authors showed as in other studies (Robinson and Pearce, 1983) that this intensity depended on managerial factors (expertise in strategy optimization and belief in a relationship between strategy optimization and performance), environmental factors (obstacles and change) and organizational factors (size and structure of the firm).

The financial indicators of optimization allow for the evaluation of the performance of companies in the banking sector in carrying out their business activities. These indicators refer to measures of productivity and profitability, such as ratios of outstanding deposits per employee, outstanding loans per employee, property insurance premiums per employee; and the following values: profit, ROI, ROA, ROS, and return on equity (ROE). These indicators can be complemented by other non-financial indicators, including customer satisfaction, service quality, and customer loyalty (Kaplan and Norton, 1999). Kargar and Parmell (1996); and Awino et al. (2013) used non-financial indicators, namely the ability of companies to solve a problem in a given situation, the ability to avoid mistakes, and the improvement of the budget implementation process. Hopkins and Hopkins (1997) in their study of banks used three measures of financial indicators of optimization: net profit or income, ROI, and increase in down payments. Awino et al. (2013) used both financial and non-financial indicators. Among the financial indicators, they chose gross profit, ROI, and ROA.

As shown above, the impact of strategy optimization on performance has still not been conclusive. Studies by authors such as Sapp and Seiler (1981); Whitehead and Gup (1985); Hopkins and Hopkins (1997); Robinson and Pearce (1983) show that the same conclusions can be made about companies in the banking sector. In the banking sector, the relationship between strategy optimization and firm performance has been well established. Sapp and Seiler (1981) in their study showed that planning banking firms perform better than non-planning banking firms. Following the same idea, Whitehead and Gup (1985) found that there were no differences in performance between planning and non-planning firms, and that planning firms also performed worse on some measures than non-planning firms. Robinson and Pearce (1983) showed that there was no relationship between strategy optimization and firm performance. Ramanujam and Venkatraman (1987); and Kargar and Parnell (1987) found that the intensity of the seven dimensions of strategy optimization ensured the effectiveness of strategy optimization and led to improved firm optimization. Following these studies, Kargar and Parnell (1996) conducted a study of small commercial banks examining the level of performance in relation to the intensity placed on the seven dimensions. They found that firms that focus on the seven dimensions of optimization perform better than those that do not. Following the same idea, Awino et al. (2013) found similar results. Hopkins and Hopkins (1997) showed that the intensity of strategy optimization by bank managers had a positive impact on the performance of banking firms. On the other hand, Gup and Whitehead

(1989) showed that banks that apply high-level strategy optimization have a significantly lower ROI than banks that apply the strategy optimization process informally. The inconsistency between these empirical results would be due to several of the same reasons mentioned above in the assessment of the causes of the inconsistencies in the empirical findings on the relationship between strategy optimization and performance.

The basic philosophy of corporate strategy is strategic fit, which implies two fundamental principles: the company must adapt to its environment to ensure its sustainability; the company must acquire a competitive advantage and then defend it to succeed. The first principle presents the genealogy of the concept of strategy optimization, from its foundation by the Harvard School to its apogee, embodied by the work of Porter, and its decline during the 1980s. This concept is structured around two key notions: SWOT and competitive advantage (Saïas and Métais, 2001). The first notion was presented in the first section, while the second structuring of strategy optimization forms the framework of the second section of our chapter. This section presents the theoretical approaches to competitive advantage that dominate the explanations of firm optimization in the field of industrial internationalization.

Porter (1993), quoted by Khamassi and Hassainya (2001), pointed out in the preface to this edition of his book that: "the paradigm of international competition has undergone a fundamental change in recent decades (...). Today, a new paradigm is gradually emerging in which international competition is actually played out at the global level. It is no longer domestic sales that enable companies to achieve economies of scale, but sales from the huge global markets. Competitive advantage no longer lies in static efficiency but in dynamism. But even if we have to think in terms of global competition, the sources of competitive advantage are primarily local. Porter's work (1980, 1985, 1990) on competitive advantage at the level of nations, firms, and clusters has motivated a great deal of interest in competitive advantage in the managerial literature, and a number of studies have built on his approach to the process of creating and maintaining competitive advantage. Thus, in the mid-1980s and much more so in the 1990s, the resource-based view (RBV) addressed the issue of creating competitive advantage. Its main authors argue that the company is made up of a set of resources, and that it is these resources that determine its performance. Thus, a firm can outperform its rivals if it owns or controls the resources that enable it to produce more economically or create more value for its customers (Grant, 1991; Barney, 1991; Conner, 1991). This section has two parts, the first reviews models dealing with the relationship between strategy optimization and competitive advantage from business portfolio models to Porter's competitive strategies. The second part of this section presents the further development of the original models of competitive advantage through the presentation of empirical results on the relationship between competitive advantage and performance; and the development of recent models based on the dynamic resources, competencies, and capabilities approach.

The basis of the strategy optimization process remains external and internal analysis and diagnosis, but gradually a number of consultancies, academics, and researchers have proposed increasingly refined analysis schemes to specify the fundamental criteria of competitiveness. This is the case of the Boston Consulting Group. Subsequently, other major consultancies, such as McKinsey or Arthur Doo Little (ADL), adopted similar methods, making them more qualitative (Chevallier, 1990). While portfolio analyses thus remain an important stage in optimization methods, other tools which make it possible to anticipate changes, to take better advantage of the conditions of the competitive field and to innovate are now being used. In particular, in the continuity of the activity portfolio models of consulting firms, Porter (1980, 1985) gives a very clear and exhaustive outline to corporate strategy, by integrating in a detailed and coherent way all the progress and contributions since the 1960s (Saï and Métais, 2001). More precisely, Porter's approach seeks to give the company a competitive advantage by analyzing its value chain with regard to the competitive forces in its market. The company's strategic positioning enables it to develop its competitive advantage and to be more competitive.

The Boston Consulting Group Matrix (1980) uses two strategic variables: market growth rate and relative market share. The basic idea behind this matrix is that it would be better for the company if a product had a larger market share and the market for goods grew faster for the company. Market growth measures market attractiveness and is based on the concept of the business life cycle. Relative market share is measured in relation to the nearest competitor or leader and indicates the competitive strength of the company as a result of a large experience effect. A higher relative market share corresponds to a higher cumulative production volume. This volume of production leads to lower unit costs resulting in higher profitability and thus a competitive advantage for the company. This model can also be used to analyze a portfolio of "country markets" in order to identify the most attractive export markets for the company. The method of the BCG 1 model consists of dividing the company into a number of strategic business areas (SBAs; a strategic business area is a segment of the company's environment within a business sector, which offers its own prospects for growth and profitability and which requires specific approach strategies on which the company has decided to act with a vision and approach of its own). For each of them, the relative share of the company in its market and the growth rate of the SBA must be assessed:

- The “stars” are SBAs that create significant financial resources to finance the needs generated by market growth. The company has a competitive advantage over its competitors in markets that still have interesting development potential. These SBAs balance or have a surplus of cash. They form the dynamic part of the business portfolio. The ‘stars’ are in the growth phase and are destined to become ‘cash cows.’ They contribute to growth and are self-financing.
- Dilemmas” are SBAs with a high growth rate, but where the company has not yet achieved dominant competitive positions. These SBAs require financial resources to ensure their development. They contribute to the growth of the company and require liquidity (LQ). The term “dilemma” refers to the strategic choice that the company has to make: either invest to make the SAD a star or divest. If the company refuses to choose, ‘dilemmas’ are likely to become ‘dead weight.’
- Cash cows” are SBAs where growth is reduced and the company has good competitiveness (strong competitive advantage). The leading position makes these activities important sources of LQ that can support other activities, but they must be replaced in the more or less near future because they no longer offer much development potential.
- The “dead weights” are aged SBAs in which the company has not been able to establish itself in terms of competitive advantages. They are characterized by a double absence of need and cash flow. They bring neither growth nor margin to the company.

This matrix provides a clear picture of the company’s position in the market. However, a certain number of factors specific to the export context (in particular the costs of entering markets, transport costs, product adaptation costs, inflation rates, exchange rates, etc.), which can influence the competitive position of companies by acting upwards on their selling prices and the dimension of risk (political, financial, commercial) are not taken into account. This is why BCG has proposed a second matrix which takes into account the competitive advantage based on the analysis of different variables and the competitive differentiation of the products analyzed. This model places more emphasis on the notion of competition because it seeks to take better account of changes in the competitive environment of companies in a more uncertain environment. The performance indicator is financial, i.e., profitability. This leads companies to adopt more targeted strategies for their production range. The adoption of a strategy is done under these conditions by analyzing the expected profitability for each project studied separately, taking into account the market share expected by the company (Lochridge, 1998). The BCG matrix is based on two variables:

- the importance of competitive advantage which represents the height of the strategic barrier that a competitor can create in relation to its rivals; and
- the number of ways to obtain competitive advantage, which represents the number of possibilities for competitive differentiation (price, advertising, service quality).

The company's activities are thus positioned in a double-entry matrix associated with the following four types of strategy according to the importance of the competitive advantage held by the company:

1. **Fragmented Activities:** In this case, competitive advantages are low, but there are many opportunities for competitive differentiation. There is no direct link between market share and the expected rate of return on the project. However, the company can exploit many opportunities for differentiation, even if these are small. The company's strategy will therefore have to be adapted on a case-by-case basis.
2. **Specialization Activities:** Competitive advantages and opportunities for competitive differentiation are strong. Activities will be profitable if the degree of specialization is appropriate given the company's competitive advantage over its competitors. Indeed, there are many substantial opportunities for differentiation. Growth in the company's market share should not always be the goal.
3. **Volume Business:** Competitive advantages are strong and opportunities for competitive differentiation are limited. It is a strategic approach based on cost domination. The higher the market share, the higher the production volume, the more profitable the investments. The company can therefore benefit from significant competitive

advantages. The company must therefore implement an offensive strategy to conquer the market.

4. **Dead-End Business:** In this case, there are few ways to gain competitive advantage and relatively little scope for competitive differentiation. Profitability does not vary regardless of the company's market share. The company's choice is very limited. The company may decide to exit the market if the observed level of profitability is below the minimum decided by the company.

The strategic choices of the company are therefore based in this model on a more pragmatic and less global approach than the first. It can be applied differently depending on the nature of the project analyzed and its expected degree of profitability. The McKinsey model is based on the approach proposed by the Boston Consulting Group but has the advantage of offering a wider range of strategic choices for the company. It is based on the representation of the company's SBAs in relation to two dimensions: medium-term market attractiveness and competitive strength or position. The establishment of its matrix requires the identification and analysis of the external factors that control the attractiveness of the sector to which the activity under consideration belongs; it also requires the identification and elucidation of the internal factors whose degree of control is the basis of the company's competitive strength. These two dimensions are assessed on a three-position scale (strong, medium, weak). The positioning of a DAS on the grid suggests priorities for the allocation of resources: invest when attractiveness and competitive strength are both high, make a profit or disinvest in the opposite case, and proceed selectively for intermediate situations; and this by seeking to concentrate resources in the most attractive sectors or the fields in which the company has significant competitive skills (Martinet, 1983). Strategor (2005) presents the attractiveness-attention matrix, composed of nine cells. This matrix makes it possible to represent the company's activities according to their attractiveness (strong, medium, weak) and their competitive position (strong, medium, weak). The strategic recommendations resulting from McKinsey's analysis then make it possible to define three main types of possible strategic orientations. The proposed strategic orientations, according to the position of the activities in the quadrants, can be grouped into three categories for the export markets and in relation to the advantage they represent for the targeted countries:

- The so-called strategic markets (leaders in a growing market) on which the company must concentrate its efforts and investments to further increase its distinctive skills. These markets are located in the "development" zone;
- Tactical" markets (good position in a mature market) for which the strategic choice is difficult (divest if the growth of the sector does not seem sustainable or maintain its position by remedying its weaknesses and/or specializing in its distinctive competence). These markets are located in the "hold" zone; and
- So-called 'losing markets' (marginal position in a low-growth market) which should lead the company to optimize the harvesting of profits in the short term until the decision is made to withdraw from these markets. These markets are located in the "abandonment" zone.

The McKinsey model, which is very complex to implement, remains a technique reserved for companies that have the necessary means to manage a large mass of information. The ADL model is structured on the basis of two variables: the competitive position (the company's assets or competitive position) and the maturity of the business (attractiveness of the sector). The method proposes the evaluation of the competitive position of a company in relation to the set of quantitative and qualitative assets that it holds compared to its competitors. To do this, the ADL method uses the notion of key success factors of the company in the sector grouped by the DAS. These are three factors, namely supply factors (integration, privileged long-term contracts, possibilities of indebtedness, cost of indebtedness and cost of short-term credit, cost of labor), production factors (capacity and flexibility of the production units, productivity of the units, level of production costs, degree of mastery of particular processes, advantages linked to the location of the production units) and marketing factors (image of the products, quality, and importance of the distribution network, extent of the range of products, facilities, and credit granted to the clientele) (Martinet, 1983). These three operations constitute three systems with different mechanisms and at the level of which the firm may or may not benefit from advantages over its competitors. The sum of the advantages or disadvantages allows it to have a level of profitability below or above that of its sector and to have greater or lesser possibilities of choice. Each key success factor is rated and can be weighted according to its relative importance in the overall assessment. The final score places the competitive position in five positions with characteristics specific to each for the company (Gervais, 1995):

- Dominance is characterized by the fact that the company is able to control the performance or strategic behavior of its competitors and therefore has a wider choice of strategic options than its competitors;
- The strong position: in this case, the company is able to pursue a certain policy of its choice without jeopardizing its long-term position;
- The favorable position is characterized by the capacity of a company to have assets that can be exploited for the implementation of certain strategies, for example internationalization, and it thus has a very good chance of being able to maintain its position in the long term;
- An unfavorable position means that the company is performing sufficiently well to warrant a continuation of operations, and is usually maintained by the tolerance of larger competitors; and
- Marginal position means that the company has an unsatisfactory performance with room for improvement, it may also have the characteristics of a better position but has a weakness and needs to improve its position to have a chance of long-term survival.

Business maturity (sector attractiveness) is an indicator for assessing the attractiveness of the sector. The concept of business maturity is a transposition of the concept of product life cycle. Each sector would go through four phases with specific competitive and financial conditions:

1. **The Start-Up Phase:** This is the period of innovation during which a new product or technology must be introduced by substitution for existing elements on the market.
2. **The Development Phase:** It leads to the choice of expansion strategies in order to maintain or improve its competitive position and image, which requires greater capacity to increase production and ensure distribution.
3. **The Maturity Phase:** In this case, positions are established and developments can only be made at the expense of existing competitors.
4. **The Decline Phase:** It corresponds to further rationalization efforts in order to lower costs while preparing for a withdrawal from the sector.

By combining the competitiveness and maturity factors, we arrive at a grid on which the company's SBAs can be positioned and from which the following four major strategic orientations can be deduced:

- Natural development, which is the need for the company to follow the development of the market;
- Selective development, which encourages the company to focus on its most competitive production range;
- The reorientation of activities based on the skills held by the company; and
- The outright abandonment of activities that the company has no reason to retain.

The ADL model is a good instrument to assess the competitive position of a company with qualitative criteria. However, it is less easy to use because of the more random positioning of activities. It refers to the life cycle, a notion that is difficult to grasp in reality. Indeed, a product in decline can be relaunched on the same market or on other international markets. Similarly, a new product may see its growth dynamic destroyed by unexpected obsolescence.

According to Ricardo's (1817) theory, trade depends on differences in labor productivity between nations, which he attributed to climatic and technological differences favoring certain sectors in a given country. The technology gap theory, which is a modern version of Ricardo's (1817) theory, argues that nations export in those sectors where their firms acquire a technological lead. Technology thus becomes a very important factor in the export process of firms and thus determines their competitiveness on the international market. Heckscher (1919); and Ohlin (1933), quoted by Lassudrie-Duchêne and Unal-Kesenci (2002), have renewed the theory of comparative advantage based on the principle that all nations have equivalent technology but are unequally endowed with factors of production: land, labor, and capital. All these factors of production allow nations to gain a relative advantage in activities that make intensive use of the factors they possess in abundance. The authors assume that countries will export products with an abundance of factors and import goods that do not use factors or that are produced in small quantities (Ohlin, 1933). The logic is that the relative costs of products are explained by the relative costs of factors, which in turn are derived from the utility and productivity of these factors, which are themselves dependent on the relative quantities of factors available to countries or regions. In a word, the quantity of factors determines the acquisition of a competitive advantage (Lassudrie-Duchêne and Unal-Kesenci, 2002).

Porter (1993) finds this theory of trade unrealistic as it is based on unfounded assumptions such as the non-existence of economies of scale, similarity of technologies, undifferentiated products, fixed factors of production of a nation and the fact that factors of production such as labor and capital are non-transferable between nations. However, there are sectors, such as the high-tech sector employing highly skilled personnel, where relative advantage based on factors of production does not explain trade. Porter's (1993) theory of competitive advantage goes beyond relative advantage and assumes that competition is no longer static but dynamic and constantly evolving, and must therefore take into account improved production methods and technological innovation as well as changes in a nation's internal and external environment.

Competitive advantage is achieved when a company discovers a new and more efficient way of working in an industry than its competitors. It will in some cases shape the industry and redefine the existing processes in it. And the company that manages to achieve this industrial revolution through its know-how has a competitive advantage (Toyo Ackah, 2010). According to Porter (1993), the main competitive advantages are built in the home country of the companies. In order to maintain a competitive advantage, a nation or a company needs to implement sustained investment and modernization policies, and this could mean abandoning some advantages in favor of other more strategic and promising ones. For the company, it is a question of acquiring a unique position thanks to a particularity that gives it precedence over any other company. This position of strength will be achieved through a distinctive element which, if perceived positively by consumers, will constitute a competitive advantage. The company must therefore seek to make the most of its competitive advantages in order to differentiate itself from its competitors in a sustainable manner. Supporting Porter's idea, Barney (1991) describes the concept of competitive advantage as a creative strategy developed by the company, which guarantees the company an advantage as long as such a strategy is not yet implemented by any other actual or potential competitor.

According to Porter's approach, the productivity of projects is determined by the structure of the industry in which it operates and by the competitive advantages obtained by the firm in a given sector (Grant, 1991; Porter, 1991). This sub-section will present the five forces that determine profitability and the related strategies, then the elements that determine the chances that companies in a given country or industry have of building a competitive advantage over competing companies in other countries, and finally the sources of the company's competitive advantage that Porter develops using the value chain. Porter's (1980) analytical framework refers to an industry, i.e., a set of companies competing directly in a market. In any industry, the competitive game results from five forces whose intensity determines an average level of profitability influencing the attractiveness of that company:

1. **Competitors in the Sector:** This is the rivalry between existing firms. It depends on structural and behavioral determinants. First of all, the size and number of competitors are important: the more concentrated the sector, the less competition there is.
2. **Potential Entrants:** This is the permanent threat of new players wanting to enter the sector. The existence of potential competitors who could enter the sector influences the degree of competition in the sector.
3. **Suppliers:** These exercise varying degrees of power over the firms in the sector. This power increases with the size and concentration of the suppliers compared to the more atomized actors-buyers in the reference sector. This power also increases with the differentiation of the inputs sold by the suppliers, and with the transfer costs that their customers (the actors in the reference sector) would have to bear in order to change input and/or supplier.
4. **Customers:** The bargaining power of the players is becoming increasingly important. This power also increases for standardized products that can be purchased from other manufacturers, especially if the transfer costs are low.
5. **Substitutes:** The existence of substitutes that buyers see as fulfilling the same functions as the reference product puts pressure on the industry in question and the value it can create. This threat is all the greater the higher the relative price-quality ratio of the substitutes and the more profitable the sectors. Innovation can often reinforce or reduce this threat. The aim of this model is to identify the forces at work in the company's competitive environment and their intensities, to develop strategies to obtain a competitive advantage and thus to achieve higher profits than the average for the sector and consequently a higher performance. In addition to identifying the sources of competition, it will be necessary to determine and prioritize their respective capacity to influence the company's performance.

Porter (1985) elaborates on the problems faced by firms in a competitive world. For him, there are three basic strategies for dealing with the five competitive forces, depending on the strategic target of the firm (whole industry or particular segment) and on the strategic advantage the firm may have (pure cost advantage or product uniqueness advantage). The three strategies are: global cost domination, differentiation, and concentration. The company has to make only one choice among the basic strategies if it wants to gain a competitive advantage. A structural change may have influenced the strategy chosen by the firm and the firm's performance:

1. **Cost Dominance:** This strategy assumes that the firm has the lowest costs, it allows the firm to fight against current and future competition and substitutes. It is only possible if the firm has a large market share and is capable of great management rigor. They correspond to volume strategies generally linked to the existence of strong experience effects.
2. **Differentiation:** This implies that the products offered are perceived by the customer as being unique, which makes it possible to charge more for them. It is a question of the company standing out with specific services, imposing its image, and building customer loyalty. This is the classic strategy of the big brands that rely on innovation, design, and advertising to make a difference. Successful differentiation leads to higher profits than competitors.
3. **Concentration:** It aims to define a narrow strategic target and to satisfy the needs of this market segment better than others. The company cannot compete on costs, and cannot differentiate itself on the whole market. It therefore restricts its target and adapts closely to a limited customer segment. It then concentrates its efforts on a limited strategic target, offering a unique advantage to this target. This is a niche strategy.

The international competitiveness of a country's industry is the result of four elements that determine a nation's competitive advantage (Porter, 1980). These determinants interact and reinforce each other in a dynamic process. This is Porter's diamond. These are the factors of production, demand, upstream, and related industries, and the strategy-structure-rivalry of the firms in the sector, to which Porter adds chance and STATE. All these factors determine the chances that companies in a given country and in a given industry have of building a competitive advantage over competing companies in other countries. These are the factors of production namely labor, land, natural resources, capital, and infrastructure. Porter (1980) distinguishes between specialized factors of production (generate competitive advantage) that are created and not easily imitated by other firms and general factors that can be obtained by any firm (do not generate significant competitive advantage). Specialized factors are skilled labor, capital, and infrastructure while general factors are unskilled labor and raw materials. The relative importance of each factor varies across sectors. Globalization has increased mobility across borders for several factors of production, including human resources, capital, and knowledge and technology. It follows that local availability of a resource is an advantage if the resource remains in the country. According to Porter (1980), resource scarcity often actually helps countries to become competitive. Resource abundance generates waste, and resource scarcity stimulates innovation. These countries are forced to innovate to solve the scarcity problem.

1. **Demand:** According to Porter (1980), a demanding domestic market is an important competitive factor. Firms facing this type of market are likely to sell superior products and close proximity to these consumers allows them to better understand the needs and desires of customers. If the nation's demanding values extend to other countries, local firms will be competitive in the international market. For companies, it will be necessary not only to seize the opportunity of demand but also to satisfy it.
2. **Upstream and Related Industries:** According to Porter (1980), for a given country, a set of well-performing related and supporting industries is important for the competitiveness of firms. This includes suppliers (upstream industries) and related sectors (downstream industry). This usually takes place at the regional, not national, level. Furthermore, the existence of related industries in a given country frequently leads to the emergence of new competitive capacities in the downstream industry.
3. **Business Strategy, Structure, and Competition:** Areas of competitiveness vary according to a long- or short-term perspective of national capital markets which affect business strategy. Thus, countries with a short-term perspective will tend to be more competitive in sectors where investment is short-term. Countries with a long-term perspective will tend to be more competitive in sectors where investment is long-term. According to Porter (1980), the best management practices vary across sectors. Some countries may be oriented towards a specific management method. These will tend to be more competitive in sectors where that management method is

appropriate. Firms that have international competitive advantages also have them domestically. Intense national competition stimulates innovation and progress as these rivals, operating under similar basic conditions (cost, local market access), will seek to develop other types of competitive advantage.

4. **The Role of STATE and Chance:** Porter also highlights the role of STATE and chance in his model. The decision-maker can positively or negatively influence each of the four determinants and thus modify the formation of national competitive advantage. For example, government policy can affect the factors through subsidies, financial policies, education. Similarly, random events can positively or negatively affect the competitive position of a company. These can include technological inventions or innovations, wars or destruction, or sharp changes in exchange rates. In sum, Porter argues that firms will be better placed in international markets the more pressure they are under from their customers (demand) and competitors (rivalry). On the other hand, in developed economies, globalization limits the possibilities of differentiation by 'factors' – low wages and abundant and cheap raw materials. The companies that fare best in the globalized economy are those that manage to secure a competitive advantage, at the national level, in at least three of the four 'determinant' areas of the diamond.

Porter (1985) analyzes the sources of a firm's competitive advantage using the value chain. The value chain is the decomposition of the different operations performed by a firm in a given industry to produce a product or provide a service, each of which contributes to the value of the product or service. It is by managing its value chain as a system that a company can gain competitive advantage. Moreover, a company's value chain is the basis of its competitiveness in a given industry. The value chain highlights those activities that have a real impact in terms of cost or differentiation from competitors.

The company is broken down into primary and support activities. The primary or main activities contribute to the creation of value for the company and include: supply logistics (reception, stock, and distribution of raw materials), production/manufacturing (transformation of raw materials into finished products), marketing (collection, stock, and distribution of the finished product to the customer), marketing, and sales (making the product known and selling it to consumers), and finally service (after-sales service, installation, maintenance, training, etc.), Assistance or support activities that allow the primary activities to function properly by increasing their effectiveness and efficiency. They include the company's infrastructure (administration, planning, quality, etc.), human, and technological resources (R&D) and economic resources (purchasing). These activities must be carried out at the best possible cost so that the profit margin is substantial. The value chain makes it possible to analyze the strategic potential of the company at the origin of the greatest value, i.e., the key factors of success which are explanatory of a competitive advantage. Porter's approach focuses on the impact of the environment (industrial structure) on the position of the firm's competitive advantage and takes very little account of the influence of the particular characteristics of the firm on its competitive advantage (Porter, 1990). It is this criticism that has given rise to another school of managerial thought based on the resources of the firm.

In this subsection, after presenting the results of some empirical studies that have shown a significant relationship between competitive advantage and firm performance, we will present the models that have been developed in the further development of the basic models of competitive advantage and firm performance. These are the resource and dynamic capability models. The resource model is firstly a reaction to the strategic models, notably the BCG, which have focused on external growth and environment-centered approaches. Secondly, it is a critique of Porter's competitive approach which considered that the structure of the industry determined the rules of competitiveness and influenced the potential value strategies of the firm (Teece, Pisano, and Shuen, 1997). Subsequently, drawing on the resource approach, the dynamic capabilities model considers the ability of the firm to renew its skills in order to keep pace with a changing environment. Several empirical studies have shown that there is a significant relationship between competitive advantage and firm performance (Barney, 1991; Ma, 1999, 2000; Fahy, 2000; Bowen and Ostroff, 2004; Wang and Lo, 2003; Rose, Abdullah, and Ismad, 2010). According to Barney (1991), the resources acquired or controlled by the firm can support the design and implementation of strategies that would improve its effectiveness and efficiency and hence its superior performance. According to Ma (1999), competitive advantage is at the root of all superior performance, and understanding the process of achieving competitive advantage is important for managers who are responsible for the firm's success and long-term survival.

Ma (2000) remarked on competitive advantage and its relation to performance: competitive advantage is not performance; it is both a relational concept and a specific context. In addition, Ma (2000) used three forms of relationship between competitive advantage and performance: competitive advantage leads to superior performance; competitive advantage can be considered without performance and performance exists without competitive advantage. Fahy (2000) argues that achieving a sustainable position of competitive advantage can lead to superior performance, as measured by financial indicators. Firms should therefore focus on their managerial strategies to achieve and maintain a competitive advantage over their rivals; hence the position of competitive advantage will lead to superior performance. Bowen and Ostroff (2004) investigated the significant relationship between the strategic human resource management system, work climate and project productivity as measured by non-financial indicators. Their aim was to explain whether the intensity in the strategic human resource management system explains how individual employee attributes can affect organizational effectiveness and firm performance. Some studies have shown a significant relationship between competitive advantage and firm performance by focusing on both financial and non-financial dimensions of optimization (Wang and Lo, 2003; Falshaw et al., 2006; Rose, Abdullah, and Ismad, 2010). Others have used only financial performance (Fahy, 2000) while others have conducted the studies with non-financial performance (Bowen and Ostroff, 2004; Neely, 2005). The majority of these studies conclude that there is a significant relationship between competitive advantage and firm performance.

In the resources and competencies approach, the company is no longer conceived as a portfolio of products and markets, but as a portfolio of resources. It is the resources and skills that the company possesses that now determine the strategy. Competitive advantage is to be sought internally. Resource theory started from the observation that performance is heterogeneous within an industry: in the same sector, why are some companies sustainably more profitable than others (Depeyre, 2005)? Developed in the mid-1980s by Wernerfelt (1984); Rumelt (1984); and Barney (1986), the RBV has become one of the contemporary approaches in the research field of sustainable competitive advantage analysis. One of the objectives in this research is to try to explain the sustainable performance of companies. RBV theory is based on the fact that firms compete on the basis of their resources and capabilities (Peteraf and Bergen, 2003). RBV focuses on firms' resources as a determinant of competitive advantage and firm optimization. The main assumption is that sustainable performance can be achieved from sustainable competitive advantage (Barney, 1997; Powell, 2001). Resources are tangible (physical, financial) and intangible (patents, brands, customer databases, supplier relationships, management systems) assets held by a firm that enable it to design and implement strategies to improve its performance. Hofer and Schendel (1978) classified resources into five categories: financial, human, physical, and technological. To these categories, Grant (1991) added a sixth, reputation. In analyzing the sources of competitive advantage, the RBV approach is based on two fundamental assumptions (Barney, 1991; Peteraf and Barney, 2003). The first assumption is that firms may be heterogeneous in terms of the resources and capabilities on which they base their strategies. Resource heterogeneity means that resources are not the same for all firms. For if they are, they cannot contribute to the achievement of competitive advantage. Resource heterogeneity is therefore a necessary condition for competitive advantage. The second assumption is that heterogeneity may persist over time because the resources used may not be imperfectly mobile across firms, due to the diversity of players in the industry.

However, the assumptions of heterogeneity and immobility are not sufficient conditions for sustainable competitive advantage. According to Barney (1991), the firm's resources must have the following properties to be a source of sustainable competitive advantage: value-creating, rare, inimitable, and non-substitutable (VRIN). Competencies refer to the organizational ability to deploy resources in combination to achieve a goal. They are resources that are not easily traded on the market because they are the result of individual and collective learning within the organization (Grant, 1991). Priem and Butler (2001) believe that the RBV approach is tautological in that it amounts to saying that a firm has a competitive advantage because it has a competitive advantage (inimitable own resources). Barney (2001) has attempted to address these criticisms and has suggested that more empirical and dynamic approaches are needed to confirm or refute the parameterization of the elements of the approach.

Dynamic capabilities model are the managerial skills needed to coordinate, harmonize, and strategically deploy available value-creating resources. These capabilities can be sources of competitive advantage (Teece et al., 1997). According to Johnson et al. (2005), the development of strategic capabilities leads to long-term competitive advantage. Grant (1991) states that resources are the source of a firm's capabilities while capabilities are the source of its competitive advantage. Winter (1982) established that a firm's know-how is embedded in organizational and strategic routines defined as the repetitive activities it develops when using specific resources. These thoughts were extended in the case of the RBV work of the 1980s and 1990s which developed an internal dynamic analysis called 'dynamic capabilities.' The element of 'dynamic' concerns the company's capacity to renew its skills in order to be in line with a context that is evolving more and more rapidly, where technologies lead to increasingly frequent innovations, accelerating the obsolescence of offers and facilitating substitutions. The dynamics of capabilities are then essential for the firm to establish a long-term competitive advantage (Cao, 2012). Building and developing competitive advantage requires the firm to exploit its current specific internal and external capabilities but also develop new ones (Wernerfelt, 1984). Teece et al. (1997) have established that, dynamic capabilities allow the firm to integrate, build, and reconfigure their resources and competencies and therefore maintain performance in a changing context. It appears that, under relatively stable conditions, it is possible to build competitive advantage from sustainable skills and resources. In this case, the RBV approach provides a relevant framework for strategy analysis (Cao, 2012). However, in a more turbulent context, it is essential to focus on the ability to change, evolve, and learn, i.e., to achieve dynamic capabilities (Johnson et al., 2005).

The models we have just analyzed make it possible to highlight interesting results concerning the links between the strategy optimization and the performance of companies in an open economy. We started from the traditional models of the Harvard school to the recent developments of the dynamic resources and capabilities approach, taking into account the current international competitive context. It appears that with the push for economic openness, firms need to internationalize to face certain challenges in order to outperform their rivals. It appears that it is important for each company to identify the factors and resources that enable it to have a sustainable competitive advantage in order to be more competitive internationally. It is a question of acquiring a unique position thanks to a particularity that gives it precedence over any other company. The company must therefore be able to renew its skills in order to keep up with the ever-changing context, where technologies lead to increasingly frequent innovations. The authors of managerial theories believe that innovation is the key to the competitiveness of companies and that it concerns everything that creates or improves its resources, processes, or value. Innovation therefore determines the ability of companies to maintain sustainable competitive advantages, to increase their competitive position and consequently to improve their performance on the international level.

KEYWORDS

- **corporate social responsibility**
- **dynamic resources**
- **globalization**
- **return on assets**
- **return on investment**
- **return on sales**

CHAPTER 4

The Role of Performance and Profitability in Optimizing Strategies

Project productivity plays a crucial role in the optimization of the firm's strategies. Indeed, performance primarily measures the achievement of strategic objectives and the organizational objectives derived from them. The strategy is assessed by comparing the strategic objectives with the actual results achieved. Recent studies have shown that financial performance is no longer sufficient to assess the performance of a company. Companies must therefore measure their progress on the basis of a more global performance that includes the economic, social, and environmental dimensions (Amedzro St-Hilaire, 2022). In order to effectively improve its overall performance and to ensure that it is superior to the competition, companies are increasingly using benchmarking. This approach consists of identifying, analyzing, and adapting, by adapting them, the practices of the best performing organizations.

In business economic and value-based management literature, value-based management is considered to be performance management and must be consistent with the overall performance sought. Managing performance means planning and implementing actions to correct a gap between the objective and the result. It means ensuring that the objectives set are achieved, but also seeking to progress. For a company, it is a question of having a performance measurement system that can help its managers to evaluate the effectiveness of the implementation of the strategy optimization process. It is therefore important to determine the tools for measuring performance. Performance evaluation can be done in relation to the objectives defined by the company. There are several performances controls tools, the most common of which are strategic scorecards and normative models. The studies carried out on this subject show that in the face of constraints, contextual threats

Value-Based Management in an Open Economy: Optimizing Strategy to Improve Business and Performance. Prof. Post-Dr Walter Amedzro St-Hilaire (Author)

to companies and the opening up of markets, value-based management and global optimization tools play an active role in the implementation and realization of the strategies defined by the company. This chapter is divided into two sections, the first presents the process of optimization, while the second section shows the role that global performance plays in the optimization of the company's strategies.

More and more companies are recognizing their social and environmental responsibility. Corporate responsibility (CSR) is now recognized as the contribution of companies to their sustainable performance. As Amedzro St-Hilaire (2011), points out, CSR is a concept that designates the voluntary integration by companies of social, societal, environmental, and governance concerns into their strategy, management, and relations with their stakeholders. CSR is a way for the company to strengthen its profits in the short, medium, and long term. Bowen (1953), one of the pioneers of the concept, defined CSR as a set of obligations leading to a set of policies, decisions, and courses of action consistent with the objectives and values of society. Friedman (1970) disagreed, arguing that: "the social responsibility of business is to make a profit" and that anything else was an undue drain on the substance of the enterprise, on the income of its shareholders and an obstacle to the proper functioning of capitalism. In 1971, the Structure for Economic Development (CED) further developed the concept of CSR by referring to three concentric circles: "the first includes the basic responsibilities for the fulfillment of the essential functions of the company, relating to production, employment, and economic growth; the second, encompassing the first, includes a broader notion of responsibility, with sensitivity to changes in society and its expectations, with, for example, consideration of issues of environmental protection, social relations or consumer information; finally, the third takes into account the exercise of emerging responsibilities, serving to improve the environment, such as targeted job creation for the benefit of particularly disadvantaged populations' (Germain and Trébucq, 2004). Carroll (1979) defines the notion of societal responsibility as 'what society expects of organizations in economic, legal, ethical, and discretionary matters.' He was the first to formulate the idea that social and economic objectives are not incompatible, since they are both part of a common goal: 'global social responsibility.' Epstein (1987) argues that CSR refers to the different consequences resulting from organizational decisions, which affect different stakeholders. Garriga and Mele (2004) classify CSR theories into four major streams (cited by Lefebvre and Lefebvre, 2012):

- Instrumental theories that consider CSR as a strategic tool to generate profits (Friedman, 1970) through the maximization of shareholder profit (Jensen and Meckling, 1976), the implementation of competitive advantages (Porter and Kramer, 2006), or the implementation of a specific business policy (McWilliams and Siegel, 2001);
- Political theories that conceptualize CSR as a way for companies to bring the social weight of their responsibility to bear (Davis, 1960), with the activity of the company seen as part of a social commitment that takes into account the social rights and obligations of the entrepreneur;
- Integrative theories that see CSR as a way for companies to respond to society's demands by acquiring social legitimacy and recognition through their management actions (Sethi, 1975), public accountability (Preston and Post, 1981), stakeholder management (Ogden and Watson, 1999), and the optimization of social responsibility (Carroll, 1979); and
- Ethical theories present CSR as an ethical requirement to contribute to a 'good society,' by protecting workers' rights (Cassel, 2001), taking measures for sustainable development (Van Marrewijk and Were, 2003), and thus contributing to the 'common good' (Velasquez, 1992).

The concept of performance has evolved from an economic to a multidimensional aspect. Financial performance consisted of achieving the profitability desired by the shareholders with the turnover and market share that preserved the sustainability of the company. In 1995, a report by the Conference Structure of Canada made several criticisms of performance based on financial indicators and concluded that strategy-oriented performance measurement systems should measure both non-financial and financial results. Thus, while financial data provide some satisfaction, there is room for improvement in the degree of alignment between strategic priorities and non-financial optimization measures (Waterhouse, 1999). In recent years, performance has tended to be approached in a more holistic way than simply assessing profitability for the company or for the shareholder. Performance now takes into account the company's social responsibility towards all its stakeholders. Corporate responsibility is no longer limited to shareholders alone, but includes other stakeholders (associations, NGOs, trade unions, customers, suppliers, etc.).

These new actors demand to be heard and this listening becomes a vital target for the performance and sustainability of companies. Based on Baret's approach (2006); Renaud and Berland (2007) consider that sustainable development highlights three objectives pursued by the company: the economic objective (creation of wealth for all through sustainable modes of production and consumption), the ecological objective (conservation and management of resources) and the social objective (equity and participation of all social groups). "The principle of sustainable development is to balance these three dimensions to avoid the pursuit of one objective being to the detriment of the other two" (Renaud and Berland, 2007). Hence the emergence of the concept of global performance. Clarkson (1995) discusses a results-based approach that focuses more on assessing social and environmental optimization in relation to financial performance. He deduces that social performance comes only from effective stakeholder management while financial performance is linked to resources, strategy, and industry structure. Baret (2006) defines overall corporate performance as the aggregation of economic, social, and environmental performance.

For Germain and Trebucq (2004), the global performance of companies is formed by the combination of financial optimization, social optimization, and societal optimization. From this perspective, Glaesner (2013) in his report refers to the notion of "global responsibility" as: "the set of economic, social, societal, environmental, and governance commitments that an organization, whether public or private, adopts, in the most concerted and open manner possible, to deploy an integrated strategy of sustainable performance, relevant, and mobilizing for its shareholders, customers, employees, and the territories in which it operates." In order to seek and maintain a global performance of companies, the social, environmental, societal, and governance dimensions can and must become a strategic driver. Furthermore, stakeholders must be mobilized in an effective, loyal, and sincere manner, within the framework of a renewed and expanded social dialog (Glaesner, 2013). In order to improve their performance effectively and sustainably, companies are increasingly embarking on benchmarking operations. By undertaking a benchmarking exercise, the company often hopes to have an objective view of its current level of performance, which should then enable it to position itself in relation to its most serious competitors.

Benchmarking is a process of systematically seeking out the best practices and innovations in a given activity with the aim of adopting, adapting, and applying them to improve the company's performance and to ensure superiority over the competition (Hermel and Achard, 2007). For Lebas

(1995), performance only exists if it can be measured and this measurement is not limited to the knowledge of a result. In this case, the results achieved are evaluated by comparing them to the desired results or to benchmark results (Bouquin, 2004). In this case, the evaluation of optimization can be likened to benchmarking. Benchmarking thus consists of identifying a standard or a paragon, i.e., a model with which to compare oneself, in order to close the performance gap. Benchmarking is defined as a continuous and systematic process for the evaluation of organizational practices recognized as representing best practices for optimization (Spendolin, 1992). According to Camp (1992), benchmarking is the search for the best-performing methods for a given activity, to ensure superiority. Brilmane (2003) defines benchmarking as the process of identifying, analyzing, and adapting, by adapting them, the practices of the best performing organizations in the world in order to improve the performance of one's own organization. Benchmarking is a lever for optimizing innovation in the company, because it is not only a method of analysis that enables one to benchmark oneself by taking inspiration from the best references, but also a state of mind and a management style (Economic Problems Review, 2006).

There are different types of Benchmarking, each company must choose the one that suits it best according to its current situation, its ambitions, and its expected results. There are several types of benchmarking, Camp (1992) classifies them into four categories: internal, functional, generic, competitive:

1. **Internal Benchmarking:** This involves comparing processes, products, or services within the same organization. This involves researching and collecting best practices and knowledge held internally by the company or held by other companies, whether competitors or not. This practice has advantages and disadvantages. The advantage is characterized by the greater ease with which confidential information can be shared within the same organization. It is limited in that the company cannot expect to highlight highly innovative practices. The expected performance improvement is about 10%.
2. **Functional Benchmarking:** This is a comparison between non-competing organizations in the same sector of activity to identify innovative techniques. The absence of competition and economic barriers allows for a thorough and detailed approach; the exercise stimulates openness, creativity, and innovation; for organizations with a high degree of similarity, good practice is more quickly

transferable. However, the benefits are limited to the sector concerned and its scale factor (capacity, size, etc.). The expected improvement in performance can exceed 35%.

3. **Generic Benchmarking:** This is the comparison between organizations in different sectors on processes or working methods. Generic benchmarking is the most powerful and beneficial, but it is more difficult to implement; indeed, generic benchmarking is the one that allows one to compare oneself with practices that are not specifically adapted to the sector to which one belongs. This benchmarking does not concern the competition and is therefore a source of innovative ideas (Costa, 2008). The advantages are multiple: partnership without confidentiality constraints, source of innovative ideas, long-lasting relationships based on a reciprocal and permanent need for information. The few remaining difficulties are linked to a greater difficulty in adapting practices that come from a different sector. Here too, the improvement in performance could be as much as 35% or more.
4. **Competitive Benchmarking:** This is a specific comparison with competitors on the product, method, or process. It is therefore necessary to compare with the best of the competitors on the market and for this type of comparison between two directly competing organizations, it is often more difficult to obtain information. Competitive benchmarking involves obtaining data of all kinds, information, and intelligence, so it can be considered a logical extension of competitive intelligence (Costa, 2008). Competitive benchmarking is fairly easy to use at the product level (it is enough to buy them), but more delicate at the functional level (it requires know-how and mutual respect between partners). On the other hand, it is for this type of evaluation that it is most difficult to obtain information because it is limited by confidentiality. The expected improvement in performance is about 20%.
5. **Other Types of Benchmarking:** These are mentioned by Lepoivre (2005): Strategic Benchmarking which consists of analyzing and adapting winning strategies with partners with whom the organization already has an established collaboration or a leading organization; cooperative benchmarking which is the logical continuation of strategic benchmarking, it is a question of a company being better than its partners and staying that way; process benchmarking consists of analyzing and adapting critical processes, knowing that each process corresponds to measurable inputs and outputs; organizational

benchmarking consists of improving the activities that have a strong impact on the organization. The aim is to highlight internal shortcomings and malfunctions. Mann and Kohl (2010) state that all types of benchmarking can be practiced formally or informally, thus highlighting two types of benchmarking: formal benchmarking and informal benchmarking.

Informal benchmarking is defined as an unstructured approach based on learning from companies in the same field. In this case, benchmarking also helps to position the company's strategy in relation to the market leaders in the industry. In this way, a company will satisfy its customers more, because it will meet the requirements of the market. Formal benchmarking is practiced consciously and systematically by companies. It is divided into two categories: "performance" benchmarking and "best practice" benchmarking. Performance benchmarking compares the level of performance of a specific process in order to identify opportunities for improvement. It can therefore be useful for a company that wants to show, for example, its dominance in a market in terms of performance compared to its competitors. Benchmarking of "good practices" for a company consists in observing, understanding, and learning from competitors (discovery of "good practices") by adapting in order to increase the performance and productivity of its company, which then becomes more competitive. It is therefore not a question of reproducing the processes of others identically. It is an opportunity for a company to learn to analyze its strengths and weaknesses and to adapt.

The aim of the benchmarking approach for a company is to bring it to have the ambition to progress, to open up to the outside world, to improve a certain number of points in its organization and to make a significant advance in its performance. There are many different methods for implementing a benchmarking approach, but they differ only in the details. In this study, we will mention four models used by four different companies, which we group together in Table 4.1.

These models have similar steps, the differences come from the order of execution of these steps or from the decomposition of certain steps into substeps. The steps of the process follow the same logic but are not as strictly determined. They follow each other according to the analysis of the situation, the search for the causes of a problem, the collection of information, the search for partners involved in the improvement operation and the implementation of the revised process. Despite the diversity of models, Benchmarking must be performed in a unique way according to well-established phases.

TABLE 4.1 Benchmarking Approaches by Companies

Steps	Description by Company			
	Motorola	**Bristol-Myers and Baxter**	**AT&T**	**Rank Xerox**
1	Deciding what to benchmark	Determine to which function benchmarking can be applied	Identify the purpose of the benchmarking	Decide on the subjects on which the benchmarking can be focused
2	Find a company for benchmarking	Identify the key performance indicators to be measured	Develop a benchmarking plan	Identify companies to compare
3	Collecting data	Identify the best companies	Choose the method of data collection	Determine the data collection method and collect the data
4	Analyze data and integrate results into action plans	Measuring the performance of the best companies	Collecting data and conducting business analysis	Identify performance gaps
5	Recalibrate and recycle the process	Measuring current performance	Choosing the best companies	Estimating future performance levels
6	–	Specify action plans to catch and surpass the best	Collecting data during the site visit	Communicating results and getting buy-in for those results
7	–	Implement and monitor results	Compare processes, identify differences, and develop recommendations	Establish functional objectives
8	–	–	Implementing the recommendations	Developing action plans
9	–	–	Recalibrate performance measures (benchmarks)	Implement specific actions and monitor progress
10	–	–	–	Recalibrate benchmarks
11	–	–	–	Achieving leadership
12	–	–	–	Integrating practices into processes

According to Camp (1992), the benchmarking process is divided into five phases: optimization, analysis, integration, action, and maturity. The following figure shows the questions to be asked before launching each phase and step in order to ensure that the productivity result achieves the desired objectives. This phase consists first of all of identifying the functions, activities, processes, products, costs, or problems in the company's activity that require a Benchmarking operation from which it can progress and obtain a competitive advantage. The company must therefore first measure its efficiency in this area. It is then necessary to select the best companies for the chosen subject. This is the Pre-Benchmarking. This selection must be made with comparable companies, i.e., with companies of the same size, or the same activity, or the same service. Finally, it is necessary to determine how to collect the data necessary for the study. Collecting information on potential Benchmarking partners is an ongoing process, the collection method(s), and criteria of which have been defined beforehand. It is important to know the company to be analyzed before contacting it and asking for information. A precise questionnaire must therefore be drawn up.

The analysis phase consists of two phases. The first phase consists of comparing the selected performance indicators to a significant group of companies (chosen for their level of comparability). These comparisons will reveal a competitive gap, positive or negative, which measures the difference between the department's performance and that of the best performers. A negative gap means that the partner is better performing and can be used as a benchmark, whereas a positive gap means that our internal functioning is better than that of the partner. The gap can be qualitative, if it suggests an opportunity to improve methods, and quantitative for performance indices. The analysis of the competitive gap should make it possible to plan the company's future performance levels in relation to those of its competitors and to provide an objective basis for action to reduce the gaps or capitalize on the advantage acquired. This phase consists of communicating and gaining acceptance within the company for the concepts, processes, methods, tools, techniques, etc., that have been analyzed and selected. Functional objectives must then be set. It is important that all changes in the organization are communicated to all hierarchical levels of the department concerned, so that they can promote and appropriate them. Acceptance of the results is an important step in the Benchmarking process, because even if it is well conducted, there can still be resistance to it. Benchmarking requires a re-examination of the objectives and their setting process. The opportunity provided by the Benchmarking process implies new strategic directions for the function concerned and/or the company as a whole, so the method followed, the results obtained, and the proposed directions, must be communicated and explained both to all levels of management and to the line staff concerned.

During this phase, it is necessary to check that all the steps have been followed and to ensure that the results of the Benchmarking are implemented. An alternative strategy may be considered during the project. It should also be checked whether the benchmarking is generating profitable results. After the action plan has been implemented, it is then necessary to estimate and measure the improvement in performance in order to adjust, if necessary, the improvement objectives or the action plans defined beforehand. Finally, feedback from staff, management, suppliers, and customers on benchmarking is expected. This allows the improvements to be realized and the Benchmarking to be promoted in order to make it habitual and thus to continuously improve the company. Maturity is reached when Benchmarking becomes an essential, permanent, and institutionalized part of the management process. It is then practiced at all levels of the organization, not just by specialists. The latter are useful for consultation on the most productive approaches, but benchmarking only really achieves its purpose when the whole organization is looking for better external methods to adopt.

The theories of strategy optimization studied in the first chapter have led to the emergence of various tools and methods for evaluating optimization and helping to manage companies. The evaluation of performance can be done in relation to the objectives defined by the company. There are several performance controls tools, but not all have the same use (objectives, levels considered, phases of use, performance criteria, etc.). The most common ones use the notions of strategy and strategic dashboards such as the Balanced Scorecard (BSC), the Skandia navigator and other normative models, activity-based management (ABC-ABM) or value-based management. In this study, we consider two performance management tools: the BSC and the Skandia Navigator. Their particularity compared to other tools is that they allow the evaluation of the overall performance of the company. The task of steering performance is devolved in part to value-based management. In order for the steering to be effective, it is necessary to first discern and prioritize the performance areas sought. This is why a value-based management system must be built specifically for each company. Value-based management can therefore be considered as performance management. Anthony (1988) distinguishes three levels of control in a company: strategy optimization, value-based management, and operational control. The basic dynamics of value-based management include two main phases: optimization and analysis of results (Giraud et al., 2004):

1. **The Optimization Phase:** These which serves to prepare for action. The role of optimization is to determine the targeted objectives (the expected level of performance) and to plan the means that the company will use to achieve these objectives. These means concern the choice of action plans to be implemented and then the determination and mobilization of the resources that will be needed for this purpose. If the optimization phase is well conducted, it will be useful for analyzing the results. The results are assessed by a deviation from the objectives.
2. **The Results Analysis Phase:** These which makes it possible to assess the degree of success of the actions undertaken and to orient future actions accordingly. If optimization is carried out in a superficial or too limited manner, it will be difficult to know whether a deviation is the result of poor performance, for which solutions must be found, or whether it is simply the result of poor planning and has no solution. On the other hand, serious optimization will provide the manager with reliable information on the level of performance achieved and enable him to focus his attention on unfavorable deviations.

The value-based management of a company must be consistent with the overall performance sought, which depends on a choice specific to each company. Its objective is to control activities and thus ensure that performance objectives are achieved. Anthony (1965), considered one of the founders of value-based management, emphasized that value-based management is the process by which managers obtain assurance that resources are obtained and used effectively and efficiently to achieve the organization's objectives. Effectiveness is measured by comparing the results achieved against predefined objectives. Efficiency is the relationship between the results achieved and the means or resources used to achieve them. The mission of value-based management is to ensure that moral commitment is translated into action through the processes of monitoring and measuring performance. Later, Anthony gives another definition that broadens the concept which is considered too restrictive. Value-based management is the process by which managers influence others in the organization to implement the organization's strategies (Anthony, 1988; Amedzro St-Hilaire, 2019). Bouquin (1990) notes that it is appropriate to call value-based management the systems and processes that guarantee coherence between strategy and concrete, daily actions. Later, in 1997, he defined value-based management as the set of systems on which managers rely to control the decision-action-result process. To this end, according to Bouquin (1997), the mission of value-based management is to:

- Ensure that the construction of action plans is in line with the operational plan and assist managers in choosing the assumptions needed to implement the action plans;
- Consolidate plans in preparation for budget negotiations;
- Draw up budgets on the basis of the plans adopted;
- Select criteria for measuring the performance of managers that are consistent with management criteria.

Originally, the value-based managementler was in charge of the management charts, which made it possible to monitor, by exception, the achievements, performance, or shortcomings of the operating process, as well as the evaluation of the profitability of optimized investments and the risks incurred. As controlling has become a strategic function, the controller is more concerned with information systems, value creation and helping to control the turbulence in the business environment. It is thus increasingly going beyond its traditional missions (Berland, 2009). The best-known approaches to monitoring the overall performance of a company are the strategic dashboards. In this study, we present the BSC and the intellectual capital scorecards. The BSC, also known as the BSC or strategic scorecard, was initiated by Kaplan and Norton (1992, 1998, 2001). It allows the evaluation of the overall performance of the company, through which elements related to financial and non-financial performance are taken into account. Kaplan and Norton believe that in the current competitive environment, financial indicators are no longer appropriate for guiding corporate strategy.

They advise managers to focus on improving customer satisfaction, quality, employee skills and motivation, and production cycle times. The BSC thus combines financial measures with banking measures, all of which are related to the overall performance of the organization. Gray and Pesqueux (1991) define the scorecard as a management tool focused on monitoring objectives. It is used to evaluate the actual performance of the company against the objectives predefined by the management system. The BSC is a tool for top management to translate strategy into operational terms, to align the organization with the strategy, and to transform the strategy into a continuous process (Kaplan and Norton, 2001). The BSC, initially designed for companies in the competitive sector to meet a need for performance measurement, has become a management tool, making it possible to explain, communicate, and implement strategy and is based on the following principles (Kaplan and Norton, 2001): translating strategy into operational terms; aligning the organization with strategy; making strategy everyone's daily business; transforming strategy into a continuous process; and mobilizing change thanks to the leadership of the managers.

The BSC breaks down the mission and strategy into objectives and measures that fall into four areas: financial, customer performance, internal processes, and organizational learning. A company will only choose one top goal to measure its long-term success. The financial axis identifies the objectives, measures, and indicators that contribute to the satisfaction of the different interests within the organization. Financial indicators are used to determine whether the intentions and implementation of the strategy contribute to improving the financial result. In general, financial objectives focus on profitability, measured for example by operating profit, return on capital employed or, more recently, by economic value added. They may also aim at rapid turnover growth. The financial goal is to raise as much cash as possible for the company. According to Kaplan and Norton (2001), the financial objectives and indicators have a dual role: they define the financial performance expected from the strategy, and they serve as the threads for those of the other TBP axes.

The customer axis of the TBP allows the company to represent its target market segments. To measure this axis, Kaplan and Norton (2001) distinguish two types of indicators: (i) key result indicators which concern the satisfaction and loyalty of existing customers, the extension of the customer base, the profitability by customer category, the market share in the targeted segments; these measure the performance achieved and the a posteriori observation does not allow for an adaptation of working methods; (ii) the determinants of customer loyalty: (iii) the determinants of customer loyalty: delivery times and punctuality, innovation in products and services, ability to anticipate their needs and to develop products and solutions that meet them. The "Internal processes" axis: The internal processes axis highlights the key processes of the company that will enable it to respond to and attract customers in the targeted market segments. They will ensure the expected long-term financial performance. The objective here is to make the best use of critical processes in order to achieve maximum customer satisfaction and to gain an edge in strategic competition. The TBP highlights completely new processes (or innovation processes). Organizational learning: This axis is about the infrastructure that the company needs to put in place to improve performance and generate long-term growth.

It must be aware that its current resources will not allow it to achieve its long-term objectives. Global competition forces the company to continuously improve its ability to meet the expectations of customers and shareholders. The BSC has been criticized in the sense that it is based on a classical conception of the value chain (Porter, 1985), the central element of which is a material production good and the outcome of satisfying the customer and ultimately the shareholder. It then consists of controlling the creation of value from the sole perspective of the customer (Bouquin and Pesqueux, 1999). The BSC then takes more into consideration the economic aspect (finances and customer satisfaction) and neglects the human resources orientation. The following figure shows the relationship between the different performance indicators and the company's strategy.

The intellectual capital dashboards were developed by Sveiby (1997); and Edvinsson and Malone (1997) and offer non-accounting indicators for monitoring social performance, to make it livelier and stimulating. These are Skandia's navigator and Sveiby's monitor. Inspired by the BSC, the Skandia Navigator was theorized by Edvinsson and Malone (1997) and implemented at Skandia AFS. It is based on an analysis of the intangible capital of the company and is designed on the basis of a resource approach to strategy (Roos and Roos, 1997). The navigator takes into account the aspect of societal optimization that relates to employees and customers (human resources). The Skandia navigator breaks down intangible capital into two categories of banks: human capital (knowledge, know-how, attitude, behavior, and agility) and structural capital (organization and partner relations). It is divided into five orientations: finance, customers, processes, renewal, and development, and human resources. Edvinsson and Malone (1997) see human capabilities as the basis for optimizing each objective for each axis. Human resources are positioned at the center of the value creation system and feed into the other four axes (Cappelletti, 2006). Human resources are therefore at the heart of the optimization of the company and constitute the primary determinant of this performance. The human resources axis corresponds to the skills of employees and the commitment made by the company to maintain their level (Germain and Trebucq, 2003).

The Monitor (Sveiby, 1997), like the navigator, proposes that people are the basis of the company's profit. According to the author, it is the people, considered as intangible assets, that enable the generation of income. The monitor develops three indicators (competences, internal, and external) of the asset that allow the competence of the personnel to be evaluated according to three criteria (growth, efficiency, and stability). This tool shows the importance of skills within a performance management system, but does not link skills management to the strategic vision of the company. Under the pressure of competition and trade globalization, it is imperative for companies to focus on improving their performance. According to this, objectives are difficult to achieve if one does not foresee its expected future results. Thus, to achieve its set objectives, the company must implement a strategy optimization. This action plan must be organized in the form of strategic projects, with objectives, means, a completion date, a manager, and a performance management system (Demeestère, 2005). Value-based management and the strategic scorecard play active roles in the implementation of the strategy defined by the company. In this subsection, we will determine the effect of each concept in optimizing the company's strategy. In the context of entrepreneurial organizations, controlling has several roles (Amedzro St-Hilaire, 2020). It is responsible for checking and controlling the efficiency of existing products and processes in order to innovate. As the market environment is changing, it will have to identify new market needs and conquer new strategic segments. Controlling should also check whether the objectives of the implemented strategy optimization have been achieved and support the implementation of new strategy optimization processes. In addition, value-based management must ensure the steering of global performance, which can allow for the maintenance and control of the company's resources and key competencies in the sense of the RBV, and oriented towards the detection and control of the company's dynamic capacities.

Anthony and Govindarajan (1998) see strategy optimizations as part of the activities of value-based management. According to Weihrich and Koontz (2003), strategy optimizations and value-based management are practically inseparable. Strategy optimizations without controlling is useless and controlling without strategy planning is irrelevant. Strategy optimizations allow companies to control their objectives and how to achieve them, and is only important if the company makes the effort to implement it. But the only way for them to check whether they have achieved their objectives is through monitoring. Monitoring therefore provides the resources for more optimization's activities. A strategy optimization ends with the start of a new strategy action. Simons (1987) built on Miles and Snow's (1978) typology with prospectors and defenders and highlighted the correlation between the control systems implemented, the type of strategy developed and performance. Strategy and control are therefore linked in that the nature of control systems and processes cannot be separated from strategic considerations. On the basis of studies on the relationship between the concepts of value-based management and strategic processes (Bouquin, 1997; Lorino, 1992; Teller, 1999), Wegmann (2001) proposes four ways of considering the relationship between strategic processes and value-based management:

- A disconnect: originally, value-based management used financial indicators, i.e., the budget and reporting. However, financial data are limited because their time horizon is short-term (one year). Beyond one year, it is necessary to use the tools of strategy optimizations which have a long-term time horizon (3 to 5 years).
- Value-based management as an aid to strategy formulation (Bromwich, 1990): in this formulation, three other types of steering measures must be combined with budgets. These are measures relating to products, customers, and the competitive environment.
- Value-based management as an aid to strategy implementation (Shank and Govindarajan, 1989): in this formulation, the traditional instruments have evolved. However, in companies, as well as in management textbooks, the instrumentation in value-based management remains mostly disconnected from that in strategic management (Wegmann, 2001).
- Value-based management as a component of strategy formulation (Simons, 1990): This formulation represents the most innovative approach to strategic value-based management, but also the most problematic, as few significant operational developments have been proposed in this framework.

Strategically oriented scorecards are systems of indicators that seek to measure overall performance in its various constituent dimensions (Gervais, 2000). They make it possible to clarify strategic objectives and translate them into concrete target values. They also ensure the deployment of the general policy within the organization and feedback on the optimizations of the strategies in order to progressively refine it. The authors of the BSC developed it according to two dimensions: a strategic dimension whose base is constituted by the vision of the managers of the general management and a value chain which is composed of strategic indicators ordered according to a determined logic and which is the mirror of the specificities of the company, its structure, and its people (Porter, 1985). By combining value chain analysis, strategic positioning analysis and competitive forces, strategic controlling and BSC highlight the cost drivers (Wegmann, 2008).

The BSC then has the role of a strategy implementation tool, it serves to translate the strategy into action (Meric, 2003). Once the objectives and the scorecard have been determined, each operational unit adapts the model to its own perimeter by defining its indicators. The BSC can thus be deployed in order to apply the company's strategy on the ground. According to Vilain (2003), the axis of learning and development strategies is the starting point for any long-term sustainable change. He distinguishes three types of objectives that lead to sustainable long-term change: strategic competencies (skills needed for staff to support the strategy); strategic technologies (information systems needed to promote the strategy); and the climate for action (motivation, empowerment of staff). Kaplan and Norton (2001) propose their BSC as a key model or tool in the effort to implement a new strategy optimizations process and a methodology for translating it to the operational level. To this end, it plays several roles, it will have to translate the project as well as the strategy to be implemented by the management. As all members of the company must be informed of the objectives to be achieved for the success of the strategy, the role of communicating the objectives and strategic indicators falls to the OCS. Furthermore, the BSC has to plan the objectives and harmonize the strategic initiatives. Finally, the BSC is responsible for monitoring the process of optimizing the implemented strategies while adapting to changes in the environment.

The BSC consists of a strategy map and a scorecard. The strategy map is the expression of the strategic proposals and determines the cause-and-effect relationships between the results measures selected and the optimizations indicators. The BSC measures in this map form a chain of cause-and-effect relationships expressing the strategic direction of the company through a set of operational objectives (Soderberg, 2006). According to the literature on competencies, Grant (1991) determines that the most important thing in a strategic approach is the ability to integrate individual competencies for organizational performance. In this sense, the navigator is similar to the idea proposed by Grant. Indeed, this performance management tool attempts to value human capital through indicators of involvement and development of individual competencies within a company. Studies increasingly show the importance of using more global measures, i.e., both financial and banking, to assess the sustainable performance of companies and institutions.

Performance is the criterion for evaluating a company's strategy because it takes into account the resources mobilized to achieve the strategic objectives. Once a strategy has been developed, the company must verify that it has been implemented by monitoring and evaluating its performance. The results obtained on the objectives allow the company either to improve performance through benchmarking or to set up a new procedure for the development of a new strategy optimizations process. The models studied in this chapter show that project productivity plays a crucial role in the search for the company's strategy.

The models studied in this chapter therefore take into account the imperfect structure of strategy optimizations to explain the existence of profitability situations. Thanks to these recent advances, it is now possible to analyze the effects of industrial efficiency on strategy optimizations, other than from a purely procedural point of view. In the models studied, when a company has a coordination and projection mechanism in an open economy context, performance, far from aggravating decisional inconsistencies, tends on the contrary to reduce the rate of procedural inefficiency. Similarly, it is now possible to analyze the effects of firm performance and project productivity on optimizations structures. In the model studied in section II, a successful firm sees its planning become more efficient due to the correlative reduction of the conceptual effect on the optimizations of strategies. Admittedly, some of these results are acquired under conditions that are too specific for them to have the robustness of models based on a measure of performance as an index of the effectiveness of the strategy optimization process. However, they show that it is possible to develop models that take better account of the interactions observed. There is a gradual trend towards a more general model that would integrate models based on a measure of characteristics as a special case. The objective of these first chapters was to show the interaction between the optimizations of strategies and the performance of firms in an open economy. The analysis of the theoretical models studied showed that the two concepts cannot be dissociated. Indeed, the optimizations of strategies play a role on the performance of firms in an open economy. Conversely, the productivity of projects plays an important role in the optimizations of strategies. The main findings are that:

- A company or institution can achieve a higher level of performance than its rivals if it possesses or controls the resources that enable it to gain a sustainable competitive advantage in order to be more competitive internationally. The company must therefore be able to renew its skills in order to keep pace with an environment that is changing ever more rapidly, where technologies are leading to increasingly frequent innovations. Innovation therefore determines the ability of companies to maintain sustainable competitive advantages, to increase their competitive position and consequently to improve their performance at the international level.

- Performance is the criterion for evaluating the company's strategy, as it takes into account the resources mobilized to achieve the strategic objectives. Once a strategy has been developed, the company must check that it has been implemented by monitoring and evaluating its performance.

The analysis of the theoretical models has thus clearly highlighted the impossibility of analyzing the two aspects of our problem separately, since the effects of the optimizations of strategies on the performance of companies depend on the role of industrial efficiency in the determination and elaboration of operational planning mechanisms. Conversely, the role of performance in the determination of decision structuring mechanisms can itself be modified by the integration or intensification of value-based management. Two elements are fundamental in the analysis of these interactions: the nature of the optimizations: The interactions between company performance and strategic planning are different, for the same model, depending on whether or not a company has a value-based management tool. This result is particularly evident in the case of the performance measurement model and in the case of the factor model (taking into account characteristics). Furthermore, these performance gains can, in a capital-abundant firm, be expressed through intensive productivity: this is the fundamental result of the integration of production theory into traditional models. Optimization structures: When optimizations and goods are competitive, these relationships are simple to grasp, but the effects of firm efficiency are limited, in terms of process, to restructuring effects, the effect on real costs being linked to the nature of the efficiency gain. Conversely, when optimizations and decision-making are characterized by non-optimal structures, be it information imperfection or design structures, then it becomes possible to analyze both process and feature effects simultaneously. Furthermore, while the models in Chapter 2 account for the non-flexibility of the process, the models studied in Chapter 1 are forced to postulate it.

KEYWORDS

- **activity-based management**
- **balanced scorecard**
- **corporate responsibility**
- **economic development**
- **mobilizing**
- **sustainability**

CHAPTER 5

Empirical Mechanisms for Identifying the Role of Strategies Optimization in Determining Comparative Advantage: Does Better Strategy Optimizing Result in Higher Levels of Company Performance and Productivity?

The search for a model appropriate to the banking sector requires an empirical study of the various aspects of the interactions between company performance and optimizing strategies. To this end, it is appropriate to begin by reviewing the studies that have examined the role of strategy optimization in determining the profitability gains of companies and to try to identify the main lessons learned. This study must be completed by a more pragmatic approach, aiming to describe the impact of economic profitability on the characteristics of optimization, strategic decision-making, and the structures of optimization mechanisms, from an empirical point of view. It this theoretical playback proves a divergence of the models regarding the usefulness and relevance of strategy optimization, it is important to verify whether the empirical results are more in favor of one or the other model. These analyzes are part of the work that aims to identify and quantify the determinants of optimization in banking firms. In an open economy context, the role of strategic performance must be situated in relation to the other determinants of optimization with which it interacts. This is why, after stressing the importance of structural factors – both from a quantitative and qualitative point of view – many empirical studies have been devoted to the analysis of its heterogeneity, in particular, to highlight the crucial role of

Value-Based Management in an Open Economy: Optimizing Strategy to Improve Business and Performance. Prof. Post-Dr Walter Amedzro St-Hilaire (Author)

process control optimization process in an open economy context (Amedzro St-Hilaire, 2022).

Many explanations have been put forward to empirically explain the interaction between strategy optimization and performance. The issue is further confused by the fact that, from an epistemological point of view, the modalities of 'scientific' verification of a theory are, to say the least, confusing. In order not to add to this confusion, we will simply recall some of the main studies and limitations that have been conducted in this respect. These studies are largely deviations and do not place themselves in the conditions of the theory. The main question behind these different studies is the following: Does better strategy optimizing result in higher levels of company performance and projects productivity? To answer this question, many researchers have tried to quantify the impact of strategy optimization on company success. In doing so, they have selected different (depending on the study) economic or financial variables and measured them quantitatively. The first empirical test of the relationship between strategy optimization and firm performance was conducted by Thune and House in 1970, who studied 36 firms in six industrial groups. Since then, the tests have followed one another, confirming, or refuting their conclusions. While some studies report a positive link, many find no quantifiable benefits, and others even find small negative effects and costs resulting from strategy optimization. Most of these studies were conducted in the United States or the United Kingdom.

The main studies proving the existence of a positive relationship between strategy optimization and project productivity are those conducted by Gershefski (1970); Thune and House (1970); Herold (1972); Karger and Malik (1975); Rhyne (1986). Other studies in the same vein were reported by Gordon Greenley in his article, Strategic Planning and Company Performance: an appraisal of the empirical evidence (1994) and grouped into three groups. While in the first group, 9 studies conclude that there is no association between strategy optimization and project productivity, in the second group, 12 studies support evidence of such an association and in the third group, 9 studies conclude that companies doing strategy optimization outperform those not doing so (Greenley, 1994). However, there are differences in methodologies, sampling techniques and variables of interest among the different studies. For example, the Ansoff et al. study uses 13 different variables to measure the performance of 93 manufacturing firms: sales, earnings, earnings/share ratio, total assets, earnings/equity ratio, dividend/share ratio, inventory value, debt/equity ratio, common equity, total earnings/

equity ratio, P/E ratio, dividend/earnings efficiency, and price/equity ratio. To ensure good reproducibility of the measures, each variable is measured 21 times. The values of these variables for companies with intense strategy optimization are compared with those obtained for companies with little or no strategy optimization. With the exception of two variables (growth rate of equity and growth of assets), companies practicing strategy optimization outperform those not practicing strategy optimization. The confidence level of the statistical tests is $\alpha < 0.1$ or $\alpha < 0.005$. Gershefski's study compares sales growth in a sample of 383 companies over a period of 5 years before the companies adopted strategy optimization, and over a period of 5 years after the introduction of strategy optimization. The results of this comparison lead the author to the same conclusion and indicate that strategy optimization is effective. Using a somewhat similar methodology, Thune, and House also come to the same conclusion and find that companies doing explicit strategy optimization outperform their own performance after the introduction of an explicit optimization system.

Numerous studies, such as those conducted by Grinyer and Norburn in 1975 and Kulda in 1980, report the absence of quantifiable benefits from the adoption of strategy optimization. Indeed, Fulmer, and Rue's 1973 study of 386 companies comparing four financial performance variables (sales growth, earnings ratio, earnings growth, and total capital) between firms that differed in strategy optimization led the authors to conclude that their results called into question most of the basic assumptions on which strategy optimization was established. Although they do not deduce that strategy optimization does not affect the final results of the company, they explain that their study does not indicate any clear relationship between strategy optimization and the variables measured. In their study published in 1980, Leontiades and Tezel analyzed 61 companies over a 6-year period. The approach used was to contact the chief executive and the head of the optimization department of each company to testify about the importance of strategy optimization on various numerical and semantic domains, in order to provide quantifiable variables that would allow the comparison of company performance. Based on five variables chosen to assess performance (rate of return on equity, rate of ROA, change in prices and earnings per unit of share growth and earnings per unit of sales growth), the study tests several hypotheses indicating a relationship between project productivity and strategy optimization. Finding that none of these hypotheses were statistically significant at the $\alpha = 10\%$ confidence level, the authors concluded that there was no evidence of such a relationship.

Although relatively few in number, some studies tend to show a negative relationship between strategy optimization and project productivity. Indeed, Whitehead and Gup found some negative effects. Some companies using strategy optimization perform worse on some measures than their competitors not using strategy optimization (Whitehead and Gup, 1985). In 1983, a correspondence survey of 316 financial firms using optimization and 133 financial firms not using strategy optimization enabled them to reach these conclusions. To ensure a good representativeness of the sample, each state is represented and contributes no more than 10% of the total responses. The distinction between planners and non-planners was initially based on the respective rates of sales, earnings, and rates of return on equity. To classify the firms, the authors considered as advanced planners those firms that had explicit and written objectives. The most advanced planners were those that had specialized departments for optimization and used econometric models and regression analysis to establish projects or analyze alternative actions. Finally, the frequency of revision of strategic plans was also taken into account for this classification. From this classification, the general trend was that the use of formal optimization was related to the size of the institution: 95% of institutions with assets of $1 trillion or more used formal optimization, while only 48% of institutions with assets of $50 million or less used formal optimization. For their analysis, Whitehead and Gup chose three variables to measure: the rate of return on equity and the rate of ROA to measure profit, and the absolute growth of customer deposits.

Using regression analysis to isolate the impact on performance, and analysis of variance to determine whether the observed differences were statistically significant, Whitehead and Gup studied the data from their sample. They found that institutions that used strategy optimization had lower rates of return on capital and assets than those that did not use strategy optimization. For the third variable, planners did not achieve significantly higher growth than non-planners. The results were obtained at the $\alpha = 0.1$ confidence level (90%). In order to confirm their results, the authors redefined the distinction between planners and non-planners with respect to market expansion, product, and service development, social development, and social, economic, and political considerations.

The results obtained by evaluating the data according to this new classification were consistent with the initial results. A series of further tests, based on a redefinition of the variables and criteria distinguishing between institutions using and not using strategy optimization, confirmed these results once again. The authors concluded that their results indicated a negative relationship between strategy optimization and project productivity in the financial and banking sector. While not rejecting strategy optimization, they questioned the quality of the optimization and the existence of any competitive advantage it provided. They concluded that optimization is negatively related to project productivity, unless it pays off in the long run. Furthermore, they hypothesized that the absence or reduction of pressure on institutions from their environment caused them to engage in strategy optimization.

Despite the diversity and number of studies conducted to evaluate the effectiveness of the strategy optimization process, a more or less blatant lack of methodological rigor has been noted in the various studies. The critical analysis carried out by Greenley (1986) highlights the methodological failings of each study. Thus, we note a bias in the methodological rigor of the authors, an absence of statistical tests (making it possible to verify whether the difference is statistically significant) or at least their omissions in the publication of the results. Furthermore, the variations between the variables used from one experiment to another, between the duration of the experiments, between the periods (bearing in mind that each period is marked by a particular conjuncture and history) and between the size and origin of the samples, do not allow them to be compared, even if they do allow a certain complementarity of the results obtained. The research parameters are fundamentally different from one researcher to another. Moreover, some studies are marked by the absence of proof of the reproducibility of the measurements made. In addition, much research has been based on questionnaires. In this respect, Grinyer and Norburn rightly note that because the optimization process is complex, and spontaneous reactions to questions are important for an adequate evaluation, mailed questionnaires are particularly inappropriate for obtaining an adequate response on the subject.

The first conclusion to emerge from this analysis of studies based on performance measurement as an index of the relevance of the strategy optimization process is that the evidence of a relationship, and the nature of this relationship, between strategy optimization and project productivity has yet to be proven. It is not possible to make an objective statement on the effectiveness or ineffectiveness of the strategy optimization process as a control tool. The methodological variability of the different studies limits their analysis and comparison. However, when grouping the different studies according to their results, it can be seen that those that found a positive, causal link between strategy optimization and project productivity are generally older than those that found no link at all and those that perceived harmful effects to the use of optimization. Is this temporal separation sufficient to suggest that the relationship between business performance and strategy planning would change over time? Although the history of strategy optimization seems to point in the same direction (decline in popularity followed by a revival of interest), the lack of methodological rigor, once again, does not allow this hypothesis to be confirmed. Rather, it reflects the bias introduced into most research by the authors, and the use of arbitrary attributes or variables. It also shows the inadequacy of the variables used for an objective evaluation of the process, and the presence of factors other than those measured.

There are, in fact, a range of other variables that can affect project productivity or the results achieved, so that the changes detected in project productivity may not have been affected, or may have been affected only partially, or may have been affected only by the strategy optimization. Higher levels of achievement are not necessarily linked to the use of strategy optimization. It may also be that an improvement in optimization gives the firm the means, resources for use, or capacity to implement strategy optimization within the firm. In addition, firms may adopt strategy optimization in order to protect a previously acquired performance without planning. In these cases, there would be a relationship between strategy optimization and performance, but not a causal relationship. Despite the conclusions have reached, a subjective assessment of results by management does not differ greatly, while an objective financial measure shows a substantial difference. It is therefore difficult to specifically define the consequences of using strategy optimization. In this respect, the analysis tends to support the model. However, there is still a category of potential benefits resulting from the use of strategy optimization. Greenley (1986) recognizes in his paper the benefits increased by the use of strategy optimization, which are inherent in it as a consequence of its use. He refers to this as the 'intrinsic values of optimization. There are therefore non-economic benefits to the use of strategy optimization. There is therefore a strong assumption that strategy optimization has a major effect on project productivity. However, does optimization affect project productivity, or does project productivity provide the latter with the resources to give managerial attention to strategy optimization? In any case, all the problems of extensive measurement of business optimization suggest that these results overstate the true relationship between optimization and performance. Models based on performance measurement are therefore not very appropriate for defining the nature of such a relationship.

In contrast to studies on the effectiveness of the strategy optimization process-related almost exclusively to financial performance as a gauge of the value of the optimization system, these models assume that the benefits of strategy optimization are more related to the nature of the process, and may or may not be a sufficient condition for optimization improvement. Strategy optimization can thus be effective as a process, despite the optimization process achieved. Hence the importance of developing models that are not based solely on economically driven performance. It is important to take into account the characteristics of the process, as well as the dimensions of the business context (notably resistance and resources) in which the optimization takes place. The study by Ramanujam and Venkatraman (1987), establishes that the dimensions of the business context have a dominant influence on the effectiveness of the strategy optimization process. Also, the models based on the evaluation of process characteristics, consider that the productivity of projects is not a sufficiently valid aspect on which to base the effectiveness of the optimization process. Thus, research by Greenley in 1983 and Dyson and Foster in 1982, among others, examined the effectiveness of the process, relative to the nature of the process itself.

The strategy optimization process is defined as the set of human interactions, formal, and informal, that take place during the generation or development of a strategic plan. This process has both a symbolic and an instrumental function. Symbolically, the strategy optimization process serves to build consensus in the company, providing simplified models for communication and understanding. At the instrumental level, the strategy optimization process serves as a performance program, absorbing uncertainties and reducing the time and cost of information retrieval that governance faces in making decisions. This process is characterized by a clarity of optimization, that is, a division of labor between different levels of management in the initiation, formulation, revision, and execution of plans. It is also characterized by an explicitness of strategy optimization. Indeed, an explicit process is a more rational system for building strategic plans. The third characteristic of the process is the diversity of the optimization. Where the diversity characterizing the optimization is high, there are many kinds of individuals. With this kind of strategy optimization process, multiple, and conflicting views are taken into account in the identification of strategic issues and in the development of proposed solutions, so that resistance is less important. Finally, the strategy system should be characterized by a particular optimization intensity. This concept refers to the level of personal resources that the participants have to dedicate to the strategy optimizations process. It indicates the involvement of everyone and their interest in the process. An additional characteristic recognized in the strategy optimization process by Glaister and Falshaw (1999) is the extent to which strategies within an organization are the result of a deliberate or emergent process. It is in relation to these different characteristics that this category of study design bases its research. Dutton and Duncan (1987) hypothesized that the strategy optimization process model systematically affects the occurrence and success of strategic change efforts through its effects on the content and form of strategic issues.

Studies based on a measure of process characteristics as an index of process effectiveness generally analyze a set of dimensions of the optimizations system and examine possible relationships with a set of dimensions reflecting the effectiveness of the strategy optimizations process. The results obtained from these different studies are also mixed and can be grouped into two categories: those recognizing the effectiveness of the process and those identifying dysfunctions related to the implementation of strategy optimizations. The study by Ramanujam and Venkatraman (1987) is a good example of research using this model. Collecting data by means of questionnaires from 600 firms selected from manufacturing and service firms, and collecting responses from 207 of them, the authors analyzed the characteristics of firms and optimization systems in relation to three main dimensions reflecting system efficiency (system capability, objective achievements, relative competitive performance). Following a statistical analysis of the results, the authors concluded that there is clearly a strong multi-variate relationship between the system dimensions and the dimensions reflecting system effectiveness. However, they were unable to determine the relative importance of the contribution of the dimensions of the optimization system to the observed relationship, i.e., the causal link.

Furthermore, Greenley demonstrated in 1986 that there can be banking consequences to strategy optimization that can provide substantial benefit to the business. Such benefits include process benefits, such as the ability to identify and exploit future market opportunities, personal benefits, such as the encouragement of a pro-change attitude, and the prospect that strategy optimization keeps the company synchronized with the external environment so that it can adapt to changes. Optimization can thus be an effective management process, despite the optimization achieved. Also, Glaister and Falshaw's (1999) survey of 500 manufacturing and service companies in the UK, of which 113 provided responses, found that the statement the strategy adopted is the result of a totally deliberate process' has a higher occurrence and is more consistent with the reality of the companies, than that – the strategy adopted has emerged over time without being the result of a deliberate plan. The perception among the sample evaluated clearly indicates that strategic formulation is in practice more of a deliberate process than the emergent and adaptive process. Similarly, the responses allow us to conclude that for the sample considered, there is little or no dysfunction due to the strategy optimization. Few studies in this category, however, elaborate on the impact of strategy optimization on the long or short term.

In their 1983 study, Bresser, and Bishop argue that explicit strategy optimization can be dysfunctional if it introduces rigidity and encourages excessive bureaucracy. In these cases, optimization results in rigidity and inflexibility of responses to the changing environment. Strategy optimization tends to increase the need for coordination and control of the usually fluid, flexible, and informal strategy formation process. The process therefore tends to stifle creative thinking and favors the maintenance of old, tried, and tested patterns or models. In other words, in the interests of maintaining control, strategy optimization tends to be an exaggerated extrapolation of the past and present into the future, rather than a quest to reinvent the future. Decision-makers generally assume that the future is a linear progression from the past. They set strategies taking into account a future that more or less corresponds to what is known, or to some accepted development. Optimizing strategies creates the illusion of certainty in a world of uncertainty, risk, and continuous change, without taking into account the contingencies of the environment. In relatively safe environments, without democratic checks and balances that allow market forces to play out, or in cases of monopolies or duopolies (such as the civil aircraft manufacturing sector dominated by Airbus and Boeing worldwide), this illusion is not a problem. There is therefore a lack of application of the process in companies. In addition to the challenges posed by the involvement of human resources, the communication and dissemination of the common company culture and the adaptation of the company structure, the lack of flexibility of optimization, as well as the limited vision of the future that it implies, constitute an obstacle to the effective implementation of the strategies that it has itself made possible to develop. The latter, when implemented, sometimes prove inadequate and lead to undesirable results. The strategy optimization is currently unable to take into account the range of possible futures, and therefore does not allow for the establishment of a long-term plan with any certainty.

Studies based on an evaluation of the strategy optimization process, although taking more account of the nature of strategy optimization, as well as the nature of the consequences of optimization within a company, also come to mixed conclusions. They confirm the existence of a link between strategy optimization and business success, but do not specify whether or not it is a causal link. Strategy optimization, when properly used, is associated with non-economic benefits, which can confer or strengthen a firm's competitive advantage, but it sometimes induces dysfunctions (notably inflexibility and inflexibility), which in turn can limit a firm's expansion and development. These studies therefore show that strategy optimization is a complex, double-edged tool, the effective use of which is not a given. The main criticism that can be made of the models measuring the characteristics of the system to assess its effectiveness lies once again in the methodology. Mainly based on mailed questionnaires given the nature of the research, these studies do not record spontaneous responses from respondents, and obtain relatively low response rates (out of 600 companies, 207 responded for the study by Ramanujam and Venkatraman, i.e., a 34.5% response rate, and 113 out of 500 companies responded for the study by Glaister and Falshaw, i.e., a response rate of 23%). This results in poor sample representativeness, since, as the various authors note, the majority of responses come from large firms, or from those performing strategy optimizations. The samples are biased towards this category of respondents.

How can we explain the differences between theory and practice? In other words, how can we explain the fact that, contrary to what is claimed in the literature, strategy optimizations do not always lead firms to success, and is sometimes associated with dysfunction? Although some firms benefit from the use of strategy optimizations, others do not recognize its benefits. Two categories of problems with the concept may explain these discrepancies. The concept of strategy optimizations, as theoretically developed, takes little account of the cognitive limits of human effectiveness, which in turn may limit the effectiveness of the process in practice. This makes it an imperfect concept that may therefore be ineffective on some occasions. A major problem is the lack of a consistent and meaningful definition of what constitutes strategy optimizations, i.e., the elements of strategy optimization. This problem is reflected in the different studies by the difficulty and heterogeneity of the definitions used to distinguish between companies that optimize strategy and those that do not. Indeed, the criteria vary greatly from one study to another to classify companies according to their practice of strategy optimization, as do the variables chosen to measure the intensity of strategy optimization within a company. All firms engage in optimization, but they differ considerably in the extent to which they implement the plans developed, grow as the environment changes, and use the optimization tools. Also, the definition of optimization varies from company to company. It is therefore difficult to say whether the definition adopted by a particular company, at a given time and in a given context, is in agreement or not with the theoretical definition of the elements of strategy optimization. The different degrees of optimization partly explain the variability of the results obtained by the empirical studies.

Effective optimization depends on the involvement and participation of all stakeholders in the company, i.e., managers, employees, shareholders, customers, and potential strategic partners, to determine the company's priorities, its strengths, and weaknesses, and to avoid damaging one sector to the benefit of another, and to avoid conflicts. However, the strategy optimization is still too often done by a small group of people. The needs (in terms of financial resources, technical resources, appropriate architecture, procurement, human resources, information, business management, etc.), and priorities of the company are only defined by this group. Therefore, optimization does not take into account a number of factors (real needs of consumers, potential, innovative, and creative capacity of the personnel employed, interest of the workers, etc.), and remains focused on the expectations of a few individuals within the company. It is therefore incomplete and sometimes inadequate, with a lack of sufficient guidance on the relative priority of core activities, especially in companies divided into multiple sectors (e.g., regional governance). Furthermore, being a long and meticulous process, the strategy optimization requires a high expenditure of energy and time for its implementation. Moreover, despite the efforts made, the process does not always lead to the expected results. In practice, it is difficult to mobilize the necessary resources for its implementation. The structure, as well as the technology, is not always suitable for a new strategy. The general structure of companies is not suitable for the use of strategy optimization. Still structured according to the needs of traditional, proven activities (and in relatively stable environments), companies are structurally resistant to planning. Furthermore, the sufficient information gathered in the traditional business, with a competitive strategy, is inadequate for developing the strategic alternatives needed for strategy optimization.

Companies do not have the complete information needed for effective strategy optimization. The information available to the company affects the way it is implemented: the more incomplete (which happens in most cases), uncertain, and unreliable the information is, the less likely the company will be to incur the risks associated with its implementation. Furthermore, in companies divided into sectors or compartments, the structure can create barriers between different sectors if it is inadequate. And, restructuring a company, or adapting structures according to changes in strategy optimization is not always easy, especially if the company is of considerable size. The strategy optimization introduces rational elements that break with the cultural history of the company and threaten the political process. Therefore, a conflict often arises in the company between traditionally profitable and innovative activities. This results in resistance, sometimes followed by an abandonment of strategy optimization, which limits the effectiveness of the process. In addition, there is an inability of managers to formulate and implement strategy. In order to be effective, strategy optimization must be combined with optimizations of leadership skills and management of the overall strategic change process. However, the many studies that attempt to establish the relationship between strategy optimizations and firm performance do not highlight efforts to optimize skills and manage strategic change and take strategy optimizations in isolation. Therefore, it is not possible to say how much strategy optimizations are actually done in companies. If in practice, this planning is carried out in isolation, companies use an incomplete tool which explains the discrepancy between the practical and theoretical models.

Another explanation for the divergence is the frequency with which plans are developed and revised. Optimizing strategies requires constant revision of plans and questioning of new strategies, as these become useless whenever the historical dynamics of a company lead it where it wants to go, or the objectives set turn out to be inappropriate. These revisions are necessary to ensure the flexibility of the process. Optimizing strategies therefore requires continuous attention regardless of the company's situation (crisis, or profitable situation), in other words, time and energy. However, in practice, firms tend to relax their attention or devote the energy needed for optimization to other activities depending on the conditions of the firm. In their article, Bresser and Bishop (1983) demonstrate, based on previous work, that both too little optimization and too much optimization led to inter-firm contradictions and threaten the viability of the firm. Indeed, intense optimization tends to increase the number of new inter-industry products. However, the latter often conflict with existing ones, resulting in an increase in contradictory activities within the same firm. These problems illustrate that the strategy optimization and the intensity of their use in a company are largely dependent on its size, resources, and field of activity.

In conclusion, it is important to note that the true nature of the relationship between project productivity and strategy optimization remains to be proven. The numerous empirical studies conducted tend to confirm one or other of the theoretical models. However, the bias introduced by the methodology limits the consideration of these different studies. Models based on an assessment of firm optimization as an index of the effectiveness of the strategy optimization process seem particularly inappropriate for such an assessment given the banking consequences of applying strategy optimization. Although more recent models based on an assessment of the characteristics of the policy optimization process itself alleviate this deficiency somewhat, they are also limited by their methodology, which prevents them from obtaining a sample and responses that are sufficiently representative of reality. However, by incorporating the dimensions of the optimization system and the internal context, these models offer a better representation of reality. Although many of the studies conducted in this respect tend to support theoretical model, some still deviate from it. Many empirical studies have highlighted the fragility of the strategy optimization process. Indeed, on the one hand, it is quite difficult to clearly demonstrate the beneficial effects, and on the other hand, these studies have shown that despite the considerable effort made by several companies to elaborate and develop coherent strategies, few of them are actually implemented and lead to the expected changes. The use of strategy optimization is not taken for granted, and the process does not systematically lead to the expected results. Opinions on the contribution of the strategy optimization process to the success of the company are still rather divided. Although there is a link between optimization and success, it is difficult to determine whether optimization leads the firm to success, or whether it is success that provides the firm with the means to implement strategy optimization which would then serve to articulate the consequences of a pre-existing strategy. However, it has been shown that strategy optimization does not systematically lead firms to success, nor does it systematically provide a comparative advantage over firms that have not adopted strategy optimization.

The models that we have just analyzed thus make it possible to highlight interesting results concerning the links between the strategy optimization and the performance of companies. In these models, two elements are fundamental for the analysis of the effects of optimization on performance in an open economy context: the nature of the competitive advantages and the flexibility of the process. Furthermore, none of these models can really account for the effects of operational optimization on firm performance in a context where strategy would be compatible with the notions of openness and efficiency. In order to define an appropriate methodological framework for the search for a model relating to the interaction between strategy optimization and the performance of the banking sector, the object of our research, it is essential to return to the context of the use of the measures of fidelity and validity that have been the basis of the different empirical models identified throughout the literature review. This is all the more significant as the different models that address the nature of the relationship between optimization and business performance are often considered as theories and socio-technical analysis grids with a very important prescriptive content.

Thus, most of these empirical studies have emerged from research carried out with the aim of creating knowledge to be applied to solve this pressing industrial problem. To do so, this knowledge must be descriptive of the phenomenon under observation, applicable to the situation and to similar situations (somewhat generalizable), in line with reality (therefore valid) and indicate how to solve the problem (prescriptive). On the basis of this knowledge, companies orient their management strategies in order to rectify the situation in which they find themselves. However, the empirical studies mentioned above base their methodology and theoretical models on constructs. And these constructs, which are neither raw facts nor the result of the application of objective criteria, cause some confusion. Indeed, how to reconcile the requirements of conformity to reality and applicability required by prescriptive research with the subjectivity, abstraction, and contingent correspondence to reality implied by the notion of constructs?

Thus, the study of the phenomenon addressed by the various models often requires the understanding and description of the cause-and-effect relationships involving the variables at stake in these phenomena. This understanding is often necessary, on the one hand, to characterize the phenomenon in question and, on the other hand, to find the remedy(s) for these management problems affecting the company. In this respect, in some of the models studied, questions of causality have been identified through correlation and survey research. Others use experimentation to study the causal links characterizing the variables of phenomena. In the case of our models, the study of causality has strengths and weaknesses, and the appropriateness of the choice depends, among other things, on the nature and objective priority. To alleviate these problems with the use of constructs, the strict application of fidelity and validity measures to solidify the models are recommended. The effect of these recommendations is to determine the extent to which these models can be relied upon, and the extent to which their results can be generalized. More specifically, these measures of reliability and validity should allow for an understanding and assessment of the influences of the use of constructs on the interpretation and consideration of data, and on the reflection of reality that these data offer. In order to better understand the scope of these recommendations, this chapter summarizes in the following sections the contexts of use of fidelity and validity measures in the empirical models that address our problem, and shows how these measures could have reinforced the results of studies with a strong prescriptive component.

Constructs are an important tool in management science research. They serve, among other things, as the foundation on which emerging theories are built that guide the formulation of management strategies. In this perspective, the criteria of fidelity and validity (logical and empirical) of constructs are of considerable importance. Generally speaking, a construct is valid if it actually measures what it is supposed to represent in theory. Validity is the degree to which the procedure really measures what it proposes to measure (Krueger, 1994). Its reliability refers to the degree of consistency between multiple measures of the reality to which it is intended to correspond. According to Herbert (1984), reliability refers to the ability to obtain consistent results in successive measurements of the same phenomenon. Reliability therefore defines the level of consistency or convergence that can exist between two sets of observations of an identical object of study – different observers arrive at the same observations for the same phenomenon under study using the same construct. Logical validity (or scientific validity), on the other hand, corresponds to the possibility of observing reality independently of the observer, under the same or similar conditions.

This will allow the process of generalization and, ultimately, the possibility of proposing laws for the functioning of the phenomenon studied. Empirical validity (or "pragmatic validity") corresponds to the possibility of making a prognosis, of taking a decision, independently of the proof of the reality of the facts concerned. In a way, it defines the usefulness of the model. There are several ways of judging the validity of a construct. For Zaltmann et al. construct validity consists of convergent, discriminant, and nomological validity. Convergent validity is represented by the correlation between two attempts to measure the same concept through maximally different methods. Discriminant validity measures the extent to which a concept differs from other similar concepts. Nomological validity, finally, validates an instrument by interpreting the obtained scores in terms of a theoretical concept and consequently generates predictions that, if confirmed, have a validating effect (Zaltman et al., 1973). Measures of validity and reliability therefore testify to the adequacy of the constructs and to methodological rigor.

Measures of validity and reliability are used throughout the research process and are employed in both qualitative and quantitative research in the studies conducted on our research problem. The phrase qualitative methodology refers in the broadest sense to research that produces descriptive data:

people's own written or spoken words and observable behavior. Qualitative research on the era is gaining in importance and is taking a prominent place in the study of strategic phenomena. Indeed, while some aspects of these phenomena are easily assimilated to numbers and quantifiable, many aspects are conceived around the internal context of strategies, emotions, feelings, thoughts, moral, and cultural values, prejudices, or ideas that cannot be characterized by numbers. They thus include variables that are difficult to control experimentally and abstract or subjective realities that cannot be measured quantitatively. Furthermore, due to their dynamic, global, and interactional nature, these phenomena need to be addressed through a systemic rather than a sectoral approach. The main advantage of qualitative research on this issue is that it allows reality to be seen from the point of view of those who experience it on a daily basis, and in a holistic way, as a set of interdependent factors. This makes qualitative approaches methods that have become essential for understanding the full reality of qualitative management phenomena. "In short, management can be rendered scientific without adopting Taylor's "scientific" principle that everyone's work should be so simplified that even an idiot could do it" (Bunge, 1998).

Thus, for the study of our phenomenon, constructs grouped under concepts, visions should be developed from the collected data patterns. Codes should also be used to develop and refine the interpretation of the data. This coding often involves the creation of categories to facilitate analysis. But to capture reality, the qualitative has not been refined and standardized like the quantitative. Different research designs are produced depending on the nature of the research and the aims it has set itself. In this regard, in the absence of a commonly accepted research standard, each process must include measures of construct and construct validity in order to make the research pathway and final designs credible. The revised models include several data collection instruments (such as interviews, participant observation, questionnaire surveys, focus groups, etc.), whose nature is likely to introduce more or less considerable biases in the consideration of the data, and which therefore require the use of validity and fidelity measures in their implementation. Concepts are the building blocks of the propositions and theories that marketers use in explaining, predicting, and controlling marketplace phenomena (Zaltman et al., 1973). On the other hand, it is through concepts and propositions that the researcher moves from description to interpretation and theory. Concepts are abstract ideas generalized from empirical facts (Taylor and Bogdan, 1984). In the case of our problem, concepts are sensitive instruments indicating a general sense of reference.

They are constructed to highlight a phenomenon or process that is not really apparent through the description of specific institutions. They are thus widely used throughout the research process, both during data collection and in the formulation of theories. During interviews, focus group discussions and questionnaires, respondents are confronted with the concepts they have chosen.

However, each person is in a particular situation and has a particular experience that is different from others. Based on this particular experience and/or situation, each person is driven to interpret a concept in their own way. Furthermore, given the changing nature of situations, the interpretation of concepts is a dynamic process that changes over time and between individuals. It would be advisable to ensure that the concept is appropriate to the phenomenon under study and specific enough not to measure anything other than the subject of interest. Respondents should understand the concept in the sense to which the research refers. In this way, potential variations observed as a result of the measurements can be considered as real or as real variations of the phenomenon under observation rather than variations of the objects being measured. In other words, in verifying the validity of the concepts, the evidence of the actual measurement of the specified aspect of the phenomenon under observation and not several different aspects should be categorical. The said phenomenon would then be more easily understood and the results can be considered as truly descriptive of the phenomenon. Most of the qualitative data collection methods (interviews, participant observation, focus group discussions, questionnaire surveys) used in the models under discussion involve interactivity between the researcher and the participants. Indeed, the presence of the researcher can lead to a modification of the usual strategy of the participants during the participant observation. Similarly, respondents may conceal part of the truth or alter it (according to the researcher's impression or personal interests) in their answers. It is also possible that the researcher's involvement and interaction with participants may affect his or her objectivity in interpreting particular data. It is almost impossible to certify 100% the veracity of a participant's statements but, the researcher also has the responsibility for imposing cross-checks on the informants' stories. The researcher should examine an informant's statements for consistency between different accounts of the same event or experience (Taylor and Bogdan, 1984). At this level, fidelity measures allow the bias introduced by the measurement instruments, and by the research process, to be assessed. High fidelity ensures that the research was conducted as objectively as possible and that the results are not derived from (or attributable to) primarily the subjectivity of the researcher or one or more participants.

Assessing the reproducibility of data often means re-examining the data presented during the research. This can be done by reconsidering the data presented or by replicating the process of data collection and analysis. However, it is often the case in our studies that the data is beyond the reach of those wishing to review it: data collection may be limited and not easily replicable (e.g., when data is obtained through interviews with elites in large companies). As Corbin and Strauss point out, unlike physical phenomena, it is very difficult in the social realm to set up experimental or other designs in which one can recreate all of the original conditions and control all extraneous variables impinging upon the phenomenon under investigation (Corbin and Strauss, 1990). This characteristic of limited replication in management science research is due as much to the dynamic and interactional (and therefore changing in time and space) nature of the phenomena under observation as to the nature of the data collection instruments. If, as mentioned above, the reliability of a construct refers to the degree of consistency between multiple measures of the reality to which it should correspond (hence its replicability), it is understandable that studies aimed at testing the reliability of a research are unlikely to be possible. In other words, it would be difficult to recreate the conditions of data collection and analysis as practiced by the researcher.

This implies that in research with a prescriptive component, companies will have to rely on the research presented to them and the results reported in the course of it, since they will not have the results of replicability studies available to them. In this case, the researcher must validate his data in order to give it more credibility. They must certify the "replicability" of their data. This would mean providing all the theoretical and practical elements, as well as the perspectives and rules that he or she has followed, which would allow analysts to assert that one is entitled to draw the conclusions or theories from the research, if one were in the same context as the researcher. Another way of explaining reproducibility is as follows: Given the theoretical perspective of the original researcher and following the same general rules for data collection and analysis, plus similar conditions, another investigator should be able to arrive at the same general scheme. The discrepancies that arise should be resolvable through re-examining the data and identifying special conditions operating in each case (Corbin and Strauss, 1990). In general, qualitative research methods in industrial economics are designed to capture a phenomenon as it emerges from observation or from what the research participants say about it. They are designed to ensure a close match between the data and what the participants actually say or do. However, they have flaws and offer freedoms that, if left unchecked, could undermine the rigor and objectivity of the research. In the case of research with a strong prescriptive component, such a nuisance could prove harmful to the operation of companies that reshape their strategies according to the results of such research. It is therefore important to introduce safeguards that are likely to bear witness to the application, know-how, and rigor implemented throughout the research process, and provide some guarantee of the credibility of the results for decision-makers.

Adopting a research model more or less similar to that of the natural sciences, these studies attempt to find cause and effect relationships through methods such as questionnaires, inventories, demographic studies, which produce data that can be analyzed by statistical tools. Although quantitative research in this field is somewhat more standardized than qualitative research, and more easily replicable, the use of constructs is likely to lead to confusion, and measures of validity and fidelity of constructs and data are just as important as in qualitative research. Indeed, the nature of social phenomena or processes means that the use of quantitative methods to study them risks reducing the data, words, or strategies to statistical equations. When we reduce people's words and acts to statistical equations, we lose sight of the human side of the social life (Taylor and Bogdan, 1984). At this level, measures of validity and fidelity would ensure that the essence of observed realities is not lost through mathematical modeling and the creation of constructs.

Another issue arising from quantitative research methods in strategic management, and related to the above, is the somewhat abusive use of statistical analysis. Quantitative methodology offers 'scientific' management expertise based on and including practical cases obtained through observation, using experience, analysis, statistics, and occasionally mathematical modeling as well, to craft and discuss policies, plans, as well as to monitor their implementation (Bunge, 1998). However, the redundant use of statistics can sometimes lead to misinterpretations. "One might even be led to believe that by the use of statistics one can somehow transform meaningless numbers into something meaningful and that the more complex and sophisticated the analysis, the more meaningful that something is bound to be" (Pedhazur and Pedhazur, 1991). Validity measures offer a certain distance from statistical analyzes. They allow us to keep in mind that the data collected are more or less complete, more, or less accurate, more or less general, more, or less specific to a given context. They thus make it possible to maintain a critical and open attitude towards the statistical results. In addition, they reinforce statistical analyzes insofar as, when the latter are based on valid and reliable data, the relationships brought to light by these analyzes have a greater impact and are of greater interest to a manager faced with a decision to be made. They give meaning to statistical analyzes.

In our case, in both quantitative and qualitative research, the interpretation of concepts by different people involved in the research can be problematic if this interpretation leads to different realities. In addition, and especially in quantitative research, a gap is created between a concept (abstract domain) and its indicator (concrete domain). This could introduce a considerable bias in the measurements made from the data. It is therefore important to check the validity of an indicator, its propensity to assess the concept of interest adequately (i.e., to capture the reality described by the concept). In some research, the importance of the issues and problems raised requires access to information that is difficult and time-consuming to collect. Therefore, the use of a published database appears to be a very accommodating option, allowing access to a wealth of data in a limited time. However, the use of published data for the purposes of a study is likely to introduce a bias, error or weakness into the study, the consequences of which may be more or less disabling for the conclusions drawn. Indeed, the data collected as part of the constitution of a database may not meet the criteria required for the secondary use that the researchers intend to make of it. In other words, the data, being taken for a purpose other than that intended by the research, are not always adequate for the rigorous measurement exercise required by the study. The researchers have no control over the data collection on which their measurements are based. Similarly, the categories created during data collection do not always correspond to the standards that the researchers have set for themselves in the studies. It is therefore important in such studies to confirm the reliability of the data on which subsequent analyzes are based. Without such checking, researchers run the danger of anchoring their conclusions on differences in the data that reflect reliability errors rather than true differences in the phenomena that they are examining (Herbert, 1984).

The rise of experimentation as we know it today, the immeasurable scientific progress it has made in all fields and the resulting craze have marked the beginnings of modern scientific research and have raised this method to the level of a standard in management science: experimentation was instituted and is still the method par excellence of research, whether in the pure sciences, the social sciences, or the humanities. With the scientific revolution, experimentation is evolving. It modifies the place of observation to make it central in the process of validating theories and argumentation. Moreover, we move from passive observation of the context to its modification in order to observe the consequences, and then to the control of external variables. Thus today, experimentation commonly involves or implies the intervention, manipulation or alteration of a given situation or factor in order to determine whether it introduces a difference from the normal. To this end, the design of experimental research is developed in such a way that differences between the control and treatment groups are due solely to the intervention. Experimental research is broadly modeled on the following approach: selection of subjects and assignment of them to treatment and control conditions, preferably using a random procedure; application of the intervention of interest to the treatment group but not to the control group (independent variable); experimental control of the research situation to ensure that there are no differences between treatment and control conditions other than the intervention; measurement of selected outcomes for both groups (dependent variables); and statistical analysis to determine if the groups differ on those dependent variable measures (Lipsey, 1990).

In applied research, and particularly in management, the identification of practical solutions to social problems is of paramount importance. Experimentation is a powerful method of investigating such solutions. It allows the cause-and-effect relationships between two or more given variables to be clearly established. Experimentation takes place in a control-variable setting, where most of the external variables that may cause the effect under investigation to vary are held constant (i.e., controlled), while the "cause" under investigation (the independent variable) is left free to vary within fixed limits. Thus, variations in the effect are attributable solely to this independent variable because the control reduces the influence of other factors and therefore reduces the plausibility of alternative explanations. Causal relationships can be complex, depending on the interaction of a multitude of factors and the presence of particular contexts. In other words, a causal relationship is not systematically expressed and requires a permissive context. At this level, data control is still a strength of experimentation as a method of investigation since by setting the experimental context, the researcher produces a particular context in which an effect may or may not be detected. When a context-conditional causal relationship between two events is known to exist, but it is not known which context allows this relationship to be expressed, experimentation is the tool of choice. When the researcher obtains a variation in the effect, he or she is able to describe precisely the context that allowed the variation to occur. Experimentation has a strong descriptive power of the experimental conditions that allowed a clear cause and effect relationship between two variables. It makes it possible to characterize the conditions that favor or do not favor the causal links, or to verify whether condition A varies the direction and strength of the causal relationship more than condition B.

As mentioned above in the general outline of experimental research, this method uses samples, the units of which are preferably obtained by randomization (i.e., chosen randomly by means involving chance, such as the tossing of unpiped dice or an unrigged coin or a randomization table). The purpose of randomization is to equalize the chances of each group, i.e., to ensure that a selected group has no greater chance than the other groups of having or experiencing an unmeasured factor that has a particular impact on the expected outcome. It thus ensures that the sample is representative of the rest of the population under study and avoids sampling bias. In so doing, it makes it possible to distinguish the variance due to the cause under investigation from that due to confounding variables (unmeasured external variables which covary with the variables of interest or on which these variables depend, and whose variation may mask the true relationship between the variables of interest or falsely lead to the belief that there is a true causal relationship between these variables of interest) which are much more present in the case where a sampling bias has been introduced. Randomization is not always carried out, however, particularly where the researcher is interested in a particular sub-group of the population, or where the presence of certain characteristics is desired, or where the aim is to reduce intra-group variability. However, one of the advantages of randomization is that it provides the study with greater external validity (i.e., the extent to which the causal relationship being studied can be said to hold over and above variations between individuals, parameters, or other characteristics).

The validity of the conclusions of such experimental research relies on the criteria of internal validity and statistical conclusion validity (or research sensitivity) to assess the reliability of potential causal links detected during the experiment. "Statistical conclusion validity is concerned with errors in assessing statistical covariation, whereas internal validity is concerned with errors in causal-reasoning errors (Shadish et al., 2002). Internal validity determines the extent to which the observed covariation between the variables of interest reflects a causal relationship. The researcher must ensure that there is a time course between the variation in the cause and the variation in the effect (that the cause precedes the effect), that the observed covariation is real, and that there is no other possible explanation for the covariation other than causality.

Sensitivity (or the validity of the statistical conclusion) refers to the ability to detect a fundamentally significant change in the effect following the change in the cause when a causal relationship between the variables actually exists. In other words, sensitivity refers to the ability to arrive at the correct statistical conclusion depending on the situation (i.e., to conclude that there is a causal relationship when in fact such a relationship exists between the variables, or that there is no causal relationship when in fact no such relationship exists between the variables). When the experimental design is well elaborated, it is possible to achieve good test sensitivity and to discriminate effectively between causal relationships and other situations, even when the effects produced are of low magnitude. Finally, experimentation uses constructs to relate the characteristics and operations used in the experiment to theories and realities of society. These constructs reflect the socio-economic and political contextual implications in which the research takes place. In doing so, they are powerful indicators of the societal situation and context of the research and allow the research to be reframed in its context. The main strength of experimentation is that it can probe the cause-and-effect relationships between a multitude of existing variables. It is thus able to weigh up competing theories that seek to identify the origins of a given social problem and to find practical solutions.

The main strength of experimentation is provided by the control of the experimental environment and the manipulation of variables. However, there are a multitude of non-manipulable variables, such as strategies, character traits, openness, religious beliefs, culture, etc. These variables are, however, known to have a significant effect on many social phenomena and on the outcomes of management policies. These non-manipulable variables are difficult to study through experimental research. This limits the scope of this method of investigation. It is, for example, impossible to study in this way whether low self-esteem is a cause of an increase in the number of resignations in a company. Furthermore, the control of variables and the resulting setting of experimental conditions can create a situation that is very different from reality where dynamism and constantly changing conditions dominate. This reduces the scope and generalizability of the results. Similarly, experimentation often takes place in a single context, among the multitude of possible contexts in which causal relationships between the variables of interest may arise, and the experiment can hardly be repeated in each of these contexts and test them all. "Few investigators are omniscient enough to anticipate all the conditions that might affect a causal relationship. Even if they are omniscient, a full solution requires the experiment to include a fully heterogeneous range of units, treatments, observations, and settings (Shadish et al., 2002).

On the other hand, the need to control several variables to reduce external influences can be enormously expensive, especially in cases where the samples studied are large in size. Each new condition to be tested or controlled is a significant additional cost. Sample selection, processing, monitoring, and analysis are also significant costs. Thus, the costs involved in experimentation often limit the size of the samples to be studied and the conditions to be implemented. Because of the budgets allocated to the experiments, they are often conducted with sub-optimal sample sizes, and a selection of study conditions is made. In general, the experimenter chooses the context most favorable to the generation of exploitable results. Thus, the experiment is reduced to a given place and time. The implementation of the experimental research design is a relatively long process that can be a problem when the sociological question under investigation requires efficient and, above all, quick answers. Developing the research design, recruiting the sample units, conditioning them, processing, and monitoring them, controlling the experimental data, collecting the data, analyzing, and comparing the data, each of these steps can take a long time. Experimentation has strong probing power, but when the social effect or problem is observed and no avenue for causation exists or is suspected, experimentation is not very useful. Experimentation requires, as a prerequisite, the existence of a set of candidate causes.

Another weakness of this research method is that identifying the effects of a cause through experimentation does not inform the mechanisms of action of that relationship (Amedzro St-Hilaire, 2018). The why and how of the relationship remain a mystery. Experimentation has a descriptive power of the relationship, but its explanatory power is very limited. Finally, experimentation is sensitive to threats to the validity of each of its components. Threats to internal validity, external validity, test sensitivity, construct validity, etc.; each can be a hindrance that limits the scope and/or validity of the conclusions drawn from the research. The researcher must be aware of the factors influencing the sensitivity of the measure, internal or external validity at each stage of the research. The researcher should also be aware that some parameters that allow for a higher level of one component may simultaneously reduce another. For example, while increasing homogeneity within a group reduces intra-group variation and thus increases internal validity, it reduces the possibility of generalization to larger, more heterogeneous groups and thus reduces external validity. The researcher must therefore make trade-offs throughout the research and prioritize the objectives to be achieved, depending on the state of knowledge in the literature and the nature of the research. The same may be true at the level of each individual component. For example, taking into account the sensitivity of the test means taking into account two types of error and determining the power of the test. This last aspect is often absent in management research, which generally uses the commonly accepted value of 5% for type I error. At the first level of analysis, the researcher might compare the relative seriousness of Type I and Type II errors. The specific issues that determine the seriousness of the respective errors will be distinctive to each study (Lipsey, 1990).

Correlation studies and surveys allow the relationship between variables or between a combination of variables to be tested. They have the advantage of being carried out in the field, in the natural context where the variables of interest occur, and of studying their variations as they occur in that context. The relationship patterns found reflect the dynamism of the context in which the relationship takes place. These research designs are designed to capture a phenomenon as it emerges from observation, or as it is perceived by those involved. They are designed to maintain a close correspondence between the data and the reality as perceived at the time of the research. They provide information on the strength and nature of the causal relationship and through the investigations on the mechanisms of this relationship. Correlation quantifies the relationship between variables and measures the association and nature of the association between variables. Correlation is a mathematical ratio measuring the direction and strength of a relationship between variables or between two sets of variables. It is used to determine how much of the variance in one variable is explained by the variation in another variable. There are three types of correlation: positive (if the variables vary in the same direction), negative (if one of the variables varies in the opposite direction to the other) and zero (if the variables have no relationship). This is a relatively simple method and can be done in a very short time. This is an advantage when the need for results is urgent. On the other hand, correlation studies can be used to explore avenues of research into the causes of an observed phenomenon whose causes and solutions remain unknown. Indeed, by determining (in the context in which the phenomenon occurs) the variables that are related to it is possible to find variables which, if studied in greater depth, could determine whether they are potential causes of the phenomenon.

Surveys include examinations, tests, questionnaires, interviews, etc., which aim to study the strategies, opinions, ideas, etc., of a sample. They make it possible to describe the characteristics of a more or less limited sample and to generalize the information obtained. They provide information on how non-manipulable and non-quantifiable variables can influence a relationship between variables of interest. Like correlation studies, they can be carried out in a relatively short time. They can also be applied on a very large scale, allowing for representative samples of the population. Surveys are very suitable for research involving non-quantifiable variables involved in a given phenomenon. For example, to study the relationship between the opinions about technical progress and job satisfaction of employees of a company seeking to implement a new technology, and the resistance to change experienced by that company. Like correlation studies, they can be useful in detecting variables that are potential causes of the occurrence of a given phenomenon. These research methodologies have the advantage of providing results more quickly than experimentation. They are easily implemented and generally use larger samples than experimental research. Unlike experimentation, these methods study relationships that are not necessarily empirical in nature, and generally do not contain a priori controls for variables or external conditions. These research methodologies are not very sensitive: the effect examined must be large enough to be detected satisfactorily. As a result, several successive studies of the same relationship will lead to mixed results. Most will fail to find a satisfactory link between variables (in cases where the effects are small or very moderate), while only a handful will find a relationship between the same variables. In the case of correlation studies, confounding variables are a problem. They induce patterns of relationships that are not systematic when the variables of interest are present. Therefore, it is important to control for relationships for potential confounding variables.

On the other hand, correlation provides a description of the strength and direction of the relationship, but evidence of a correlation between two variables does not indicate that the correlation is due to a causal relationship. Thus, the use of correlation studies to study causality implies that one already knows that there is a causal relationship between the variables, but wants to test the strength and direction of that relationship in a given context. Surveys are very sensitive to the subjectivity of participants and interviewers and may reflect ingrained beliefs or prejudices, rather than reality. They are also very sensitive to the ambiguity of concepts or constructs, and require particular care in the definition of the concepts used and in the use of each of them. Interviewers must constantly ensure that the concepts used are interpreted in the same way by respondents, interviewers, and even readers. On the other hand, they require interaction between the researcher and the participants, an interaction that can alter the reliability of the data if careful attention is not taken at this level as well. Unlike experimentation, the absence of data control makes the results obtained by these methods subject to the vagaries of the situations in which they take place. They are not very reproducible and require a rigorous description of the context in which they were conducted in order to validate them. Validation here implies that a researcher in the same conditions would logically arrive at the same results if confronted with the same data. They do not establish clear cause and effect relationships since other possible explanations have not been ruled out.

This chapter aimed to highlight the inadequacy of partial approaches in analyzing the interactions between strategy optimization and company performance. In the first part, we see the progressive evolution of the results concerning the role of performance in determining the effectiveness of management practices, leading to underline the increasing importance of management optimization in order to gain and maintain a competitive advantage. The chapter therefore returns to the need to restructure the methodological content of studies addressing the issue of strategy optimization in the face of performance. Thus, according to Taylor and Bogdan (1984), whereas qualitative researchers emphasize validity, quantitative researchers emphasize reliability and replicability in research. Nevertheless, measures of validity and reliability are necessary for both types of research to strengthen their results. The list of contexts in which these measures are used is not exhaustive, but some cases can be highlighted. In this respect, ensuring the adequacy of the research methodology becomes important. Do the instruments used allow the research goals to be achieved? In other words, does the methodology used allow for the achievement of the objectives set. If the methodology does not lend itself to the aims of the research and the context, the results obtained, however interesting or true to reality, cannot be considered valid. At this level, validity depends not only on the research procedure but also on the context in which the research is conducted. Validity measures allow us to assess the care taken in developing the research design. In general, these measures allow us to keep a foot in the door of reality, in the consideration of research results in the decision-making process, and to judge the quality of the research. Applied to certain methods, these measures strengthen the credibility of the results by compensating for the method's drawbacks. Decision-makers should view the results provided to them, in relation to measures of reliability and validity, with faith but also with skepticism.

In this respect, since the nature and purpose of the research determine the most appropriate methodology for conducting the study, when a cause-and-effect relationship must be clearly established and empirical data can be obtained, experimentation is more relevant. This is the case when a study is commissioned by a manager who wants to know the cause of an industrial problem and several causes are suspected. Experimentation makes it possible to probe among these potential causes and find a practical solution to the problem. It has a strong descriptive power. Also, when samples are not too large in size, it is possible to subject them to experimental conditions. For studies done for generalization purposes, experimentation is less adequate. Correlation studies and surveys are feasible on larger samples and allow access to groups that experimentation could not cover. These methods are most appropriate when an explanation of the relationship is required, or when the study budget is limited. They allow explanations of causal relationships to be obtained. Correlation is particularly appropriate when a manager is faced with several alternatives and wants to know whether it is worthwhile to devote more attention to one or the other. When the correlation between variables is low, it is preferable to look for an alternative variable with a stronger relationship to the dependent variable. They can also be used to highlight unknown variables in relation to the dependent variable in order to investigate each of them further experimentally. In sum, the empirical approaches to the analysis of the interactions between strategy optimization and firm performance are in a sense incomplete since none of them can answer the other side of the question. This is why it is now necessary to go beyond these partial approaches and find an explanatory model specific to the banking and financial case regarding the nature of the relationship between optimization and performance.

KEYWORDS

- **empirical mechanisms**
- **empirical studies**
- **generalization**
- **heterogeneity**
- **hypothesis**
- **strategic management**

CHAPTER 6

Economic and Management Theories for Controlling Opened Business Context

As the open economy context is one of the elements of this book, it is important to understand its ins and outs. Although economic and management theorists, as opposed to practitioners, have been slow to discover and take into account its importance for business management, several disciplines are interested in it. Each one gives one or more different versions of it. Moreover, the variety that exists, even within the same disciplinary field, is surprising. We have chosen three models which we believe are useful in addressing the relationship between strategy optimization and the performance of companies and institutions. For each one, we will see how it treats the object "open business context," and we will propose a metaphor which summarizes its meaning.

Economics was the first to address the question of the relationship between context and business performance from the perspective of market openness. The center of gravity of economic approaches is efficiency. Logically, they approach the business context as a market, but as an imperfect market. Welfare economics is an important starting point for understanding economic models. Welfare economics is a good starting point for understanding economic models. It favors the market, but considers that the market is 'imperfect' and suffers from 'failures' that it is unable to correct naturally. In this way, it justifies the intrusion of socio-politics into the economic context, in the form of public intervention, to correct these failures. It brings us very close to the classical economic type of reasoning, which still occupies a central place in the study of economic decisions, whether they be economic policies or private strategies. For many authors, the classical economic model, revised or not by Welfare Economics, is unsurpassable. For others, it is the paradigm to avoid.

Value-Based Management in an Open Economy: Optimizing Strategy to Improve Business and Performance. Prof. Post-Dr Walter Amedzro St-Hilaire (Author)

Considering the imperfections of the economy, Welfare Economics advocates the regulation of economic contexts. The regulation model considers that this regulation makes it possible to compensate for the insufficiencies and weaknesses of the industry context and to correct certain forms of imperfections for the good of the actors. Antony Ogus (1994) exposes the reasons, economic, and non-economic, linked to the interest of the agents, which support the regulation of the market. He also illustrates situations of failure and inability of unregulated contexts to generate an efficient distribution of wealth and allocation of resources, and to ensure the interest of the actors. Ogus (1994) reminds us that in reality, several conditions necessary for achieving efficiency through the interaction of forces in an unrestricted context are not met. These conditions are, among others, the perfection of information, competition, the absence of externalities. They reflect the imperfections of the market, which, coupled with the imperfections of private laws supposed to remedy those of the context, make regulation necessary, and this in the interest of the actors.

In order to demonstrate the need for regulation, the author demonstrates the absence of the conditions put forward in the model supporting the deregulation of economic contexts. In particular, he points to the existence of 'natural' monopolies, which arise when it is cheaper for a firm to have all production carried out by one institution than by several (as in cases where there are economies of scale and fixed costs are very high relative to demand). In an unregulated context, competition, and free-market entry would make it less profitable to serve the poorest, smallest, and hard-to-reach areas (Ogus, 1994). They would not be able to benefit from the services offered by the various companies and institutions. Another illustration of the inability of deregulated contexts to ensure the general good is related to the nature of public goods (goods that are accessible to all and whose consumption by some is not at the expense of others). Ogus (1994) demonstrates that the supply and demand of these goods do not reflect their true value and lead to a misallocation of resources and an underproduction of these goods (this would be the case in particular for the supply of electricity and for health care services for the population). Ogus (1994) also provides evidence that the prerequisites for the functioning of a context of pure and perfect competition are not present. In an unregulated context, externalities (cost imposed on a third party, not necessarily directly involved in the exchange operations) are not considered at their fair value. Unregulated contexts do not generate optimal information for a particular decision. Individuals cannot afford to move around the entire market in search of the best deal whenever they want. Full mobility implies that individuals are fully autonomous, independent, and uninfluenced. In reality, they are held back by their work, peer influence, community ties, what people say. Furthermore, the mobility of supply and demand elements, necessary for the achievement of the optimum, is not entirely possible (full information being restricted by the impacts of false advertising, by those of information withholding…).

In the face of these imperfections, regulation facilitates coordination, by setting standards, limiting the costs of information and adjustments required by a lack of coordination. It also balances distributive justice and efficiency, and takes into account the less well-off. Regulation is paternalistic, applying uniform control over certain activities where it is assumed that many individuals 'instinctively' make the wrong decisions. It also reflects the shared values of the context, promotes their application, and takes into account the welfare of future generations. The foundations of regulation must be laid by decision-makers. It is up to them to overcome the imperfections of the context for the good. Based on the welfare implications of regulation, many models have been developed to account for the potential impacts of market deregulation and to explain the economic strategy of agents as a result of the economic context. As banking institutions have been deeply affected by deregulation, we will see to what extent these models are able to answer the problematic posed in this research.

The deregulation of American economic markets, following various reform movements, is one of the most important economic policy issues of our time. Winston (1993) investigated whether the theoretical models of microeconomics developed before the reforms (and supposedly at the origin of some of the reforms) were good enough to explain the mechanisms of regulation and to predict the consequences of deregulation on the environment and firms. Welfare economics criticizes the idealistic nature of the classical economic model and raises awareness that, beyond economic interests, there are non-economic criteria that play an important role in the consideration of agents' interests. However, many theories have questioned the validity of the theory of regulation, arguing on the one hand that regulation is imperfect, that it favors well-organized groups (such as producers) more than more diffuse ones (consumers) and creates more than it reduces industrial inefficiency. Moreover, the costs of maintaining regulation are likely to cause marginal costs or a considerable loss that reduces profits, thereby preventing the mechanisms of competition from working to their advantage (to offer a wider range of products to consumers, and at potentially lower prices).

Regulation prevents firms and institutions from suffering from competition but reduces their possibility.

The proponents of deregulation, who initiated the reform movements, predicted that unregulated competition would replace the service costs of maintaining regulation and reduce productivity imperfections and the effects of overcapitalization. From models, they suggested that deregulation would increase profits in some industries and have little or negative effects in others. This is because, on the one hand, lower costs would increase profits (positive effects for some) and, on the other hand, increased competition would disadvantage (or eliminate) some firms and institutions (negative effects for others). Comparing the observed effects of deregulation in different sectors with predictions made on the basis of theoretical models, Winston (1993) concluded that the general predictions of theoretical microeconomic models were correct but that these models were limited in predicting the details of the distribution of changes. He observed that following deregulation, price changes followed in response to the new cost and competitiveness realities, as expected. However, some unexpected differences were also observed (e.g., due to the adoption of new operating modes and/or new technologies). Furthermore, the effects of deregulation were not shared equally among consumers.

He also noted that the methodology used to make these predictions did not predict the effects of deregulation on the non-monetary attributes of change and did not anticipate the importance of government policies in ensuring the success of deregulation. Early predictions of deregulation were based on empirical studies of the effects of regulation. However, the analysis of deregulation is counterfactual since regulation and deregulation do not occur systematically in the same context. The comparison between deregulation and regulation needs to take into account contemporary changes that might favor or disfavor the outcomes of one or the other (Winston, 1993). A full counterfactual analysis must consider that industry performance is influenced by internal technological changes and external economic events, as well as by policies not directly related to reform. Because of the difficulties in accounting for all the factors influencing market structure, the models did not pay much attention to the effects of deregulation on labor. The main predictions concerned consumers and producers. On the other hand, since the process of deregulation is a continuous one, some effects may not have occurred yet, and some unanticipated effects may still occur. Therefore, the comparison between the observed effects of deregulation and those anticipated may be biased.

Counterfactual analysis requires a high degree of intuition to predict the effects that would occur if the situation were different. Winston deduced from his observations that the accuracy of the predictions depended more on the intrinsic characteristics of the industries than on the methodology employed, because of the difficulty of anticipating the operational changes adapted by firms in response to deregulation in each sector. The speed of technological progress in each industry is an important factor that the models failed to incorporate. This severely limited the scope of economic models (specifically microeconomic models) to predict or consider the operational and technological changes adopted by institutions as a result of regulatory reforms, as well as the response of consumers to these changes. Winston (1993) concluded that the economic models considered allow for the anticipation of aggregate effects, but the predictions remain limited, as they are based on schemes that do not assume any technological change in the activity of firms and institutions.

Within economic approaches, two so-called 'institutionalist' schools attempt to move beyond the Welfare Economics model. Both acknowledge that Welfare Economics has contributed to a better understanding of the socio-political elements present in the economy. However, both criticize it for being prescriptive, prescribing roles for actors without taking into account the contextual reality: according to Welfare Economics, when the context impacts badly, the state must simply act to prevent spillover because there is no other recourse. Institutionalist analyzes claim to offer a more refined explanation of the presence of the socio-political within the economic. They are also more ambitious, since they extend their analysis to all contextual institutions, not just the state. Their name is based on the word institution alism, as they both seek to explain the presence of economic institutions (rather than just exchanges). However, they are at loggerheads because they offer radically different explanations. The controversy between these two 'institutionalist' economic currents is often acrimonious, with each accusing the other of scientific quackery.

Institutional economics dates back to the end of the 19th century. Once known simply as institutionalist economics, it is now referred to as orthodox institutionalism or Old Institutionalism, to differentiate it from neo-institutionalism. It is politically social-democratic (in contemporary terms) and was the basis of the New Deal in the United States in the 1930s. For a long time, this current has been a competitor of neo-classical economics, which it criticizes for its reductionist character and its visions based on the erroneous assumption that economic 'facts' (institutions in particular) are simply the result of the aggregation of economic calculations by actors. He takes it for granted that the socio-economic reality is far more complex. Quite eclectic, he borrows a lot from the social sciences, and in particular from sociology, political science, and social psychology. Its concern for realism leads it to develop fairly complex descriptive models, and to reject the canonical scientific method, and in particular mathematical modeling. Institutional economics considers that institutions (established systems and/or prevalent rules structuring interactions), by acting on the agents' vision of things, modulate their habits and way of thinking. It seeks to build an economy based on concepts and metaphors borrowed from evolution (biology). Emphasis has been placed on more or less artificial evolutionary processes (economic, institutional, and technological) and on the way in which actions are shaped by circumstances.

In contrast to classical economic models, the lack of a precise, consistent, and formal theoretical support for the development of this current has considerably hindered the evolution of institutional economics, and has often hindered its analysis. Thus, in the return of Institutional Economics, Hodgson (1994) attempts to reconstruct the history of the original (or critical) institutional economics, to identify its essential theoretical building blocks and to distinguish it from orthodox economic currents and particularly from the new institutional economics (NIE). Hodgson attempts to reconstruct its original theoretical foundations by looking at the major contributions of the founders of institutional economics, notably John Commons and Thorstein Veblen. Institutional economics places particular emphasis on the importance of routines in cognitive processes, on the limits of efficiency in taking into account the entire context and on the role of institutions and routines in selecting and understanding the context. As actors evolve in a constantly changing context, their conduct overtime is influenced by institutional relationships and institutional change. The historical process and the interaction between agents and institutions must be the basis for a systemic study. This results in a rejection of some of the foundations of orthodox economics that have subsequently been adopted by neo-institutionalism: the atomic view of actors, reductionism, and utilitarianism.

From this rejection arise the two sources of continuing controversy between old and NIE: on the one hand, the utilitarian theory of rational choice to explain industrial strategies, on the other, the consideration of a world existing independently of any structured contextual reflection. Orthodox institutionalism thus criticizes the mechanistic and uncreative view that classical economic theories offer of the efficient economic agent, marginally adjusting to circumstances in a closed system. Because of this static view, these models are inadequate for understanding the economic context. The interest of the NIE in the mechanisms for achieving equilibrium is also criticized. Faced with the inadequacy of classical economic theories, Hodgson proposes to rehabilitate traditional institutionalism by drawing the necessary lessons from the reconstruction of the history of institutional economics. But as he points out, it is important to determine the extent to which economics and management science can employ and rely on biological metaphors.

The NIE (or neo-institutionalism) adopts the insights of old institutionalism but rejects its methods and political overtones. It has a triple parentage: the political philosophy of liberalism, classical microeconomics, and scientific empiricism (operating on the basis of mathematically testable hypotheses). This trend dates back to the 1960s. It explains the presence of socio-economic institutions by the fact that these institutions offer collectively less costly solutions to certain tasks that the multiplication of interacting exchanges would make outrageous. It forms the basis of what some call 'neo-liberalism.' It seeks to generalize microeconomic theory by retaining all the essential elements (the core) of the economic approach, namely stable preferences, the efficient choice model, and equilibrium. Neo-institutionalism is thus distinguished from critical institutionalist contributions that questioned one of the essential components of the economic model: the efficient choice model. It reframes the essential elements of the economic approach in a context of information and transaction costs and constraints on governance rights. Neo-institutionalism thus seeks to generalize the economic approach in the light of these new considerations and apply it to new areas. Thus, exchanges between and within firms are re-analyzed.

Transaction costs are opportunity costs that arise when agents exchange governance rights and are distinct from the costs of executing or implementing a contract. While information costs represent the expenses or losses incurred in acquiring the information necessary to carry out an exchange, transaction costs result from imperfect information about the context, which requires risk-taking and protective (as well as preventive) measures against opportunistic strategies. On the other hand, uncertainty about the quality of goods and about agents' strategies influences production and trade. This uncertainty leads to the need for measures, which also entail costs. The allocation of resources depends on costs, and these transaction costs alter the balance between costs and benefits. Therefore, they must be included in the calculations of economic agents.

Constraints, on the other hand, come from the context in which the actors act. They influence the calculations and optimization decisions. The latter are only optimal in the context (with the constraints) that have modulated them. Thus, actors adapt to changes in exogenous parameters inherent to their environment. They are in a naturally balanced system which, following changes in its parameters, evolves towards a new balance. Central to neo-institutionalism is the efficient choice model, according to which agents maximize an objective function subject to constraints. Governance rights increase the constraints. Under these conditions agents not only seek to maximize their utility in a context characterized by a set of rules and constraints, but also seek to modify these rules to achieve a regime more favorable to their performance. This changes the equilibrium and leads to new rules and a new distribution of wealth, hence to a new equilibrium. Thus, when stability is reached and there are no more modifications in the constraints and rules, the sequence of decisions of the different economic actors to achieve the optimization of the satisfaction of their interests in the new circumstances, lead to this desired equilibrium. Similarly, competition between companies and institutions leads to an industrial (contractual) or institutional equilibrium.

By extending the neoclassical model and adding new variables, the NIE makes the analysis of industrial strategies in specific situations more complex. It implies new realities, such as the anticipation and correct identification of constraints. In sum, at this point in the research, we can already make an initial observation. We have tried to conceptualize the business context of the company by limiting ourselves to a single discipline, economics, a science that claims to be more 'scientific' than others. We might have thought we would find some certainties. This is not the case. The field is crisscrossed by multiple controversies, where each defines the business context in his own way, and follows an analysis that presents profound divergences with those of his competitors.

Economic models shed light on the exchange relationships between different actors, but these exchanges are too often based on purely economic terms (profit, costs, externalities). The business context is understood as a market where more or less formal, more, or less institutional rules compensate for the imperfections arising from the nature of this market. The business context as a market is analyzed by considering almost exclusively the confrontation between supply and demand, and the factors that redefine the nature of this confrontation. However, developments in the economic theory of the firm show us that the context cannot be reduced to such an overly simplistic description. Many aspects are not taken into account by economic approaches. The economic view suffers from the reductionism of orthodox economics, despite the remarkable advances made over the years. It fails to consider elements other than the market in the context of firms (Pasquero, 1990). This context is therefore reduced to the economic actors (producers and consumers) to whom the concept of efficiency is attached. These actors form large, relatively homogeneous, and undifferentiated groups that pursue their interests solely through market mechanisms. The actors maintain mainly contractual relations and the state intervenes in these relations only as an economic agent. Socio-politics is excluded from this perspective.

The second analysis chosen for this research with regard to the open business context is that of political science. Its center of gravity is power. It sees the context as a field of power struggles. It sees the exercise of power as a game in which actors with very different interests pursue their interests through specialized structures and through power relations (understood here as the different means of pressure deployed by the actors to meet their objectives). It also focuses on the actors of this strategy, the pressure groups, by examining how they exercise their power. Political models incorporate into the analysis the strategies of actors to influence the mobilization of contextual resources to their advantage. Thus, there are power relations underlying the adoption of different strategies. In the case of the Banking sector, these approaches make it possible, beyond the economic issues, to consider the process underlying the establishment of different industrial strategies. The state is not only considered for its regulatory power but as a political actor deploying strategies. In the context of public policy, Gerston (1997) introduces the concept of strategies and outlines the factors that contribute to their establishment, in order to remove the confusion and contradictions surrounding the field of industrial strategies. Strategy-making is first and foremost a dynamic and continuous process operating in an open economy. Strategies are governmental actions or provisions responding to economic or technological issues on the agenda of the actors. The main elements affecting the process of strategy setting are the nature of the issues contained in the agenda, the actors presenting, interpreting, or affected by these issues, the resources available to deal with the issues, the institutions dealing with the issues and the levels of government to which the issues are addressed.

Gerston (1997) highlights the variety of strategies adopted. There is variability in each element, so the strategies that emerge are distinguished from each other by their impact: the changes they bring about, the number of people they affect, the length of time they are implemented, etc. As a result, not all issues are treated equally. Gerston (1997) emphasizes the continuous nature of the strategy adoption process. Through the presentation of the main elements involved, he presents the interconnectedness that causes socio-political dynamics and variability. It demonstrates the interdependence between context and institutions, and the need to analyze strategies as a continuum. Strategy-making is not only the domain of government, but also a global phenomenon. Political models allow for a view of the business context that is not possible with economic approaches. They consider the strategies of actors to change conditions to their advantage and to constrain the actions of other actors. They therefore encourage consideration of the political factors and influences that constrain industrial action in a given context. However, like economic models, they are unable, on their own, to account for the reality of the Canadian case that interests us.

There are many approaches applicable to the conceptualization of the economic context, such as interactionist, empiricist, structuralist, phenomenological, and many others. In sociology, the controversies are perhaps even more numerous, more abrasive, and more enduring than elsewhere. We will retain institutionalist sociology because it seems to us the best alternative to the two previous perspectives. Being institutionalist, this sociological perspective defines the business context as a web of institutions as a set of collective rules of operation. It includes several variants that complement each other more or less well. We can thus note:

- Economic sociology, whose ambition is to explain economic facts through the concepts of sociology. As opposed to institutional economics, which seeks to explain the context through the concepts of economics, economic sociology is rooted in sociology;
- Functionalist sociology, which considers that any collective of agents, market, company (public, private, non-profit, pressure group, network, etc.), is ultimately an institution, or has a very high proportion of so-called institutional characteristics, and must therefore be treated as such. In order to explain the economic context, it uses only concepts from sociology. This approach is closest to the theory of the firm;
- The socio-political stream, which is a separate stream in sociology, and would be considered marginal in a department of economics and management. It tries to explain contexts by a combination of sociological concepts and concepts borrowed more from political science (power) than from economics (efficiency).

These three models are in fact largely complementary. Each one starts from the same conceptual bases, enriched or not, of sociology, and explores different facets of the reality of companies and institutions. We note that certain currents of institutional sociology (notably economic sociology and the socio-political current) are attempts at hybridization. In contrast, the second analysis is indigenous. It is based on a long tradition in sociology, functionalism and epistemology It is based on a long tradition in sociology, functionalism, a once glorious epistemology that has fallen somewhat out of favor since the 1970s, and which contemporary institutionalists have successfully revived by modernizing it. This brief overview shows that the term 'institutionalist' is used in very different contexts, and by authors belonging to distinct, even radically opposed, schools of thought. In our view, there is no 'institutionalist school' as such. There is only a common interest from one discipline to another in explaining the formation and functioning of firms and institutions. But the similarities end here.

Everything else is subject to controversy. From this flexibility and generality of the notion from this flexibility and generality of the concept, there is a great deal of diversity in the work on institutions and firms, a diversity that is confusing because of the inconsistencies generated by this work. In order to take stock of the currents related to institutionalism, Scott (1995) recognizes that there are three distinct models of institutions (i.e., regulatory, normative, and cognitive), which may operate at different levels of strategy (some more restricted than others) and which may be promoted or generated by different host systems (culture, social structures, or routine). From this analysis of institutions, three institutional building blocks emerge: a regulatory, a normative and a cognitive. The different theories do not give equal weight to each but consider one or the other as central. Institutional theories differ in the level to which they apply:

- Thus, models based primarily on the regulatory pillar place greater emphasis on the processes of regulation, on the ability of firms and institutions to set rules, to ensure and monitor compliance and to impose sanctions as necessary to contain and regulate strategies. The costs of regulation and monitoring performance are also addressed. This conception implies a realist perspective assuming that agents have natural and innate interests that they pursue efficiently, according to a cost-benefit assessment approach. Explicit rules and references are needed to preserve stability. Firms affect strategies mainly by altering cost-benefit speculations.

- Where the emphasis is on the normative pillar, it is the systems that define the goals and appropriate methods for achieving them and that introduce a prescriptive, evaluative, and obligatory dimension into the context, through a system of values and norms, that are addressed. Companies structure choices and exert their stabilizing influence in the first place, through the establishment and reinforcement of norms and values granting rights, responsibilities, privileges, duties... Actors do not only pursue their interests but act to meet norms;
- Finally, the cognitive model emphasizes the mediating power of symbols. Symbols modulate the subjective meaning that is attributed to the constituents of the context. It considers that agents construct and conceive the context in the light of pre-existing cultural systems, which provide the necessary orientation for dealing with new stimuli. They also determine which aspects should be given greater importance. A cognitive element of importance is represented by the constitutive rules that involve the creation of categories and the construction of a typology of actors. This construction also defines their interests, skills, utilities, and identities. Once these characteristics are established, they serve as cultural models to modulate the similar forms subsequently encountered. These socially constructed characteristics vary over time and space, which explains the variety of actors' interests across time and cultures. Cognitive processes provide a common framework for considering the context in a coherent way. The actors' strategies are explained by routines, habits, contextual identity, and the consideration that certain types of strategies are inconceivable within the constituted framework. These strategies retain a certain consistency through various situations since they derive from the identity constituted from the cognitive processes.

Whatever model of institutions is chosen, they are embedded in a context whose emphasis depends on the model of institution. This framework can be cultural (based mainly on accepted rules and patterns that reinforce strategies but can be changed by those strategies), structural (emphasizing the role of systems and position) or routine. A firm or institution is supported and constrained by a context, to integrate institutional forces within it. Firms and institutions are commonly seen as a multi-facial system incorporating symbolic systems (cognitive constructs and normative rules) and regulatory processes that generate and shape strategy.

These symbolic systems are mutually reinforcing and contribute interdependently to the constitution of a powerful context that provides the institution with its guiding force. In the face of this integrative conception, Scott (1995) considers that institutional theory would benefit from distinguishing these constituents from each other and identifying the specific characteristics of each. Indeed, the analysis of these models from an integrative conception of firms and institutions the analysis of these models from an integrative view of firms and institutions brings out apparent contradictions between theories, which would be considerably limited if this analysis were carried out in the light of the different institutional models. On this basis, he identifies the differences between the considerations and conceptions of the different schools of thought and demonstrates the existence of a three-level controversy: on the emphasis placed on each of the 'three pillars,' on the emphasis placed on each of the systems that generate institutions and on the level at which each institutional model operates on strategies. The different possibilities that emerge from the combination of these factors add a new dimension to the flexibility of the concept and offer a multitude of levels of analysis to the different fields of institutional and industrial study. For Scott (1995), understanding contemporary models is firstly a matter of identifying the level of analysis at which the theories of interest are situated. The conception of reality and its impact on strategies determine the type of institution studied by each. The framework of each analysis must be analyzed before comparing the models, in order to place them in their context.

The context for the optimization and performance of companies and institutions is complex because it includes many aggregates. The different analyzes discussed show us that there are several ways to account for each of the specific aspects. Each one offers us a particular, unique way to analyze part of the complexity of the interaction. This interaction is intrinsically linked to the very nature of the open economy context. To understand the context as an imperfect market (like the economic perspective) is to assume the productivity of projects and institutions in purely economic terms. Indeed, it would mean explaining the different strategies in terms of the costs, benefits, and resource allocation associated with each action. This also implies assuming that these strategies are the result of more or less exact calculations carried out efficiently by essentially maximizing agents. On the other hand, the context as a locus of power advanced by political models, while allowing for a wide range of strategies to be considered and analyzed, provides a rather limited view. It leads us to consider the company or institution as a punctual, homogeneous actor, coherent according to the distribution of powers in place.

The participation of actors in the evolution of companies and institutions is conceived here under the political aspect (establishment of power relations). The omnipresent regulation in the life of companies and institutions makes all the actors intervene by means of pressure means increasing their capacity to intervene in the institutional context and to influence their strategies. However, this perspective does not have the capacity to take into account the redefinition, the evolution of the different agents over time, apart from the relations of power and control is rather limited. Many aspects would hardly find an explanatory framework if only one or the other of the different perspectives discussed were considered. An integrative model to judge the different faces of this interaction is to be sought. Sociological approaches consider that the context is inseparable from economic life. For these approaches, the market is above all an institution defined by rules, formal or not, evolving, and not arbitrary, determined by the powers that be. It offers interesting bases for the pursuit of our research objectives, but a number of difficulties remain. The multidisciplinary characterizing these approaches has the advantage of freeing the analyzes from the constraints limiting the definition of institutions to partial models, but also the disadvantage of the complexity of this structuring.

KEYWORDS

- **classical microeconomics**
- **methodology**
- **neo-institutionalism**
- **new institutional economics**
- **scientific empiricism**
- **welfare economics**

CHAPTER 7

An Operating Framework for Clarifying the Relationship Between Strategy Optimization and Improve Performance in Digitalized Business Context

The main objective of this chapter is to present the operating framework, in order to clarify the principles and foundations that have guided our study. This clarification includes, of course, the methods we used to isolate the empirical facts, as well as the techniques used to formulate them. The chapter discusses, among other things, the issues at stake in the choice of paradigm, which we believe constitutes a system of beliefs about the world and the way in which it should be understood and apprehended (Denzin and Lincoln, 1994), thus providing answers to the questions we ask ourselves about our reality and its scientific knowledge. In this respect, from the diversity of existing paradigms, the choice of a paradigm to define the relationship mentioned, must be based on two different considerations: either consider that, being by nature not demonstrable, the best is to orientate oneself towards the paradigm that lends itself most effectively to the study or to retain the one that seems most likely (Amedzro St-Hilaire, 2018).

We have chosen to consider each paradigm not only as a unique opportunity but also as an alternative way to approach a given reality. The quality of the paradigm is therefore considered in terms of its exclusive correspondence with reality and incorporates the way it helps us to think about it. The chosen paradigm is therefore an objective anchor that takes into consideration relative issues. Hence, considering the theoretical models mentioned above, the objective is not so much to arrive at a functional model, but to elaborate an operational model at the end capable of accounting for the reality specific

Value-Based Management in an Open Economy: Optimizing Strategy to Improve Business and Performance. Prof. Post-Dr Walter Amedzro St-Hilaire (Author)

to the banking sector with coherence (Amedzro St-Hilaire, 2018). In this perspective, we consider a criterion of exhaustiveness to be a necessary condition for the scientific validity of the knowledge produced in the framework of this research, while at the same time attaching equal importance to its acceptability to the actors and to the usefulness it represents for the sector in terms of management (Amedzro St-Hilaire, 2018).

The paradigm used in this research thus mixes empiricist-inspired statements with others of a more substantial nature recognizing objective realities. Indeed, the study of the interactions between strategy optimization and performance from a perspective of observation of meaning is, in our view, resolutely situated in an empiricist paradigm in the sense of Guba and Lincoln (1994). In this view, reality is an experiment that takes multiple, often intangible forms and results from experiences that, although shared, remain essentially specific. This position is close to subjectivism without considering that all reality is only a projection of the actors (Morgan and Smircich, 1980). Rather, it is a conception of the world as a continuous process of creation from the meaningful actions and interactions of actors (Berger and Luckman, 1996). Knowledge, from an empiricist point of view, therefore remains contextual and relative, the object of research being called upon to redefine itself in the interaction it has with reality (Allard-Poesi and Maréchal, 1999). Admitting the existence of a constructed reality, multiple positions can be envisaged on this basis. Empiricism does not seem to us to constitute a unique and coherent form but appears to us to be a set composed of forms, some of which are difficult to reconcile. We distinguish, at least, two distinct forms of empiricism based on the idea that actors construct the world: some think they can give an objective account of the phenomena at work; others resign themselves to interpreting them in a plausible and coherent way (Amedzro St-Hilaire, 2018).

The former envisages a rather positivist-inspired epistemology, based on the principle that scientific research could, in a way, 'transcend' the analysis of actors. Through specific methods, it could access substantial or constructed realities without having to go through subjective representations, thus being able to produce objective knowledge. We opt for the second alternative. Our approach to reality considers the existence of an objective substantive reality while integrating constructed realities. We admit and assume the representational character of scientific knowledge. In this respect, we are in interaction with our study environment, in contrast to the pure positivist position, which assumes a neutral researcher who is external to his or her research object. Thus, we do not attempt to establish absolute causal relationships, nor to predict the strategies of actors, but simply try to increase the degree of understanding of the interaction between strategy optimization and firm performance in an open economy context through multiple explanations (Lincoln et al., 1985). In this way, we recognize and integrate objective and symbolic realities simultaneously. Facts can be analyzed in these terms, but in practice, the interpreted data often mix different levels of analysis whose boundaries remain uncertain. Similarly, these criteria are mixed at the level of the resulting explanatory model. We consider its validity around at least three criteria:

1. **Refutability:** This model must first of all be able to account for the entirety of the facts on the basis of our interpretation within a homogeneous and coherent logic. Given the steps taken throughout the research, this criterion is in fact established. The model, on the other hand, has only acceptable external validity, but some of the statements are subject to assumptions in the determination of the quantitative data;
2. **Plausibility:** The meaning of the explanation produced must be plausible. Stakeholders must be able to recognize the facts of the case in the discourse. This plausibility is achieved when the model is reported to the stakeholders in such a way that they feel it is obvious;
3. **Usefulness:** The representation of reality introduced by the models must also be useful. This refers to the relevance of this research to the issues of our stakeholders. The model must also introduce an order that integrates the concerns of the Banking sector, while giving it control over its management models. It must allow the sector to situate its issues and to envisage possibilities for dealing with them (the utility thus defined does not, however, imply an instrumental model). The evaluation of the model thus combines explanatory, comprehensive, and, to a very lesser extent, operational capacities, each of which takes on greater or lesser importance depending on the evaluating bodies to which the various reports are made (Bournois, 1993).

The methods we have used to collect our data and the techniques used to interpret them allow us to operationalize and put into practice the relatively general statements we have developed so far. In this respect, even if some people believe that by mobilizing a certain category of methods, we are necessarily moving in the same direction, we believe that this convergence at the level of the tools can be misleading from the moment that mixed methods bring together a set of alternatives from very different paradigms. The meaning and relevance of the tools are not given. Data collection methods and interpretation are therefore two vectors which gradually shape the analysis model and which we consider to be moving in the same direction: the methods collect data which are then interpreted. They capture fragments of the reality being questioned, but it is the interpretation that effectively links them to the key concepts. This sequence applies to some extent to the quantitative approach, but it does not fully account for the process followed in the qualitative approach. The explanatory model that we propose to develop must, in time, be able to account for all the facts interpreted on the basis of the empirical data collected. Our interpretation therefore confronts the data collected with theoretical concepts.

To understand the complexities of the relationship between strategy optimization and business performance, it is important to identify the elements of this relationship but also the symbolic richness of these elements in the context in which they exist. It is also important in the case of the banking sector to pay attention to the representations of the actors involved in the shaping of this relationship: reality is first and foremost evidence that presents itself in multiple forms. Consequently, two approaches were available to us in the implementation of our research: the hypothetical-deductive approach and the empirical-inductive approach. Both approaches allowed us to identify a large number of fragments of reality. But in the hypothetical-deductive approach, we had to know precisely what we were going to observe. We had to establish precise indicators that allowed us to consider exactly the same facts from one situation to another. In contrast, in the empirical-inductive approach, the approach is to report as much detail as possible about the situation under study without regard to a pre-established categorization. We do not impose any censorship on ourselves, the selection being made during the interpretation. Given the conceptual framework used, we have opted for a mixed approach (Amedzro St-Hilaire, 2018). This choice responds to research aiming to produce relatively in-depth knowledge about banking and financial institutions, some of the experiences of which it is the scene. This approach allows us to collect the empirical data necessary to determine the explanatory model. This allows us to shed light on the logic and meanings at work in the environment studied: it gives an account of objective reality while bringing out the actors' constructed reality. The method chosen thus does not deny the existence of an objective reality, but allows it to be contrasted with that constructed by the actors.

Therefore, only the empirical-inductive approach is detailed in this chapter on research methodology as the arrangements in place required special attention. Hence, data was collected by adopting the case study as a mode of investigation because we were not able to control the different elements involved in the processes under study: the boundary between the phenomenon and the context was not well delimited (Lessard-Hébert et al., 1995; Robson, 1993). The data that it allows to be collected is considered by many authors (Gillham, 2000; Maxwell, 1997; Robson, 1993) to be the most appropriate way to explore a field in depth. In addition, it favors a rigorous analysis of the multiple aspects of the situation, while generating rich and useful data in an approach such as ours. With this in mind, our research focuses on a specific sector, allowing us to limit ourselves to variations within a single context. Performance is the numerical result obtained at the end of a competition. At the company level, it expresses the degree of achievement of the objectives pursued (Amedzro St-Hilaire, 2017). A company is successful when it is both effective and efficient. Effectiveness refers to the achievement of the objectives it has set itself, while efficiency seeks to minimize the means used to achieve these objectives. The performance of a company can be understood as a multidimensional phenomenon, since the phenomena that measure it are so multiple. The strategy optimization represents the conduct to be followed by the company if it wants to achieve its objectives. It enables the company to achieve its medium and long-term vision. It is therefore a very important tool if the company wants to be competitive because it forces the organization to examine its internal and external contexts, to make projections into the future and to determine the strategies that will enable it to achieve its mission and vision (Amedzro St-Hilaire, 2018).

In the context of open markets, mastering the interaction between these two concepts becomes an advantage for any company that wants to stand out in terms of efficiency and productivity. Thus, to understand the complex aspects of the relationship between strategy optimization and business performance, it is important to determine the elements of this relationship but also the symbolic richness of these elements in the context where they exist. It is also important in the case of the Banking sector to pay attention to the representations of the actors involved in shaping this relationship. In order to take this reality into account, especially in the context of the open economy in which we position Canadian financial institutions, several sources of data and techniques were mobilized to bring out a common denominator of the various situations and points of view that would make it possible to explain this interaction. The plurality in favor of the methodologies used will produce one type of representation of reality among others possible. An inductive approach, the use of various types of qualitative analysis, expressing processes at work, and discourse analysis are all methodologies that make it possible to construct representations of reality (Groulx, 1997).

Indeed, methodological pluralism and data triangulation refer to a strategy of objectification and validation of data. This approach is similar to the work of an engineer where the combination of methods in the collection of data aims to increase "the internal consistency and reliability of the instruments while increasing the power of the analysis." Thus, the multiplication of methodological points of view increases objectivity because it reduces or even cancels out the biases inherent in each particular point of view. This multi-method approach, by diversifying the angles of observation, makes it possible to correct measurement errors and increase the validity of analyzes. Consequently, two approaches were available to us in the implementation of our research: the hypothetical-deductive approach and the empirical-inductive approach. Both approaches allowed us to identify a large number of fragments of reality. But in the hypothetical-deductive approach, we had to know precisely what we were going to observe.

We had to establish precise indicators that allow us to consider exactly the same facts from one situation to another. In contrast, in the empirical-inductive approach, the approach consists of reporting as much detail as possible about the situation under study without regard to a pre-established categorization. We do not impose any censorship on ourselves, the selection being made during the interpretation. Given the conceptual framework used, we have opted for a mixed approach. This choice responds to research aiming to produce relatively in-depth knowledge about Banking and financial institutions, some of the experiences of which it is the scene. This approach allows us to collect the empirical data necessary to determine the explanatory model. Moreover, we have to admit that our study is best understood from a rational observation perspective. Indeed, reality is first and foremost evidence that presents itself in multiple forms.

For the descriptive and analytical purposes of the research, the method chosen to conduct it is a quantitative-qualitative mix that was introduced by post-positivism, a paradigm that assumes that there is an external reality, but one that is impossible to fully grasp. This reality can only be approximated. Thus, the combined use of these two epistemological approaches is in line with the dynamics of research as a process and as a product, which implies many different and complementary approaches in order to capture the characteristics of this reality as much as possible. The data collected is mainly based on secondary data, interviews, and questionnaires. This makes it possible to highlight the logics and meanings at work in the environment studied: it gives an account of objective reality while bringing out the actors' constructed reality. The method chosen does not deny the existence of an objective reality, but allows it to be contrasted with that constructed by the actors. In order to achieve our research objectives, we adopt a methodological approach that is both descriptive and analytical, resorting at times to statistical and econometric analysis of the data in its hypothetico-deductive aspect. For the statistical and econometric analysis of the data, multiple regressions are used. These are intended for the manipulation, analysis, and representation of data.

The productivity of projects does not only depend on the dynamism of the company's own products but also on competitive advantages: infrastructures, public services, education, research, living environment, local subcontracting. This conception of the company requires behavior that is attentive to its effects in organizational terms and not only in economic terms. The performance of companies would then be the result of many factors, including management practices, their capacity to innovate, investments in decision planning and cost control. This is why a mixed approach is favored here. The elements of an open versus a closed economy are specific. A closed economy is an economy that has no economic relations with the rest of the world, unlike an open economy. An open economy exports and imports goods and services to and from the rest of the world. In an open economy, the balance of trade, i.e., the difference between exports of goods and services (EX) and imports of goods and services (IM), i.e., EX – IM, can be positive or negative. Conversely, in a completely closed economy, there would be no exports and no imports. But there is no country whose economy is totally self-sufficient or exclusively outward-looking.

Some countries, such as Canada, are very open to foreign trade. Thus, in 2014, there was EX = 462 billion dollars for a gross domestic product (GDP) of 1,723 billion, i.e., (462/1723) × 100 = 26.81%. This proves that Canada is a country whose economy is very open to the outside world and in particular to the NAFTA countries and more widely to the industrialized countries. The same observation can be made for imports, which in 2014 amounted to 621 billion, or (521/1723) × 100 = 30.23% of GDP. We can see that this is even more important than exports. This is a country that is open to the outside world. This openness manifests itself in 2014 by a deficit equal to EX – IM = 521 – 462 = 59 billion. To understand the relationship between strategy optimization and company performance, it is first necessary to take into account the key concepts underlying the relationship between these two entities. The first concept is that of efficiency. The efficiency of economic actors is at the heart of business theory and has supported its expansion over time. Economic efficiency is a hypothesis of individual behavior that provides a set of canons for judging the actions of an individual or group of individuals. It is often assessed in terms of the agent's goal, the means employed and the outcome of his or her actions. However, this concept has undergone an astonishing evolution over the course of the various theories of business economics, each of which has contributed to bringing it closer to reality.

Constructs are an important tool in applied management. They serve, among other things, as the foundation on which emerging theories are built that will guide the formulation of strategies by management professionals. Reliability defines the level of consistency or convergence that can exist between two sets of observations on an identical object of study; different observers arrive at the same observations for the same phenomenon studied using the same construct. Logical validity (or "scientific" validity), corresponds to the possibility of observing reality independently of the observer, under the same or similar conditions. This will allow the process of generalization and, ultimately, the possibility of proposing laws for the functioning of the phenomenon studied (Amedzro St-Hilaire, 2018).

Empirical validity corresponds to the possibility of making a prognosis, of taking a decision, independently of the proof of the reality of the facts concerned; it defines, as it were, the usefulness of the research conducted. There are several ways to judge the validity of a construct. For Zaltmann et al. construct validity consists of convergent, discriminant, and nomological validity. Convergent validity is represented by the correlation between two attempts to measure the same concept through maximally different methods. Discriminant validity measures the extent to which a concept differs from other similar concepts. Nomological validity, finally, validates an instrument by interpreting the obtained scores in terms of a theoretical concept and consequently generates predictions that, if confirmed, have a validating effect (Zaltman et al., 1973). Measures of validity and reliability reflect the suitability of the constructs for the research in which they are used, and the rigor of the researcher's methodology. Strategy optimization is a strong concept used in the corporate world, in the context of decision making, to address strategic issues. The latter are defined as developments, events, orientations, or trends likely to have an impact on the company's strategy.

In a firm, strategic issues translate individual concerns into organizational actions. They can thus be seen as having both political and informational consequences. These consequences can in turn influence decision-making activity and strategic change in the company. In the face of changing internal organizational conditions, and external environmental variables, the outcome of strategy optimization is, in theory, a viable alternative to ensure that the plan continually realigns the firm's objectives and strategies with changing conditions. Thus, by identifying and exploiting future opportunities, the use of strategy optimization would allow major business decisions to be made more efficiently and to be more closely linked to established objectives. It would also allow for a better allocation of time and resources to identified opportunities, and avoid the time and resources wasted in correcting erroneous or ad hoc decisions. It also helps to create a framework for internal communication between staff and allows for the identification of priorities within the timeframe of the plan. Ultimately, strategy optimization provides a competitive advantage over the firm's competitors. Strategy optimization appears to be a tool for managing environmental turbulence. Stakeholders can be defined as any group or individual that influences or is influenced by the business. They are therefore an integral part of business life and are involved in the running and direction of the business. They are the source of social problems. The notion of interest rate parity highlights the importance of the two variables that explain the formation of short-term exchange rates.

The limits of qualitative methods, in terms of external validity, refer to the law of large numbers, a reality that we admit without difficulty. On the other hand, they are associated with other limitations of a more uncertain nature, such as the fragility of their scientific validity, basing this objection on the subjectivity of the data collected. Faced with these remarks, Denzin (1994) recommends the implementation of different forms of triangulation, with the aim of reducing subjectivity as much as possible. The paradigm adopted in this research integrates the subjectivity of the methodological approach as an inherent and necessary element of any knowledge, even scientific. The frames of reference constructed on this occasion define scientific validation criteria adapted to this characteristic. From this point of view, the only problem with this subjectivity lies in the partial nature of the perceptions we develop in this context. In order to remedy this, the principle of triangulation to which we refer does not aim to eliminate the subjectivity of the research, but to multiply the sources of data in order to access the situation through the greatest number of aspects and visions possible. In other words, the use of triangulation in research does not aim to increase the objectivity of the findings but is a means of multiplying approaches to the situation and integrating its various components. Thus, the elements of triangulation that we have implemented, if they contribute to reducing the partiality of the production, aim at taking into consideration and relating different levels of the studied realities. The forms of triangulation thus implemented within the framework of our methodological device to reduce the partial character are:

1. **At the Level of Data Interpretation:** The selected model was fed back to different actors in the field research. In order to assess the plausibility of the model, this triangulation process aimed to see to what extent its actors found their own experience of the situation in the model;
2. **At the Level of Data Collection:** A triangulation of the actors considered: different stakeholders with different interests and characteristics, were considered and approached in the framework of the data collection. This triangulation is a direct consequence of the problematic as the situation is not exclusively substantial. The understanding of our problematic is only possible if we integrate the point of view of the actors with whom it is confronted and refers to. At this level, triangulation allows us to identify the elements of consensus on which the actors agree. It also contributes to the evaluation of individual contributions in terms of performance. As Bournois (1993); and Igalens (1994) point out, we cannot limit ourselves to the points of view expressed by the management. New potentialities and avenues of research to be explored within the framework of our problematic have thus been revealed to us.

Triangulation in terms of modes of insertion in the field allows us to position ourselves in relation to the situation and to make the most of it, according to the constraints and opportunities it presents. As observation was not always possible in our case (particularly because of the authorities' demands, fearing that our presence would encourage the leakage of strategic information, the exposure of malpractice and the choice of the actors to remain anonymous), we had to resort to the study of documents (even if we were formally forbidden to take out these documents, which were classified as secret), interviews (the parties opted for anonymity) and meeting sessions. These methods of insertion allowed us to access ground that we would not otherwise have been able to cover. The most relevant data sources, which collect information to quantify the relationship between strategy optimization and company performance, are the Multilateral Investment Guarantee Agency and the Bank of Canada. Both institutions have multiple objectives and the data is structured around a number of modules corresponding to the areas of investigation according to the objectives. However, we also solicited data from Finance Canada, Statistics Canada, Industry Canada, DataStream, SIRCA, and the websites of the institutions studied. Four types of data were used to study the relationship between our two key concepts. The first is the questionnaire survey. This was conducted with the Canadian Bankers Association and focused on management breakdowns and dysfunction. More than 30 stakeholders at all levels had to comment on and validate the different versions of the questionnaire, especially with regard to its content. The dysfunctions surveyed met two criteria: they corresponded to reality (validity), and they were noticed in a fairly common way, without this implying that all of them had noticed them (variance).

The respondent, i.e., the head of the planning and control department, was asked to note the presence (yes) or absence (no) of the malfunctions. The survey also aims to collect some indicators of logistical breakdown that are not available at the Bank of Canada. It should be noted that all variables used here are measured at the design level and refer to the year 2014. The survey was conducted in the summer of 2015. Of the 68 questionnaires mailed out, 50 were returned. The return rate was 74.4%, which is higher than what is generally observed in studies of this kind (Becker and Huselid, 1998). The second consists of various databases and constitutes the majority of secondary sources. On the one hand, the Bank of Canada has provided several pieces of information on the competitive advantage of firms in the Banking and financial sector. This information covers 37 banks and financial institutions in Canada. In addition, it, and Finance Canada provided financial reports and market information. This second database and the data updated by Statistics Canada covers a sample of 712 institutions.

The composition of the panel data is made up of firms primarily from Nunavut, British Columbia, Ontario, Quebec, Alberta, and Manitoba. The data includes 4550 observations for the period 1997–2013 with an average of 6.39 observations per institution. A minimum of 4 observations is required for effective use. In addition, annual reports published by the Multilateral Investment Guarantee Agency on the financial and banking sector between 2005 and 2011 were sought. We also used DataStream and SIRCA databases to complete the secondary data. Finally, information on the age of institutions was collected from Industry Canada. The sample consists mainly of North American banks and financial institutions. The third type of data is the desk research. It was used to collect indicators of optimal governance regarding accounting standards and the legal system from the Canadian Institute of Chartered Accountants and Industry Canada. The fourth type of data is the document review. To collect data on management optimization and financial data for the period 2005–2011. This method was used to complete the information on the age of the institutions. This was best done by consulting the websites of the financial institutions studied. Again, the sample consists mainly of North American banks and financial institutions.

As mentioned above, optimization, if done well, can give a company an advantage over its competitors. Understood as the main variable explaining the performance of a company, we need to highlight the different aspects of optimization that contribute to its effectiveness or ineffectiveness. The achievement of a company's objectives and the maintenance of its position over time is influenced by several factors including the human factor. Thus, the individual will within the organization influences or is influenced by the management mechanisms in place both at the level of the company and within the economy, which in our case is open. To study the impact of strategy optimization on the competitive advantage of companies, a multitude of variables were mobilized. Two axes have been identified to understand this interaction. First, we have the effect of the adoption of optimization on competitive advantage by introducing the time factor. We use the effect of time T, the optimization of the P and TP strategies (the product of the time variable and the P variable) in addition to turnover, the GDP growth rate, and other variables. We also studied the impact of the quality of the optimization on the competitive advantage. For the purpose of this second analysis, we used an economic model to measure the lack of competitive advantage, which is reflected in the lack of value-added. This choice is based on studies where this dimension is the main dependent variable studied as well as on the existence of proven models that can be used to accurately estimate the role of performance in the lack of competitive advantage in the Banking and financial sector, holding other determinants constant. The model has four components: lack of competitive advantage in the banking and financial sector (dependent variable), other determinants of comparative advantage (control variables), dysfunction, and logistic breakdown (independent variables).

Logistical breakdown refers to process difficulties and indiscriminate strategic practices and refers directly to the concept of industry practices. Dysfunction, on the other hand, generally refers to the norm in banking and financial institutions. These variables constitute the two dimensions of inefficiency in operational optimization. In addition, three dimensions of effectiveness are examined here: lack of productivity, inadequacy, and unprofitability. To operationalize the analysis model and compare the aggregates, three economic estimation models are used: a production function, a cost function, and a profitability model. The operational model is adapted to the Banking and financial sector. The control variables used are specific to each of the dimensions of the concepts under study. The main advantage of using three different models to operationalize the conceptual model is that each model employs indicators that are conceptually linked to the other models. For example, unprofitability can be seen as the difference between the revenues from production and the costs incurred in generating those revenues. The estimation results for each model can be interpreted in relation to each other because of the complementarity between the indicators specific to competitive advantage. In addition, each estimation model includes its own control variables and two separate indicators.

To capture the effect of strategy quality on competitive advantage we first used factor analysis with orthogonal rotation to group the indicators of logistic break in order to measure it. The method used is the varimax method because this orthogonal rotation method minimizes the number of variables with high contributions on each factor. It thus simplifies the interpretation of the factors. Given the effect of time in the relationship between the logistical break and the postulated costs, we have chosen as an indicator of each logistical break the number of years since it was identified in the system. This measure allows us to take into account not only the presence of these mismatches but also the "maturity" of the mismatch, which is likely to vary its effect on the dependent variable under study. The dysfunction variable is measured by an index composed of four indicators: turnover, failure, and reliability important to the optimization process. The grouping of these indicators is mainly aimed at increasing the content validity of the measure, which thus covers various facets of the dysfunction concept.

The logistic break index created adds up the values of the four indicators previously transformed into a centered and reduced form (Z scores). This sum was then divided by the number of valid values for each of the cases; this way of treating the data makes it possible to consider each indicator as having an equal weight in the index and at the same time to settle the issue of missing values. Finally, the composition of this index is interesting because the measures retained all point in the same direction: the higher the value of this index, the worse the dysfunctions will be considered to be, which should have a negative effect on productivity, accentuate production costs and reduce profitability, leading to inefficiency in the short, medium, or long term. Subsequently, multivariate regression analysis is the statistical tool used to test our hypotheses. Despite its limitations (Becker and Gerhart, 1996), we found this tool to be the most suitable as it allows us to control for the effects of various other determinants of competitive advantage. The variables dysfunction and logistical breakdown were introduced into each of the six basic estimation models of the three dimensions of competitive advantage (lack of productivity, inadequacy, and unprofitability) in order to test our hypotheses. The period covered is the year 2014.

In order to understand the link between competitive advantage and the strategy optimization following their adoption, two methods will be used. A first evaluative approach consists of testing the effects of optimization on competitive advantage using non-parametric tests. This first step allows us to test the static advantage. To do this, the average and median of practices during the three pre-planning years [–3/–1] and the three post-planning years [+1/+3] are calculated for each institution. Subsequently, tests for differences in the median (Wilcoxon test) are applied to the two series of averages. In order to ascertain the significance or non-significance of the values found, additional tests were used: Mann Whitney, sign test, Wilcoxon signed-rank, Kruskal Wallis, Friedmann, and the Kolmogorov Smirnov test. In order to locate the effect of strategy optimization on competitive advantage over time (pre- or post-optimization plan), we use one-tailed tests on the left and right. The second approach concerns the dynamic effects of strategy optimization on competitive advantage. It tests three models. In a first regression model (model 1), the effect of time T, strategy optimization P and TP (the product of the time variable and the P variable) on efficiency and competitive advantage is estimated. Note that time T covers the years of observations; the variable P is a dummy variable that takes the value 0 when the institution does not use optimization and the value 1 when it adopts optimization. The variable TP is a variable that captures the combined effect of time when the institution has optimization tools. The coefficient of P measures for each institution the threshold effect of optimization mechanisms. However, this coefficient is static and unable to show the dynamic effect of planning. To overcome this problem, we use the TP coefficient, which measures the effect of the dynamics of optimization during the strategy optimization.

In order to determine the impact of strategy optimization on banks and institutions, we first looked at internal and external strategy optimization systems or mechanisms as they have a significant influence on the ability of banks to manipulate their earnings. Earnings here are understood in the accounting sense. Internal strategy optimization mechanisms include investment decisions, dividend strategy and optimization structure. On the external side, these mechanisms include the competitiveness of the banking system and the legal and regulatory systems. Among the independent variables, we have those that are directly related to the internal systems (capital structure, dividend optimization, and optimization structure) and those that are usually called control variables and are essentially variables that need to be introduced into the model in order to reduce sub-identification problems. The control variables are size, profitability, default risk, dummy variables, and time. The last two variables are necessary because of the panel data structure of econometric models. The variables can be divided into two groups depending on whether they are used to study policy optimization at the institutional level or institutional optimization in digital economies.

We use two alternative measures for the capital structure: book value optimization leverage (BLOV) and market value optimization leverage (MLOV). The variables used for dividend optimization correspond to two alternative payout ratio measures. The first one (Div1) is calculated on the basis of dividends and earnings per share and the second one (Div2) is calculated according to Pindado et al. (2006). For the optimization structure we also use three other measures for robustness purposes. The first one includes the governance held (PctShClHeld) which is represented by the assets. The second variable (Herf) corresponds to governance practices as measured by the Herfindahl-Hirschman index. The last measure relating to the optimization structure (Own1) is represented by the proportion of shares. Among the control variables, we included size, profitability (ROA), and default risk. For default risk, we chose two alternative measures (Z1 and Z2). The first (Z1) is the Z-score. We decided to also use the version of the score (Z2) for open economy developed later (Altman, 2005).

For the strategy optimization in open economies, we have the following institutional variables: accounting standards, legal system, and banking efficiency. For accounting standards, we use a dummy variable (IFRS) that takes the value 1 if the institution reports its performance under IFRS and 0 otherwise. Following Kaufmann et al. (2011), for the legal system we include variables resulting from the six dimensions of governance: (i) voice and accountability (VA), which is the process by which governments are chosen, controlled, and replaced; (ii) political stability (PS), which measures perceptions of the likelihood that the government will be destabilized or overthrown by unconstitutional means; (iii) government effectiveness (GE) corresponds to the quality of public and civil services, and the degree of its independence from political pressures, the quality of policy formulation and implementation, and the credibility of the government's commitment to these policies; (iv) regulatory quality (RQ), which measures perceptions of the government's ability to formulate and implement sound policies and regulations that enable and promote private sector development; (v) rule of law (RL), which reflects agents' confidence in and ability to abide by the rules of society, and in particular the quality of contract enforcement, governance rights, police, and courts; and finally (vi) control of corruption (CC), which measures perceptions of the extent to which public power is exercised for private ends, including grand and petty corruption, as well as lobbying by elites and private interests. Although these six original indicators range from about –2.5 (low) to 2.5 (high) in governance productivity, the values are not at their extreme.

For the banking system, we used six indicators to measure its efficiency, three of which measure the level of efficiency of the banking sector, and three for the capital markets: (i) Central bank assets to GDP (CBAGDP); (ii) Other financial institutions assets to GDP (OIAGDP); Private credit by deposit money banks to GDP (PCGDP); stock market capitalization to GDP (SMKGDP); International debt issues to GDP (IDGDP); Private bond market capitalization to GDP (PBGDP).

Secondly, we have linked management mechanisms to management efficiency. By uncovering the nature of the relationship between the optimization variables at the level of management practices, we will be able to understand its impact on the efficiency of Banking and financial firms. The factors explaining managerial efficiency identified could be grouped into six major groups: decision structure, decision duality, optimized value-based mechanisms, optimized mechanisms control incentives, capital structure, and external audit. Here financial efficiency is the dependent variable measured by ROA, return on investment (ROIC), and Tobin's Q. The independent variables (the factors of optimized mechanisms control) include the size of the board SSIZE, the independence of the structure SIND, the decision-making forecasts SMEET, the decision-making duality DUALITY, the independence of the audit committee ASIND, the audit frequency ASMEET, the autonomy of the remuneration structure RSIND, equity INSID, institutional contribution INS, public financing SATATE, foreign investment FORGN, management incentives SALARY and audit quality QAT, and ΣX is the vector of the other explanatory and control variables (firm size, leverage, growth prospects, risk, age of institutions and liquidity (LQ) ratio).

Optimization mechanisms can impact the accounting performance of firms via discretionary governance behavior. The method of generalized moments has been used in the econometric analysis here to study this link. The panel data methodology, which covers the period 1997–2013, allows us to control for two fundamental problems in this kind of study: unobservable heterogeneity and endogeneity problems (Arellano, 2002; Arellano and Bover, 1990). The relationships between the characteristics of institutions and their impact on the management of productivity optimization should be interpreted with caution because of the possibility of observing spurious relationships that would favor endogeneity problems. These models could also suffer from unobserved heterogeneity, where the identified relationships are symptoms of an unobservable factor driving the dependent and independent variables.

In such problems, because the independent variables are endogenous and correlated with the residuals of the regressions, OLS estimation is both biased and inconsistent (Brown et al., 2011). As a result, we address the problems of endogeneity and unobservable heterogeneity by using the GMM estimator (Blundell and Bond, 1998; Bond, 2002) which provides us with efficiency gains. Several statistical tests are used. The Hansen/Sargan test assesses the specification of the validity of the model (Hansen, 1996). This test examines the lack of correlation between the instruments and the error term. The AR1 and AR2 statistics measure the correlation of the series. The Wald test is also used to assess the significance of all independent variables in the sample. The Lind-Mehlum contrast is used to study the statistical significance of the proposed non-monotonic relationships (the special case of the optimization structure).

In addition, two models were estimated to measure productivity optimization. Among the measures to assess performance management, discretionary adjustments were estimated following the categorization of Leuz et al. (2003); and Shen and Chih (2007). Since discretionary behavior in performance management can be used to either increase or reduce payoffs, we follow Gabrielsen et al. (2002) and calculate the absolute value of DACC to measure the magnitude of this discretionary behavior instead of its direction. The second model is a cross-sectional model of discretionary regularizations described by Dechow et al. (1995); and based on the Jones (1991) model. Like Cohen et al. (2008), while we calculate non-discretionary accruals, we adjust the reported revenues on the sample firms for the change in accounts receivable to capture any potential accounting discretion resulting from the sale of credits. We then investigate the possible relationship between management optimization mechanisms and management efficiency in the Banking and financial sector. Similar to Himmelberg et al. (1999) and Schultz, Tan, and Walsh (2010), panel data analysis is applied in this study using a panel of banks and financial institutions over the period 2005–2011.

The study uses a comprehensive and holistic approach to examine the relationship between the chosen optimization mechanisms and management efficiency. Firstly, the study uses OLS estimates: although this regression procedure allows causal links to be postulated, OLS results are not always reliable and are often biased. Secondly, the study applies the panel technique to confirm the causal links between the optimization mechanisms and the management performance measures. However, although the panel effect model is used here to account for unobservable heterogeneity, it is unable to correct for endogenous causality problems. In line with the literature, the optimization mechanisms identified to assume managerial variables are exogenous factors to efficiency (Klein, 1998; Mehran, 1995). Also, several empirical researches Demsetz (1983); Demsetz and Lehn (1985); Himmelberg et al. (1999); Denis and Kruse (2000); Demsetz and Villalonga (2001); Dilling-Hansen (2005); Omar Farooque et al. (2007); Wintoki, Linck, and Netter (2010); Schultz et al. (2010) argue that strategy optimization and management efficiency are endogenously determined: optimization can be one of the endogenous variables. The regressions are performed using GMM (generalized methods of moments).

In order to validate our assumptions, the mixed approach to data collection was used in this section. The subjective and objective character of our database was attributed to the longitudinal and cross-sectional aspect of the data, as it was collected over a certain period of time and simultaneously from different entities. The first method of data collection used was the field survey conducted at the Ministry of Finance. However, the data obtained after the survey was supplemented by the collection of information from the companies concerned. The field survey and information gathering provide so-called primary data, which can be quantitative, qualitative or both, depending on the researcher's objectives. The collection period was from January first to July 30, 2015. It allowed us to build up a panel of data over a period of 6 years (2006–2012). For the purposes of confidentiality and relevance of the analysis, the companies were grouped by category of activity banks according to the national accounting nomenclature. These are: investment banking, savings banks, corporate banking, online banking, bancassurance, private banking, deposit banking, universal banking, electronic banking, free banking, cooperative banking, direct banking, network banking and general consultancy.

There are several approaches to measuring the performance of a company. The first is the use of accounting indicators. These are calculated from the company's operating account or balance sheet and are associated with its economic and financial performance. They include turnover growth (Gauzente, 2011), strategy optimization, value-added, gross operating surplus, gross operating profit and net profit (Nalwango, 2009). The second concerns financial ratios. Some authors such as Louizi (2011); Adams and Mehran (2005); Charreaux (1997) have used ratios such as: profitability (the ratio of realized profit to sales), capital turnover (the ratio of sales to investment), ROI (the ratio of realized profit to investment), ROA (the ratio of net profit to total assets), return on equity (the ratio of net profit to equity) and Tobin's Q. These accounting indicators and financial ratios are frequently used to understand performance. They provide an effective summary of the business, but they represent only a narrow view of optimization. They do not offer a multidimensional and global measure of optimization, as do the synthetic efficiency indicators. Since performance is a construct (Lenz, 1981), the use of composite indicators has become popular. These are the productivity index (the ratio of outputs to inputs) and the Malmquist index (Färe et al., 1994).

For the purpose of this book, the following variables were selected:

1. **Output (ptv) in Value:** This is the dependent variable to be estimated. In fact, the initial objective was to determine the impact of optimization on growth. However, given the difficulty of capturing growth for the categories of activity banks, we considered "output," available for all categories of banks, as a proxy for growth. Indeed, growth is a variation of GDP. Now, GDP is the sum of value added of production units, value-added being itself a difference between production and intermediate consumption;
2. **Performance (perf):** This is the synthetic performance measurement indicator obtained from the Data Analysis Control Optimization. This indicator is expressed here as a percentage and ranges from 0 to 100. It was used by Djimasra (2009) who demonstrated that the evolution of the growth rates of the factors was due to gains in technical efficiency, and therefore to optimization;
3. **Malmquist Productivity Index (malm):** This is a synthetic indicator for measuring overall productivity of production factors. This index is measured between two successive years. It is not calculated for the first year of operation. It was constructed using software. The

link between growth and productivity is the work of many authors, notably Solow (1957);

4. **Technological Change (techn):** This is an indicator that quantifies the change in technology between two consecutive periods that a given category of business bank has had to face;
5. **Other Expenses (OTH):** This variable represents all expenses (expressed in millions) that do not fall under any of the expenditure items reported by enterprises. It represents a major component of the operating expenses of enterprises. Also, because of the diversity of elements that can enter into this variable, it has not been considered as an input. It would therefore be interesting to see its influence on the production of the units considered.

The multidimensional aspect of performance allows us to assess its impact on the strategy optimization depending on whether we are interested in its accounting measures or its financial measures. We focused on accounting measures but used several methods to capture the role of firm efficiency in strategy optimization. First, we assessed performance. This assessment is a prerequisite for factor analysis and econometric regression. To measure performance, we use an efficiency frontier method which consists in estimating an efficiency function (Farrell, 1957). This efficiency frontier performance measure has the advantage of providing a multidimensional measure of optimization. Given the deterministic interest, the absence of assumptions on the functional form of the efficiency frontier and taking into account the absence of factor prices taken into account by allocative efficiency, we use technical efficiency to measure the performance of firms through performance data analysis control. The purpose of this evaluation is to determine the performance measurement index of the production unit from the technical efficiency score. Thus, the performance data analysis control is introduced as an input-output ratio represented by the maximization program in order to determine the optimal input and output quantities that maximize the efficiency of each production unit.

Multiple factor analysis technique is applied in the case of multiple tables where the same group of individuals is described by several groups of variables. It allows a characterization of the Banking sector companies according to the relevant variables. Since in our case the variables are all quantitative, we use principal component analysis (PCA) instead of multiple correspondence analysis (MCA). A multiple linear regression model on panel data that would detect the impact of optimization on production is introduced after the construction of the performance index and the characterization of firms. However, before any estimation, it is necessary to specify the appropriate model. The use of the Fisher test has allowed us to reject the hypothesis of homogeneity of the banks' behavior. In order to model the total heterogeneity of individual behavior or effects, a Hausman specification test is applied. It follows that the most appropriate model is the random effects (RE) model whose estimator is the GCM. Indeed, the Hausman test shows an absence of correlation between the individual effects and the explanatory variables.

As noted above, our study is based on a case where data was collected over a period of 12 months on the basis that the case study is particularly suited to research in a real context. If your main objective is to understand what is happening in a specific situation, if you are able to access the setting and are able to get the co-operation of the people involved, then do a case study (Robson, 1993, p. 168). In this respect, case determination involved a number of trade-offs to be kept in mind:

- We could have worked on two sectors. Indeed, it would have been perfectly conceivable to limit ourselves to the variations existing within such a restricted and reasonable case, which would have favored an in-depth study of each of them while allowing for comparative analysis. This choice could not, however, be considered as representative from a statistical point of view; they would constitute nothing more than two possible cases;
- We could have included a larger number of banks. Canada has nine domestic banks, 50 foreign bank subsidiaries and 37 foreign bank representative offices. Canada's major banks, which account for approximately 90% of banking assets, are the heart of Canada's banking system. They provide a wide range of financial services across the country, and this range will continue to grow. Canada's major banks have significant international operations in the United States and most other parts of the world. Canada's banks are well-capitalized, far exceeding Bank for International Settlements standards. By international standards, Canada applies a prudent definition of bank capital. The major banks have well-developed branch networks (approximately 8,000) and automated teller machine (ATM) networks (over 12,500). Retail deposits, a relatively stable source of funds, account for two-thirds of bank deposits. Branch structure and employment levels have been streamlined to some extent, partly

as a result of technological change, and Canada's banking system is considered to be highly efficient. With increasing competition from traditional lending and deposit-taking businesses, non-interest income is an increasingly important source of bank revenue. Bank profitability has improved as the economy has recovered and loan losses have declined, although many small foreign bank subsidiaries have not fully benefited from the recovery. Since the financial sector reforms of 1987 and 1992, there has been a process of concentration in the Canadian financial services system, with banks moving into the securities and trust business, acquiring, or starting up companies. Recently, the government has been working with the banks to improve the delivery of financial services to the small business sector. This would have changed the research considerably: the focus would no longer be on the characteristics of the interaction but on how the context shapes the interaction. Indeed, since we could only integrate a small number of representations for each sector, we would at best have been able to identify a certain trend. On the other hand, it would probably have been possible to identify a more generalizable pattern;
- We finally chose to conduct our research on the basis of a single case. This case allowed us to limit ourselves to the variations existing within a single context: banking and financial institutions.

Fearing that a choice based on a draw, for example among the various forms of banking, would not result in such a satisfactory set, a single-case principle prevailed in the choice of the banking sector. However, the quality of the choice does not lie in the intrinsic quality, but in the variety of aspects found there and the range of possibilities that this variety allows us to study. Our relationship with the companies was established with ExpertActions Group who put us in touch with the managers and the different stakeholders. We worked with these people, supervisory board members and institutional staff for seven months (almost 711 hours of desk research and over 149 hours of interviews). We met with them every week or so, and the rest of the time we conducted desk research. At each visit we reported our findings, which were then used as a basis for discussion. The meetings with the various officials were conducted in parallel. We only met them once, when our explanatory model was ready on 11 December 2015. The purpose of the meeting was to validate the results obtained. Apart from the exchanges with the managers, we took advantage of these visits to hold various discussions with the employees of the services and the unions. We met with the parties and submitted our results to them for validation. We also attended various meetings.

Intended to validate the results of the research, the exchanges that took place there are among the richest in our corpus. It has been shown that the written output of companies, such as annual reports, stock exchange prospectuses and financial accounts, is a primary source for understanding companies (Atkinson and Cofey, 1997). The data sources were written documents: annual reports, company archives, press reports and other company literature, laws, and regulations. These data were important for our triangulation processes. Document analysis is usually introduced as a data collection technique in its own right. We consider it, like Venkatesh, as a fragment of thought in the same way as the interview. It is an already transcribed text in which the company expresses itself on a given subject. The key to analyzing this document is to understand the context in which it is written: who wrote it, on whose behalf, for whom, in what context and for what purpose. On this basis, it can be used in the same way as an interview. Some texts claim to be from the company and are therefore not signed. We believe, however, that apart from certain elements of consensus, a common background emerging from the confrontation of several texts, the official discourse always emanates from the company.

The interview, and more specifically the semi-structured interview, is the method we have chosen to collect 'fragments of thought' about how the participants see the relationship under study. The questionnaire is often contrasted with the non-directive interview, and the semi-structured interview is introduced as an intermediary between these two methods. However, the difference between these three modes of data collection is, in our view, not so much a question of nature as of degree. It lies in the number and size of the fragments collected, in the degree of standardization of the questions asked and in the degree of pre-structuring of the spaces given to the answers. The questionnaire collects relatively small fragments within the framework of more or less closed questions. The non-directive interview, on the other hand, seeks to collect the broadest possible fragments which it then reduces in the context of interpretation. In both cases, however, we are dealing with "fragments of thought." The interview grid we used tends to classify us as a semi-directive interview. However, at the outset, we ask only one question, formulated as follows: Do you believe that the current management practices in which companies in Canada are immersed allow a sector such as yours to be effective? The initial question, which is open-ended, to say the least, is intended, among other things, to limit the halo effect as much as possible, since it is, a priori, sufficiently broad to "drown out" our concerns. A second major objective presiding over its formulation was to obtain fragments of thought opening up access to content beyond that which concerned us at the outset. The space that this question leaves to the participant allows him to express himself freely. Once this question has been asked, we improvise our questions by bouncing off what our interlocutor has said, the aim being to help him or her to clarify and put his or her thoughts into shape and order.

Traditional techniques tell us that the researcher is involved in an informal conversation with the respondent. Thus, he/she should maintain a friendly tone, chatting while staying close to the guidelines of his/her research topic. The researcher starts by breaking the ice with general questions and gradually moves on to more specific questions, while also discreetly asking questions designed to verify the veracity of the propositions made by the respondent. The researcher, still following the traditional techniques, should avoid getting involved in a conversation in which he/she answers the respondent's questions or provides a personal opinion on the subject discussed. He/she avoids this by saying that his/her opinion does not matter or by feigning ignorance. The researcher can reject these traditional techniques and get down to the level of the respondent and engage in a real conversation with "given" and "taken" and empathetic understanding. This makes the interview more honest, more ethical, more reliable because it treats the respondent as an equal, allows them to express personal feelings and in this way presents a more realistic picture than that presented using traditional methods. The use of language and specific terms is very important to create shared meaning (Fontana and Frey, 1994).

The interview is therefore usually described as a feigned conversation, punctuated by reformulations aimed at showing the interviewer that the researcher is following what he or she says. The challenge is not to influence the interviewee. We adhere to this point of view, insofar as we carefully avoid showing our interlocutor the issues that concern us, thus avoiding making him say what we want to hear. This bias threatens the researcher, but it also affects the respondent: the latter may tend to agree with the researcher. The relatively broad question we have chosen does not prevent us from remaining vigilant as to what we can say and do afterwards. But beyond managing this bias, the researcher's absolute neutrality seems to us to be a difficult objective to achieve and does not necessarily seem desirable.

There is a large body of literature that describes at length the rules that must be followed if the data obtained from interviews is to be considered reliable. One of them is to reduce the interaction to achieve a neutrality so that one can access what the respondent really thinks. Like Fontana (1994), we believe that the subjectivity of the relationship is inescapable and that it is better to acknowledge and manage it than to try to reduce it at all costs. Generally speaking, we do not believe that good interviewing practices alone can guarantee the quality of the interview, and if they are misused, they can even be detrimental. If we are too preoccupied with technique and neutrality, we could introduce a cold and uptight atmosphere. The respondent may sense this distance and react to the fact that we are not really present in the relationship.

In contrast to these practices, Venkatesh (1995) invites us, for example, to accept the invitation of the respondent. This involvement in the relationship: sharing something and relaxing, contributes to trust and allows both protagonists to feel comfortable in the relationship. Our belief is that the interview is first and foremost an interpersonal relationship. What is at stake is our ability to enter into a relationship and above all to listen. Techniques such as rephrasing are generally not enough to establish a relationship; moreover, as they are now widely known, they can be confused.

On another note, we believe it is better to improvise than to systematically try to formulate exactly the same questions at the same time. The interview is an undetermined process that we manage without trying to control. What counts, from our point of view, is a truly sincere listening on our part, a genuine interest (Rogers, 1961) in the experience of the other. This sincerity is reflected in non-verbal behaviors that naturally invite the respondent to confide in us. However, despite these remarks, this relationship requires that a number of practical conditions be met:

- In our opinion, it is totally impossible to conduct an interview using a tape recorder and taking notes. This would make the spontaneity necessary for this relationship impossible;
- Isolation and the certainty of not being disturbed. The relationship is a subtle thing that is built up little by little. It is also very fragile and can be destroyed by the slightest disturbance. The interview often leads the researcher and his interlocutor to disconnect from the present situation in order to invest themselves completely in the ideas exchanged. On several occasions, we have experienced relationships ruined by an unexpected visit.
- In addition to the necessary materials, we add a notepad. This is used to note down the ideas that it would be interesting to come back to in order to submit them to the respondent once he has finished speaking.

> This notepad also makes it possible to retain a certain number of elements relating to the interview situation, the material and subjective context in which and in relation to which the respondent expresses himself.

In the facto-empirical approach, interpretation is the foundation of the research activity and its central pillar. It is the interface between field data, i.e., reality, and theoretical models. The main objective of the research is to understand how the explanatory models selected apply to the interpretation of the case of the Banking sector. Therefore, the data analysis grid chosen determined and allowed us to apprehend all the possible phenomena implied by the field data. It allowed us to observe the theoretical and empirical data from the same point of view, or rather to define and characterize them using the same concepts in order to be able to compare them. The set of empirical data collected was analyzed on the basis of the different theoretical approaches chosen, in order to determine how each lends itself to the interpretation and understanding of the case of the Banking sector. The ability of each approach to interpret, fit into the model, and integrate each of the phenomena and relationships implied by the empirical data allowed us to judge the strengths and weaknesses of each approach, and their inadequacy to coherently interpret the case at hand. Finally, a model has been identified that is able to interpret the interaction between strategy optimization and performance in the Banking sector in a consistent and realistic manner.

This interpretation activity involved a constant process of back and forth between the theory and the field data. Our objective at this level was to reach a level of data saturation that would eventually allow us to obtain a model representative of all the data collected. Thus, this process of going back and forth between theories and data continued until the model was saturated, in other words, until all the empirical data found a place in it without the need to modify it. Beyond the saturation reached by our back and forth between theories and field data, the quality of the model lies in its ability to capture the most significant aspects of the reality of the Banking sector. This model is, in the long run, able to account for all the facts interpreted on the basis of the data collected throughout the research.

Miles and Huberman (1991, 1994) denounce the fact that interpretation remains an activity whose paths are rarely made explicit. Based on this observation, they consider data analysis as an activity aimed at reducing the data to a manageable volume, taking into account the cognitive capacities of the researcher and those who will read it. They insist on the need to make the activity explicit and formalize it, and propose a number of tools (matrices, indexing systems, etc.). Miles and Huberman's contribution is interesting insofar as it perfectly explains the questions that arise for us in the context of this activity. The answers provided are, however, from our point of view, rather uncertain. The techniques and tools that the authors propose certainly contribute to the management of data, but they do not integrate the interpretative activity as such: they do not specify the criteria for selecting the statements studied, nor the modalities according to which we must attribute meaning to them. This aspect of data processing seems to us to be difficult to make explicit and formalize in the sense that Miles and Huberman mean it. In a relatively logical way, since interpretation is a matter of understanding the statements, it is not possible, in our opinion, to give an account of them in the form of a causal diagram. Thus, rather than trying to explain how we are to determine the meaning of statements, the reader can try to understand the mode of interpretation to which we refer. For this to be possible, the latter must set out:

- the questions he asks himself;
- the units of analysis to which it refers, i.e., its concepts;
- the principles on which its interpretation is based, the nature given to the data collected and its relationship with the reality under study.

From Miles and Huberman's work, we retain the invitation to rigor in data management. Thus, the explicitness of the different steps avoids implicit treatments. On the other hand, standardization of this process does not seem desirable to us: it would certainly make it easier to compare research between them, but as the Handbook of Qualitative Research emphasizes, methods are plural. The diversity of paradigms and techniques leads to significantly different modes of interpretation. The standards that Miles and Huberman claim should, in our view, be limited to the post-positivist mixed-methods research in which they are engaged. Depending on the nature of the research project and the status attributed to the discourse, the interpretative activity may vary significantly. Different approaches are thus possible. We define our way of interpreting as an activity consisting in attributing meaning to the data studied. In doing so, we refer to our own categories. We seek, for example, to characterize the parameters of the optimization of the strategies faced by the banking sector; this is not necessarily the meaning intended by the representative at the time of the interview, but it corresponds to a level of the reading grid according to which the comments collected will be analyzed. The aim of our interpretation is to establish facts, which we classify into two groups:

- We refer to facts established on the basis of fragments of substantial (or objective) realities as 'substantial facts;' and
- We call the meaning attributed to the words of the Leaders "symbolic fact."

In order to better explain the choices made at this level, we feel it is important to go back over a number of possible modes of interpretation. The interviews studied can, first of all, be interpreted according to a comprehensive approach as defined by Max Weber. This type of interpretation grasps the text at the first level; it seeks to understand the meaning intended by the speaker; the objective being to understand the messages he wanted to convey, the meaning he intentionally wanted to transmit. The reduction of the interview is then done with reference to the categories used by the person. Understanding what the speaker meant is, above all, an exercise in communication that puts our qualities of empathy to the test. This exercise requires an effort of decentration in which we try to put aside our own questioning, our personal categories and try to put ourselves in the place of the speaker. What differentiates us from the ordinary person is very thin here: it is only a capacity to be open to the other's thinking.

A second step in interpreting the meaning intended by the speaker consists in integrating the elements of context (the meaning of an utterance depends, in fact, very largely on the latter). Generally speaking, the approach consists in grasping how the speaker perceives the context of the interview, what relationship he establishes with us. Discourse can, in this perspective, be conceived as a socially situated symbolic action. The third level of interpretation of the actor's meaning is linguistic analysis. This is concerned with the way in which the speaker uses language to produce the meaning of his or her speech. It assumes that meaning is not only in the signified but also in the signifier: the way an idea is expressed is in itself meaning-making. Sociolinguistics can contribute to the interpretation of discourse in any approach, although its aims are different from ours. It is specifically at this level that we refer directly to the meaning intended by the actor in our research. We seek to understand the stakeholders in the Banking sector in order to interpret the data collected in relation to their issues. In general, the way we analyze the interviews is reflected in the statements we select throughout the analysis to illustrate the different arguments. Most of the interpretations we made seemed relatively obvious: the statements could be assigned, relatively unambiguously, to the different categories. Some of the interpretations are based on elements of meaning that are initially more implicit and indirect, in that they emerge as much from the manner of expression as from the content.

The validity criteria of our study are related to our paradigm. A mixed study in a real context cannot claim to meet the positivist criteria of replicability and generalization of results. According to Lincoln et al. (1985), in empirical research, the criteria for quality are rather credibility, i.e., likelihood in our view – and authenticity. Miles and Huberman (2003) indicate that, in general, the quality of mixed studies should be assessed in terms that are specific to this methodical field. Although empirical research is made up of subjective interpretations and so on, the fact remains that studies take place in a real social world, that they can influence people's lives, and that in any particular situation there is a grounded view of what happened – including what was believed, interpreted, etc. (Miles and Huberman, 2003).

Stressing that the empirical world does not tolerate all interpretations equally, Laperriere (1997) shows that questions of internal or external validity are important whatever the paradigm in which the researcher finds himself, and that it is the criteria that must be different. Internal validity "which resides in the correctness and relevance of the link established between empirical observations and their interpretation" Laperriere (1997, p. 404) will be assessed on the basis of a correspondence criterion in a positivist framework, whereas an empiricist position will lead to a search for concordance. Similarly, external validity, which refers to the degree to which the results of a research project can be applied to other populations and situations with the same characteristics, will be assessed according to the capacity to generalize the results in a positivist framework and transferability in an empiricist framework. Looking at the evolution of validity criteria in mixed research, Whittemore, Chase, and Mandle (2001) found that there are still many tensions among scientific researchers where poles such as rigor and subjectivity, creativity, and scientificity are in conflict. Since the recognition of the validity of research is closely linked to the sharing of and adherence to criteria recognized by a community, we have chosen the framework proposed by Miles et al. (2003) to present the elements linked to the validity of our research. Based on both the scientific literature and empirical data from a wider community of researchers working with diverse approaches, these authors have grouped the validity criteria around five main questions (Miles et al., 2003, p. 502), namely:

- Objectivity/compliance;
- Faithfulness/seriousness/audibility;
- Internal validity/credibility/authenticity;
- External validity/transferability/integration; and
- Use/application/prescription.

Their classification meets the criteria generally recognized by scientists as a whole, but incorporates widely used modifications derived from naturalist research (Lincoln et al., 1985). The authors state that this is not a set of rules, but rather a guide with overlapping elements to reflect on the quality of a proposed work, taking into account its particularities, both in terms of the objectives pursued and the context in which the study took place. The presence of a researcher in the field always raises ethical issues. In order to respect the members of the company under study as much as possible and to ensure that they could give their free and informed consent to the study (Robson, 1993), our presence was explicitly mentioned to the participants, as well as the aims and implications of the research, and they were assured that the anonymity of the individuals would be preserved in the publication of the results. There were a few occasions during our visits when people questioned what we had just written. Whenever a participant asked, the sentence that had just been written was reread and explanations were given to the participant on the relevance of this point of view for the research. On each occasion, we also offered to stop the note-taking if the participant felt uncomfortable with the situation. However, this request was never made.

In this respect, ensuring the adequacy of the research methodology becomes important. Do the instruments used allow the research goals to be achieved? In other words, does the methodology used allow for the achievement of the set objectives. If the methodology does not lend itself to the aims of the research and the context, the results obtained, however interesting or true to reality, cannot be considered valid. At this level, validity depends not only on the research procedure but also on the context in which the research is conducted. Validity measures therefore allow us to assess the care taken in developing the research design. In general, these measures allow us to keep a foot in the door of reality, in the consideration of research results in the decision-making process, and to judge the quality of the research. Applied to certain methods, these measures strengthen the credibility of the results by compensating for the method's drawbacks. Decision-makers should view the results provided to them, in relation to measures of reliability and validity, with faith but also with skepticism. They can also be used to highlight unknown variables in relation to the dependent variable in order to investigate each of them more deeply experimentally.

KEYWORDS

- **book value optimization leverage**
- **government effectiveness**
- **gross domestic product**
- **market value optimization leverage**
- **methodology**
- **political stability**
- **voice and accountability**

CHAPTER 8

The Quality of Strategies and the Structural Effect of Competitive Advantage

After having analyzed the theoretical and methodological aspects of the integration of strategy optimization into the central conceptual body of value-based management theory, it is appropriate to question the empirical as well as the theoretical scope of these developments. The models analyzed so far clearly give the impression that the interactions between strategy optimization and business performance, far from being negligible, are on the contrary multiple and complex. The two issues we have distinguished appear to be very closely intertwined. Indeed, each model studied offers its own answer to these two questions. In this respect, it seems difficult to answer one without implicitly answering the other.

This distinction seems to dominate the empirical work, most of which is explicitly devoted to one or the other of these problems. But partial approaches, despite their obvious interest, can be outdated even at the empirical level. In this second part, the inadequacy of partial empirical approaches will be shown without questioning their often-decisive contribution. In particular, it is necessary to go back to the empirical work, and especially to that which subsequently insisted on the importance of the heterogeneity of the processual factor in determining competitive advantages. It is also necessary to test the interest of partial models regarding the effects of strategy optimization on efficiency (productivity and profitability). This will be undertaken by applying to the Canadian case analytical and statistical methods largely developed by previous studies, while trying to identify their limitations. This is essential to move towards a more systematic approach, and will lead to our own conceptualization that we will apply to the study of Canadian

Value-Based Management in an Open Economy: Optimizing Strategy to Improve Business and Performance. Prof. Post-Dr Walter Amedzro St-Hilaire (Author)

banks (banks in Canada are well capitalized and efficient by international standards. Banks in Canada operate in markets that are highly competitive despite the presence of a relatively small number of large players. The trend towards innovative product lines is expected to continue. Until recently, the large banks have been evolving in parallel.

The increased commercial powers conferred after the recent reforms, recent corporate acquisitions and technological progress have allowed and encouraged them to take somewhat different paths): since there must be consistency between the analysis of interactive effects, the main aim will be to empirically propose the most appropriate model to interpret the reality of the Canadian case. Also, it gathers our empirical results: each of these results remains interesting in itself, but their reading, one after the other, allows the reader to get an impression of the universe addressed, before we propose our scientific interpretations in the discussion part. The specific objective method has been directly integrated into each chapter to facilitate the analysis and scope of the results. The section also discusses the empirical results by comparing them with the explanatory power of the different theoretical models. The models analyzed have shown above all that the strategy optimization has an impact on the competitive advantage, performance, and efficiency of companies. Naturally, these impacts, while complex, interact in a specific direction. For example, it can be shown under certain assumptions that productivity costs are a proxy for the profitability costs of internationalization of firms. A complete empirical analysis must, all other things being equal, explicitly take into account all of these interactions by observing, on the basis of facts and statistical data, which model is best suited to explain the reality of the banking sector.

Such a study implies an analysis of the role of the structural factor and more generally of practices in determining competitive advantage. In contrast, this chapter presents, at the forefront, results based on a method developed independently of any elaborate theoretical considerations. The first section will be devoted to the analysis of the effects of optimization quality and operational efficiency in the banking and financial sector in an open economy context. The methodology used will be outlined and the results presented. The limitations of this approach and new developments arising from them will then be analyzed. Also, the effect of optimization on performance structures will be analyzed from a partial point of view, using, where possible, empirical data specific to the banking and financial sector.

Concerning the quality of strategy optimization and competitive advantage in an open economy context, a method is developed here to understand quantitatively whether strategic quality control plays a role in the competitive advantage of firms? Data were collected to estimate the role of two dimensions of inefficiency in operational mechanisms: process flexibility and logistical disruption on the competitive advantage of the banking and financial sector. The method is then applied to the Canadian case. Indeed, the search for a link between strategy optimization and the competitive advantage of the banking and financial sector in an open economy context is an important concern for decision makers. This question (The main question behind the different studies is the following: Does better operational planning result in and contribute to better industrial competitiveness?) is fundamental to industrial economics and strategic management as it touches on the very credibility of these disciplines. While all seek to highlight the value to firms of properly managing the core processes of optimization management, demonstrating the role of strategic boundaries on indicators of competitive advantage is undeniably a key argument for management science and the application of its principles and methods in practice. This section attempts to provide some evidence of this significant contribution.

Also, the optimization process and its indicators are considered by some theorists as the cornerstone of corporate strategy and a relationship between the latter and the competitive advantage of companies is widely assumed. Given the skepticism of some about the importance of optimization in corporate strategies, it is legitimate to ask what is the actual role of optimization in operational competitive advantage in an open economy context. Does the optimization process as the central pivot of corporate strategy empirically have an impact on the competitive advantage of the company? Can these shortcomings affect competitive advantage in an open economy? How much of the lack of industrial competitive advantage can be attributed to process limitations? What specific features negatively influence the quality of the Banking and financial sector?

In order to answer these sectoral strategy questions, this chapter empirically investigates the effect of optimization failures on competitive advantage by attempting to find the relationships between competitive advantage and the characteristics of the strategic flaw (a first step in the appreciation of the interaction between strategy optimization and industrial performance). The aim here is therefore to provide a clear and synthetic assessment of the situation that still prevails in this Canadian sector of activity. In this context, it seems particularly relevant to consider two weak points: dysfunction (failure and reliability) and logistical breakdown. The objective is to find out whether these elements affect the competitive advantage of the Banking and financial sector. The approach integrates the actors, using objective data, as well as dimensions that affect the process of operational quality in an open economy. Therefore, estimates are made using models that are recognized and proven in the literature in order to control for the influence of other variables that may influence competitive advantage: something that studies on strategy generally do not do adequately. The analytical model articulating the various concepts relevant to the study of the problem and allows the research hypotheses to be deduced. The model has four components: the lack of competitive advantage of the banking and financial sector (dependent variable), other determinants of comparative advantage (control variables), dysfunction, and logistical breakdown (independent variables).

With regard to the lack of competitive advantage, the multiplicity and often conflicting nature of the criteria for the notion make it a difficult concept to grasp: measuring competitive advantage means making a judgment on the quality of the banking and financial sector, based on a number of criteria, which are undesired, undesirable, undesired outcomes. This chapter identifies four views of the representation of the lack of competitive advantage in banking and financial institutions: the economic, social, systemic, and political models. Here we stick to the economic model of failure, which highlights a fundamental aspect for the whole sector, i.e., the lack of added value. This choice is also based on studies where this dimension is the main dependent variable studied, as well as on the existence of proven models that can be used to accurately estimate the role of performance in the lack of competitive advantage of the Banking and financial sector, holding other determinants constant. Three dimensions are examined here: lack of productivity, mismatch, and unprofitability. There does not appear to be an explanatory model of competitive advantage that encompasses the complexity and diversity of the factors involved.

However, we identify two categories of determinants from the theoretical models: dysfunction (failure and reliability) and logistical breakdown. These determinants are considered as control variables to assess the effect of optimization flexibility on the competitive advantage of the banking and financial sector. Logistical disruption refers to process difficulties and indiscriminate strategic practices and is directly related to the concept of industry practices. The second aspect retained is dysfunction, which generally refers to the norm prevailing in Banking and financial institutions. For our purposes, dysfunction refers to context, measured by quantitative indicators of attitudes and strategies. As for its place in the model, we had to decide between two ways of conceptualizing it. On the one hand, it could be considered as a moderating variable which interacts with certain other variables to influence competitive advantage, or on the other hand, it could be considered as a full-fledged determinant of non-competitiveness, in the same way as the control variables of our model or the logistic break. We have opted for the second option, i.e., to consider it as an independent variable and to stick to its direct influence on the competitive advantage. This choice seems reasonable, given the literature and our desire to simplify the demonstration.

To operationalize the analytical model and compare the aggregates, three economic estimation models are used: a production function, a cost function, and a profitability model. The operational model is adapted to the Banking and financial sector. The control variables selected are specific to each of the dimensions of the concepts under study. The main advantage of using three different models to operationalize the conceptual model is that each model employs indicators that are conceptually linked to the other models. For example, unprofitability can be seen as the difference between the income from production and the costs incurred to produce that income. The estimation results for each model can be interpreted in relation to each other because of the complementarity between the indicators specific to competitive advantage. The profitability estimation model identifies the net effect of the variables, considering their effects on the two dimensions of competitive advantage (non-productivity and non-costs). This approach has not been used in empirical studies, which generally rely on a pair of indicators without any link between them to corroborate the results. These models do not claim to be complete descriptions of the determinants of competitive advantage and it is certainly not the aim of this chapter to develop such models. The aim is to use models that are sufficiently well specified so that the estimated effects of the variables are not biased.

The logistic break' variable is the subject of three assumptions in this chapter, all other things being equal:

- **Hypothesis 1:** Logistical disruptions lower productivity.
- **Hypothesis 2:** Logistical disruptions increase production costs.
- **Hypothesis 3:** Logistical disruptions reduce profitability.

Regarding the dysfunction, three hypotheses are formulated, all other things being equal:

- **Hypothesis 4:** Dysfunction decreases productivity.
- **Hypothesis 5:** The malfunction increases production costs.
- **Hypothesis 6:** Dysfunction decreases profitability.

The data collected covers 37 banks and institutions. By limiting ourselves to a single industry as outlined in the methodology section of this research, we eliminate various influences that could bias our estimates, thereby increasing the validity of the thesis. The first source of data is the Bank of Canada, which itself collects a variety of information on the competitive advantage of firms in the Banking and financial sector. The second source is a questionnaire survey we conducted with the bankers' association, which focuses on management breakdowns and dysfunctions. More than 30 stakeholders at all levels were asked to comment on and validate the different versions of the questionnaire, especially with regard to its content. The malfunctions surveyed met two criteria: they had to correspond to reality (validity) and they had to be noticed in a fairly common way, without this implying that they had all been noticed (variance). The respondent, i.e., the head of the planning and control department, had to note the presence (yes) or absence (no) of malfunctions. The survey also aimed to collect some indicators of logistical breakdown that were not available at the Bank of Canada. It should be noted that all variables used here are measured at the design level and refer to the year 2014. The survey was conducted in the summer of 2015. Of the 68 questionnaires mailed out, 50 were returned. The return rate was 74.4%, which is higher than what is generally observed in studies of this kind (Becker and Huselid, 1998).

The bank association includes: B2B Bank (1980 McGill College Street, Montreal), Bridgewater Bank (10310 G.A. MacDonald Ave. 8180, Edmonton), CFF Bank (2020 Winston Park Drive, Oakville), Canadian Tire Bank (3475 Superior Court, Oakville), CIBC (Commerce Court West 199 Bay Street, Toronto), Canadian Western Bank (10303 Jasper Avenue, Edmonton), Canadian Hollis Bank (44 King Street West, Toronto), Citizens Bank of Canada (401–815 Hastings Street West, Vancouver), CS Alterna Bank (165 Atwell Street, Toronto), BMO Financial Group (First Canadian Place, Toronto), Equitable Bank (30 St. Clair Avenue West, Toronto), Canadian Imperial Bank of Commerce (10303 Jasper Avenue, Edmonton), (30 St. Clair Avenue West, Toronto), HomEquity Bank (1881 Yonge Street, Toronto), Laurentian Bank of Canada (1981 McGill College Avenue, Montreal), President's Choice Bank (27 York Street), Manulife Bank of Canada (500 King Street North, Waterloo), National Bank Financial Group (600, de La Gauchetière Street West, Montreal), Rogers Bank (333 Bloor Street East, Toronto), Pacific, and Western Bank of Canada (140 Fullarton Street, Talbot Centre, London), Royal Bank of Canada (200 Bay Street, Toronto), Zag Bank (6807 Railway Street SE, Calgary), TD Bank Group

(C.P. 1 Toronto-Dominion Centre, Toronto), Scotiabank (44 King Street West, Toronto), Tangerine (3389 Steeles Avenue, Toronto), Amex Bank of Canada (2225 Sheppard Avenue East, Toronto), BofA Bank Canada (1595 Telesat Court, Ottawa), Bank of China in Canada (50 Minthorn Boulevard, Markham), Canadian Habib Bank (918 Dundas Street East, Mississauga), HSBC Bank Canada (885 West Georgia Street, Vancouver), ICICI Bank of Canada (150 Ferrand Street, Toronto), Industrial, and Commercial Bank of China (333 Bay Street, Toronto), International Bank of Commerce (4950 Yonge Street, Toronto), J.P. Morgan Chase Bank (200 Bay Street, P.O. Box 80, Toronto) 80, Toronto), Korea Exchange Bank of Canada (4950 Yonge Street, Toronto), National Bank of India (200 Bay Street, Toronto), Bank of Tokyo-Mitsubishi UFJ (P.O. Box 42, South Tower, Royal Bank Plaza, Toronto), Sumitomo Mitsui Bank of Canada (222 Bay Street, Toronto), UBS Bank (154 University Avenue, Toronto), Walmart Canada Bank (1940 Argentia Road, Mississauga), BNP Paribas (1981 McGill College Ave. Montreal), Citibank Canada (123 Front Street West, Toronto), Société Générale (1501 McGill College Avenue, Montreal), Bank of America National Association (181 Bay Street, Toronto), The Bank of New York Mellon (320 Bay Street, Toronto), Barclays Bank PLC (333 Bay Street, Toronto), Capital One Bank (5140 Yonge Street, Suite 1900, Toronto), Citibank N.A. (123 Front Street West, Toronto), Comerica Bank (200 Bay Street, Suite 2210, Toronto), Deutsche Bank A.G. (199 Bay Street, Suite 4700, Toronto), Fifth Third Bank (70 York Street, Toronto), J.P. Morgan Chase Bank (200 Bay Street, Toronto), Maple Bank GmbH (79 Wellington Street West, Toronto), Mizuho Bank, Ltd. (100 Yonge Street, PO Box 29, Toronto), PNC Bank Canada Branch (130 King Street West, Toronto), The Royal Bank of Scotland plc (79 Wellington Street West, Toronto), Société Générale (1501 McGill Avenue, Montreal), State Street Bank and Trust Company (30 Adelaide Street East, Toronto), UBS AG Canada Branch (161 Bay Street, Toronto), Wells Fargo Bank (40 King Street West, Toronto), Credit Suisse, Toronto Branch (1 First Canadian Place, Toronto).

Effectiveness/non-effectiveness and control variables: two separate indicators of productivity, efficiency, and cost-effectiveness were used in the estimations. Each estimation model has its own control variables.

Logistical breakdown – we measured logistical breakdown, which required the grouping of logistical breakdown indicators. Factor analysis allowed us to group these indicators based on the latent dimensions of the system. We have presented the results of the factor analysis with orthogonal rotation (varimax), which identified three factors that explain almost 46%

of the variance. In the estimates, three indicators measure, for each institution, the degree to which the logistical break in place emphasizes each of the following three dimensions: mobilization, mechanism, and involvement (contribution to the achievement of objectives). Given the effect of time that appears to play a role in the relationship between logistical breakdown and the costs postulated in Hypothesis 2, we have chosen as an indicator of each logistical breakdown the number of years since it was identified in the system. This measure allows us to take into account not only the presence of these mismatches but also their "maturity," which is likely to vary its effect on the dependent variable under study. This is an interesting way of exploring the role of time in the effect of logistical breakdowns on competitive advantage, while at the same time allowing Hypothesis 2 to be verified.

Malfunctions: Finally, the dysfunction variable is measured by an index composed of four indicators: turnover, failure, and high reliability with respect to the optimization process. These indicators are concrete, easily observable characteristics of dysfunction for which the banking and financial sector compiles data. The main purpose of grouping these indicators is to increase the content validity of the measure, which thus covers various facets of the concept of dysfunction.

The logistic break index created adds up the values of the four indicators previously transformed into a centered and reduced form (Z scores). This sum was then divided by the number of valid values for each of the cases; this way of treating the data makes it possible to consider each indicator as having an equal weight in the index and at the same time to settle the issue of missing values. Finally, the composition of this index is interesting because the measures selected all point in the same direction: the higher the value of this index, the worse the malfunctions will be considered to be, which should have a negative effect on productivity, accentuate production costs and reduce profitability, leading to inefficiency in the short, medium, or long term.

Multivariate regression analysis is the statistical tool used to test our hypotheses. Despite its limitations (Becker and Gerhart, 1996), we found this tool to be the most suitable as it allows us to control for the effects of various other determinants of competitive advantage. The variables dysfunction and logistical breakdown were introduced into each of the six basic estimation models of the three dimensions of competitive advantage in order to test our hypotheses.

The estimates on productivity or non-productivity show that all dimensions of logistical breakdown have a positive and statistically significant relationship (or close to it in the case of one of the six estimates) with the non-productivity indicators. We also find that the results are more mixed for production costs, where none of the estimated relationships are statistically significant; the coefficients on two dimensions ('mobilization' and 'involvement') are negative, while the coefficient on the 'mechanism' dimension is positive. Finally, the estimates for unprofitability indicate that each of the dimensions of the logistic break has a positive but insignificant effect on unprofitability. By virtue of the logic linking these criteria together, a net positive effect of the dimensions of logistical breakdown on unprofitability was expected given their positive and significant impact on non-productivity and their mixed impact on costs. These results comfortably corroborate Hypothesis 1 to the effect of a positive impact of logistical disruption on non-productivity. Hypothesis 3, however, is not verified by virtue of the results obtained with non-productivity. It is not so surprising that the effect of logistical breakdown remains marginal since this measure of non-efficiency is influenced by a myriad of other factors and also because logistical breakdowns are likely to have a more direct influence on intermediate outcomes. As for the mixed effect of logistical breaks on costs as posited in Hypothesis 2, it is corroborated. This means that, when the presence of logistical breaks alone is taken into account, the estimated effect is positive for all dimensions of the system and significantly so for the "mechanism" dimension. Taking into account the number of years since the observation thus mitigated the effect of logistical breakdowns on costs.

With regard to malfunctions, the estimates reveal that they are associated with lower productivity. As for the cost estimates, the dysfunction variable is positively related to costs: dysfunction is significantly associated with higher costs. Hypotheses 4 and 5 are thus corroborated. As for Hypothesis 6, given the negative effect of malfunctioning on productivity and its positive effect on costs, its net impact on profitability was expected to be negative. The estimated coefficients have the expected sign, but the relationship is not significant at conventional levels. Hypothesis 6 is therefore not supported. In summary, the results show that dysfunction has a generally positive role on competitive advantage: positive on the competitive advantage of the banking and financial sector in Canada.

Two conditions for the influence of logistical disruptions on practices optimization in an open economy context are identified: their constitution as a system and their "duration." Firstly, it is found that it is entirely appropriate to study logistics disruptions from a system perspective in order to fully measure their effects. The configuration approach used here provides a better explanation of the effects of these disruptions on competitive advantage than the universalist approach. A practical implication of this finding is that one should not think that sectoral competitive advantage is substantially influenced by identifying a logistical break. In order to have a substantial impact on the lack of competitive advantage, it is necessary to play on several dimensions and to identify mutually reinforcing and complementary logistical breaks. The dimensions of the logistics disruptions in place in an open economy context also appear to have been well identified by the factorial analysis, which allows a detailed analysis of the specific effects of each of these dimensions of the system.

The first dimension, 'mobilization,' is probably the most likely to have a positive role on the lack of competitive advantage, among others on non-productivity. The second dimension, the 'mechanism' dimension, consists of typical breaks in the strategic process, which can be found even in several economic sectors. This may explain the significant positive role on non-productivity and the positive effect on costs. As for the last dimension, the "involvement" dimension, the results are somewhat more ambiguous and less easily interpreted. If there are no 'strong' weaknesses, are there dimensions that are more likely than others to lead to a lack of competitive advantage? It must be admitted that the chapter does not allow for a definitive conclusion on this point. It must be emphasized that there are various links between these dimensions in logistical breakdowns that statistical tools do not necessarily allow to differentiate. For example, would mobilization or involvement be as effective without practices linked to mechanisms? We doubt it, given what the results tell us about the need to consider breakdowns in a systemic rather than specific way.

The second condition for the influence of logistical breaks concerns their "duration." The results obtained here with costs are particularly interesting in this respect. Thus, considering the number of years, logistical disruptions appear to have a mixed role on costs. By comparing these results with others, it seems that the longer the malfunctions have been in place, the less they contribute to cost increases. Considering that they involve various costs (implementation, running-in, etc.), this chapter shows, however, that these seem to be amortized in time. It may be that malfunctions can take some time to play a major role in the lack of competitive advantage in the sector and that we should therefore be patient before making a judgment on the value of one or more malfunctions. Moreover, the results demonstrate the interest of considering dysfunction in an explanatory model of the effect of the limits of operational optimization on sectoral competitive advantage in an open economy context. First of all, dysfunction appears to be an important explanatory variable in the determination of competitive non-advantage. Our results indicate that dysfunction plays a non-negligible role in the inefficiency of operational optimization. In order to measure the role of operational and strategic performance on the lack of competitive advantage, one should not rely on a single characteristic such as logistical breakdowns: dysfunction must be taken into account. Furthermore, conceptually, we note that dysfunction and logistical breakdowns are quite distinct features of the strategic management system. Thus, one could have breakdowns and dysfunctions and vice versa. One thing is certain: no matter how well one solves logistical breakdowns, the fact remains that dysfunction is a factor that can undermine competitive advantage. All other things being equal, in practice, we must also act on dysfunction if we want to obtain all possible benefits from the operational management system.

This chapter, which is essential to the mastery of the interactions between strategy optimization and efficiency in the Banking and financial sector, shows, among other things, that the quality of optimization plays a role in the problems of competitive advantage in the context of an open economy and contributes to the current literature from various points of view. First, the model allows for an analysis based on features rarely considered together in the theoretical models reviewed earlier in the first part of this research: dysfunction and logistical breakdown. The influence of several of the many determinants of competitive advantage is likely to bias the measurement of the precise effect of boundaries in this kind of study. Here, this influence has been reduced by a variety of means, including the use of proven models of competitive advantage and the restriction to a single sector (the banking and finance sector). Another contribution is the use of three interconnected dimensions of competitive advantage that further corroborate the findings here.

The chapter thus highlights certain limitations that our explanatory model should make it possible to overcome. The first limitation is that the sector studied is composed of elements that are limited to a single context. It would be interesting to take up the conceptual model and possibly apply it to other sectors. A second limitation relates to the problem of simultaneity, i.e., the situation where a variable can be both a cause and an effect. Although we have assumed that optimization failures affect competitive advantage, our cross-sectional estimates do not allow us to rule out the possibility that the relationship between these variables runs in the opposite direction or is even circular. The strategy adopted here remains quite appropriate because it is based on the consistency of the analytical model we have developed. The issue of causality between strategy optimization and competitive advantage can be examined using a system of simultaneous equations or multi-year databases. Although longitudinal studies do not solve all methodological problems (Huselid and Becker, 1996), this type of study may be preferred in the future.

For the models supporting the lack of outcome, it is possible that there are still variables that could influence it that have not been taken into account. It would have been interesting, for example, to examine the effect on management effectiveness. Qualitative studies with actors in the field would also be needed to further examine the mechanisms and factors involved in order to corroborate the results obtained here. The evidence provided in this case makes an original contribution to the advancement of knowledge on this complex but important issue. However, to account for this relationship between strategy optimization and competitive advantage, a more complete model would need to be specified, including a mechanism to account for expectations regarding the future level. The use of the notion of competitive advantage in a context in which no optimization or interest rate adjustment process takes place may seem surprising. However, the notion is given a rather vague meaning here: an anticipation is a forecast of the level of a variable based on the past distribution of the levels of that variable. This is a restricted definition since it does not imply, for example, the idea of conditional probability, in which the realization of a given level of the variable depends not only on the distribution of the variable but also on the past and current levels of the other variables.

KEYWORDS

- **conditional probability**
- **hypothesis**
- **mobilization**
- **optimization**
- **statistical methods**

CHAPTER 9

Optimization Mechanisms When Confronted with the Variability of Sectoral Performance Factors

This chapter mainly focuses on the effects of OMR mechanisms on sectoral performance. While over the years, OMV has become a popular topic among academics and practitioners (Dilling-Hansen, 2005; Leng, 2004) in applied economics and management, the recent financial crisis has refocused the importance of this issue on the banking and financial sector in open economies. It is well accepted in this respect that optimal decision-making plays an effective role in the management of firms, the production of reliable financial information and the enhancement of investor confidence. In addition, this form of strategy optimization would play a key role in enhancing transparency while mitigating conflicts of interest. The objective of this chapter is, at its core, to uncover the nature of the relationship between the variables of management practice optimization and its impact on the efficiency of Banking and financial firms. This chapter thus provides a more in-depth examination of the relationship between optimized mechanisms value-based mechanisms and the financial efficiency of firms in today's globalized market environment.

A number of models have addressed the structural effect of strategic decision-making mechanisms on the effectiveness of firms. Jensen (1993) has shown that size is important in the effectiveness of optimized mechanisms control structures. This was confirmed by Cheng, Evans, and Nagarajan (2008), who states that there is a significant relationship between small optimized structures and better management efficiency. For their part, Eisenberg, Sundgren, and Wells (1998); Jensen (1993); and Yermack (1996) have shown a significant negative relationship between market value and

Value-Based Management in an Open Economy: Optimizing Strategy to Improve Business and Performance. Prof. Post-Dr Walter Amedzro St-Hilaire (Author)

structural size of optimized mechanisms value-based mechanisms. They note that smaller size can improve the effectiveness of the firm's structural optimization. These results are consistent with Hermalin and Weisbach's (2001) assertion that management structure is endogenously determined as one of the variables of management optimization in firms. These authors found a negative relationship between the size of management structures and the efficiency of firm optimization, even though the composition of the structure has no influence on financial efficiency. Also, similar results are reported by Drakos and Bekiris (2010); and Filbeck and Lee (2006), arguing that there is a negative correlation between size and the effectiveness of optimized practices.

Bhagat and Black (2002) examine (them) the management optimization variables (independent mechanism and agency size) that affect the effectiveness of practices. The results highlight that the size of the arrangement has no significant impact on the effectiveness of optimized mechanisms control. However, a negative relationship is found between the independence of the mechanism and the effectiveness of the practices. These results are confirmed by Bonn et al. (2004) who indicate that the ratio (specific planners, external decision-makers ratio, and practice diversity) has a significant positive relationship. Also, Florackis (2008) argues that small structures are less effective than large ones and therefore large mechanisms are essential for large firms. Bonn, Yashikawa, and Phan (2004), find a positive and significant correlation between size and structural effectiveness of optimized mechanisms control in large companies.

Elsayed (2011) examined the effect of the size of management planning mechanisms on firm effectiveness. The results of the study found a significant positive relationship between size, managerial effectiveness, and planner quality in the context of non-duality. However, a negative relationship is found between size and managerial quality on firm effectiveness. More recently, Uadiale (2010); Kim et al. (2012); and Fauzi and Locke (2012) report a positive significant relationship between the size of management mechanisms and firm efficiency in the situation of strategy optimization. This result differs from that reported by O'Connell and Cramer (2010) showing a negative significant relationship between firm size and efficiency. However, they also found a significant positive relationship between the ratio of external strategists and practice effectiveness. As well as structural size, the independence of mechanisms plays an important role in the process of reducing conflicts of interest between the majority investor (the capitalist) and governance (strategists). The fit between the mechanisms and the optimized practices of the firm suggests that the management structure must be independent of decision making in order to reduce the risk of moral hazard. The findings of Hermalin and Weisbach (1988); and Bhagat and Black (2002) showed that lower-performing firms were more likely to increase the proportion of external strategists than higher-performing firms, suggesting a tightening of supervision to mitigate agency problems.

However, some studies have found that improved effectiveness of optimized mechanisms control is more a result of these changes. Brown and Caylor (2004) found that companies with a higher proportion of structural independence had higher returns on equity, higher profit margins, higher dividends, higher returns, and greater share buybacks. External stakeholders cannot always add value to the optimization of a company's management as not all external structures are truly independent. Bhagat and Black (2002) have suggested in this respect that many external structures are involved in business relationships with the company. Even if these structures should be truly independent, they would not be able to make appropriate decisions. These independent structures may simply not have the qualities required to be effective. Pass (2003) points out in this vein, that external structures do not always have the industry knowledge to adequately support or guide the optimization of the company's strategies. Roberts et al. (1998) state in this sense that the dynamics of optimized mechanisms control practices tend to fail as long as these major reforms are more control-oriented than strategy-oriented. If the objective of considering independent structures is to mitigate agency costs, risks may indeed be reduced, but the effectiveness of optimized control will not necessarily be improved (Kapopoulos and Lazaretou, 2007).

There are other studies that suggest that there is no relationship between management quality and practice effectiveness. Paul and Polytechnic (2011); Kiel and Nicholson (2003); Dalton, Daily, Ellstrand, and Johnson (1998) confirm no significant association between the quality of optimized mechanisms control arrangements and the effectiveness of firms under strategy optimization. On the other hand, Choi, Park, and Yoo (2007) find a significant positive relationship between the independence of optimized mechanisms control structures and the effectiveness of practices. This result is consistent with the contributions of Schellenger, Wood, and Tashakori (1989) who show a positive effect of structural independence on firm effectiveness. In contrast, other studies show a significant negative relationship between independent structures and the effectiveness of strategy optimization (Agrawal and Knoeber, 1996; Yermack, 1996; Klein, 1998).

The frequency of decision forecasts also plays a crucial role in the optimization function. Empirical studies show that companies with planning activities such as management decision forecasting have more efficient management functions (Nikos Vafeas, 1999; Yatim, Kent, and Clarkson, 2006). For Finegold et al. (1998) this is because the optimization of management practices plays an important role in improving the effectiveness of the management structure. Although the presence of independent structures is essential, it is important to consider the level of optimization in the effective management of predictive decision tools (Brick and Chidambaran, 2007; Khan, 2006). Other previous studies have shown that the frequency of forecasting is positively associated with the effectiveness of business management practices (Brick and Chidambaran, 2010; Conger et al., 1998; Liption and Lorsch, 1992; Rosenstein and Wyatt, 1990; Nikos Vafeas, 1999). Based on these empirical studies, our hypothesis regarding the management structure is the following:

- **Hypothesis 1a:** There is a significant positive relationship between firm effectiveness and the size of the optimized mechanisms structure in Banking and financial institutions.
- **Hypothesis 1b:** There is a significant positive relationship between corporate efficiency and the independence of optimized mechanisms structures in Banking and financial institutions.
- **Hypothesis 1c:** There is a significant positive relationship between business efficiency and the quality of management practices in Banking and financial institutions.

The quality of decision making is an important part of the management optimization mechanism. Decision duality exists when decision making is directly associated with the management structure (Boyd, 1995) because differentiation plays an important role in the proper functioning of strategic planning. In order to avoid any ambiguity in their respective roles, the management structure and the decision-making structure of a company must not be fully dissociable. This is confirmed by Jensen (1993); and Fama and Jensen (1983). Since one of the main tasks of optimized mechanisms control is to evaluate practices, if the structures are not decoupled, the ability to evaluate practices may be affected (Jackling and Johl, 2009). In the same sense, Jensen (1993) argues that decision-making concentration can lead to partisan decision-making to the detriment of stakeholders. A company seeking to increase its overall efficiency should diversify its strategic decision-making.

On the other hand, there are authors who defend decision-making duality as something positive for the effectiveness of optimized mechanisms control (Bradbury, 1990; Gendron and Bédard, 2006): they argue that if the overall power of the company is concentrated within the same structure, there will be fewer conflicts of interest and management is facilitated, thus achieving greater efficiency in practices. In this sense, the literature is contradictory as to the results: some find a positive relationship (Bradbury, 1990; Willekens, Bauwhede, and Gaeremynck, 2004), and others a negative one (Akeel and Dennis, 2012), or even consider that there is no optimal structure, but various influential factors (Elsayed, 2007; Vineeta, Vic, and Barry, 2009). Therefore, we will test the hypothesis that:

➢ **Hypothesis 2:** There is a significant negative relationship between practice efficiency and decision duality in Banking and financial institutions.

Optimized audit mechanisms have become one of the main elements of management optimization (Verts, 1994). Several characteristics of effective auditing practices have been discussed in the previous literature. Keung, Robin, and Tessoni (2007) determine the effective role of the optimized audit mechanism by highlighting the quality of the audit and its impact on improving the efficiency of firms. They conclude that optimized audit mechanisms play an important role in management optimization. Collier and Zaman (2005) indicate the importance of optimized mechanisms and experience in judging the effectiveness of firms. To get the most out of optimized audit practices, it is imperative that the audit mechanisms are effective and active. Optimized audit is not the only factor influencing management effectiveness as audit standards also affect its effectiveness.

Also, Zoort et al. (2002) suggest that one of the main functions of audit mechanisms is to monitor the integrity of performance statements and reports. Other responsibilities of optimized audit practices include the supervision of external controls and activities (Krishnan, 2005). Anderson, Mansi, and Reeb (2004) report evidence that companies with better optimized audit mechanisms are more successful with shorter financial statement monitoring times. Thus, Aechambeault and DeZoort (2001); and Vafeas and Waegelein (2007) found that many grants are given to encourage external mechanisms.

Based on these empirical studies, our hypotheses regarding optimized audit practices to be confirmed or refuted are as follows:

- **Hypothesis 3a:** There is a positive relationship between business efficiency and optimized audit practices in Banking and financial institutions.
- **Hypothesis 3b:** There is a positive relationship between business efficiency and the frequency of audit optimization in Banking and financial institutions.

Empirical evidence on the relationship between optimized mechanisms control incentives and management efficiency is mixed. Jensen and Murphy (1990) provide evidence of a strong relationship between the quality of optimized mechanisms control premia and stock returns. They suggest that optimized mechanisms control premiums play an important role in mitigating agency problems. A similar result was found by Kren and Kerr (1997). They reported a positive relationship between firm efficiency and advance compensation, but found no significant association between optimized mechanisms control premiums and firm efficiency. Vafeas and Theodorou (1998) report that this form of optimization has no impact on firm efficiency.

While Calleja (1999) finds that firms with an optimized pay structure have higher returns to shareholders than firms without pay optimization mechanisms. Laing and Weir (1999) confirm the existence of a positive association between optimized compensation and firm efficiency. Klapper and Love (2004) investigate the relationship between optimized compensation mechanisms, including decision structures, and the operating efficiency of firms. They report that better planning is positively correlated with firm performance.

Similar results were reported by Bozec (2005). Christensen, Kent, and Stewart (2010) find that optimized compensation mechanisms are positively related to financial performance (in line with expectations from agency theory). These results are confirmed by Benson, Hutchinson, and Sriram (2011) showing a positive relationship between optimized practices and performance. More recently, Malik (2012) reveals that the existence of such compensation mechanisms is positively related to prices. On the other hand, the findings of Lam and Lee (2012) show a significant negative relationship between optimized compensation mechanisms and managerial efficiency, but a significant positive relationship between compensation practices and managerial efficiency.

Therefore, this section tests the impact of optimized remuneration mechanisms on management effectiveness through the following hypothesis:

- **Hypothesis 4:** There is a positive relationship between business efficiency and optimized compensation arrangements in Banking and financial institutions.

The relationship between capital structure and the effectiveness of optimized practices is mixed or inconclusive. Studies show that there is either a positive or a negative or endogeneity relationship. Demsetz (1983) argues that the composition of capital is endogenously determined and that there should be no systematic relationship between capital structure and changes in business practices. This is confirmed by Demsetz and Lehn (1985); and Demstez and Villalonga (2001). Their studies show that the capital structure does not modify the value and practices of the firm: confirming its endogenous character. A similar result is reported by Welch (2003). The endogenous character of the capital structure is also found in small firms. Dilling-Hansen (2005) reports a non-linear endogenous relationship between the composition of capital and the firm's planning practices. Farooque et al. (2007) examines the relationship between the variables of mechanisms optimization and business performance through capital structure. They conclude that capital composition is negatively related to the efficiency of management practices.

A second category of studies on the relationship between capital structure and the effectiveness of optimized mechanisms control does not take into account the endogenous character of capital structure but its relationship with the practices of companies. Shleifer and Vishny (1986) find a positive relationship between capital concentration and management efficiency. This is corroborated by the studies of Alonso-Bonis and Andrés-Alonso (2007). In this respect, they report a positive relationship between systematic and high capital concentration and firm practices. This result differs from the Demsetz and Villalonga (2001) studies. Also, Wei, Xie, and Zhang (2005) found that the share of equity and institutional capital has a significant negative impact on management efficiency. In addition, foreign capital has a strong and positive impact on the efficiency of optimized practices. Although, Wu and Cui (2002) found that firms with high capital concentration have better accounting profits, but this realized performance is poorly represented by the ratio of market to book value and costs to profit. In other words, they show a non-linear relationship between the capital structure and the contribution to the firm.

Following, Shleifer and Vishny (1986); Morck, Shleifer, and Vishny (1988) examined the relationship between internal capital structure and market valuation as measured by Tobin's Q. The results of the study showed that there was a non-linear relationship between capital concentration and management efficiency. Other authors such as Morck et al. (1988) show similar non-linear or curvilinear relationships (McConnell and Servaes, 1990; Hermalin and Weisbach, 1991; Hyeon Cho, 1998; Himmelberg, Hubbard, and Palia, 1999). Hyeon Cho (1998) showed that there was no relationship between capital structure and management efficiency. Rogers, Dami, Ribeiro, and Sousa (2007) report the lack of influence of capital composition on the effectiveness of optimized mechanisms control. Chiang (2005) investigates the relationship between management optimization variables and operating performance. The results indicate a significant negative correlation between the proportion of capital held by governance and the effectiveness of practices. However, the relationship between various forms of capital and management efficiency is significantly positive. Gugler and Yurtoglu (2003) indicate a significant positive relationship between management effectiveness and capital at about 21.5%. However, they find that the ROI is negatively associated with capitalization between 21.5% and 63%. They also find a positive relationship between the investment ratio and capital equal to 100%. Farooque, van Zijl, Dunstan, and Karim (2010) examine the co-deterministic relationship between capital concentration and optimized mechanisms control effectiveness. The results indicate that there is a significant relationship in both positive directions between capital concentration and management effectiveness. Thomsen and Pedersen (2000) indicate, in parallel, a significant positive effect of capital structure on shareholder practices. The study by Gurbuz and Aybars (2010) reveals that foreign capital improves firm performance, while (Bai, Liu, Lu, Song, and Zhang, 2004) indicate that the issuance of shares to foreign investors has a positive and significant impact on market practices. Jensen and Meckling (1976) argue that capital diversity has different effects on firm efficiency. Except that in their case, considering the conceptual parameters invoked, it is important to analyze the effect of various types of capital on performance.

Based on these empirical studies, we will try to test the following capital structure hypotheses:

- **Hypothesis 5a:** There is a significant positive relationship between the effectiveness of optimized practices and equity in Banking and financial institutions.
- **Hypothesis 5b:** There is a significant positive relationship between the effectiveness of optimized practices and institutional input in Banking and financial institutions.

- **Hypothesis 5C:** There is a significant negative relationship between the effectiveness of optimized practices and public financing in Banking and financial institutions.
- **Hypothesis 5D:** There is a significant positive relationship between the effectiveness of optimized practices and foreign investment in Banking and financial institutions.

Jensen and Murphy (1990) show a negative relationship between optimized bonuses and managerial performance. Similar results are reported by Core, Holthausen, and Larcker (1999); Chen and Jermias (2012). These results differ from those described by Leonard (1990); Gregg, Machin, and Szymanski (1993); Conyon and Gregg (1994); Conyon (1997); Andjelkovic, Boyle, and McNoe (2002); and Banghøj, Gabrielsen, Petersen, and Plenborg (2010) as they find a weak and insignificant relationship between managerial effectiveness and optimized bonuses. Although the above studies show negative evidence or no relationship between optimized bonuses and managerial performance, other studies corroborate the agency theory: bonus optimization policies provide incentives for better management practices (Fama and Jensen, 1983). Thus, a positive relationship between monetary incentives and firm efficiency is found by Jensen and Meckling (1976). Previous studies confirm a positive result (Coughlan and Schmidt (1985); Abowd (1990); Lewellen, Loderer, Martin, and Blum (1992); Janakiraman, Lambert, and Larcker (1992); Mehran (1995); Bruce and Buck (1996); Elston and Goldberg (2003); Elayan, Lau, and Meyer (2003); Sun, Cahan, and Emanuel (2009); Ozkan (2011); Farmer, Archbold, and Alexandrou (2013). These empirical studies support the following hypothesis:

- **H6:** There is a positive relationship between management efficiency and optimized bonuses in Banking and financial institutions.

It has been shown that external audit is an important factor that can affect company performance. In the literature, external audit plays a key role in improving the transparency of performance statements (thus increasing audit quality and optimization practices) (Mitton, 2002). Audit quality as a mechanism for external management optimization could reduce agency costs and information asymmetry: thus, audit quality has a significant impact on management effectiveness (Adeyemi and Fagbemi, 2010; Willenborg, 1999). Previous empirical studies argue that companies with a high external audit requirement have more industry expertise and de facto more chances of discovering irregularities in performance. These conclusions are supported by numerous studies (Francis and Krishnan, 1999; Willenborg, 1999; Lennox, 1999; Jagan Krishnan and Schauer, 2000; DeFond, Francis, and Wong, 2000; Ferguson, Francis, and Stokes, 2003; Weber and Willenborg, 2003; Fan and Wong, 2005; DeFond and Francis, 2005; Francis, Reichelt, and Wang, 2005; Hay, Knechel, and Ling, 2008; Wahab, Haron, Lok, and Yahya, 2011; Fooladi and Abdul Shukor, 2012).

From this, our hypothesis to be confronted is the following:

- **H7:** There is a positive relationship between the effectiveness of optimized mechanisms control and the quality of external audit in Banking and financial institutions.

The data is collected from several secondary sources and from company document reviews. The primary data is collected from the annual reports published by the Multilateral Investment Guarantee Agency and covers the financial and banking sector between 2005 and 2011. We also used DataStream and SIRCA databases to complete the secondary data. To collect the data on management optimization and financial data for the period 2005–2011, it was preferable to consult the websites of the financial institutions studied as well as that of Industry Canada to collect information on the age of the institutions. The sample consists primarily of North American banks and financial institutions.

The main objective of this chapter is to investigate the possible relationship between management optimization mechanisms and management efficiency in the Banking and financial sector. Following the example of Himmelberg et al. (1999) and Schultz, Tan, and Walsh (2010), panel data analysis is applied in this study. Specifically, using a panel of banks and financial institutions over the period 2005–2011, the chapter adopts a comprehensive and holistic approach to examine the relationship between selected optimization mechanisms and management efficiency.

Firstly, the study uses OLS estimates: although this regression procedure allows causal links to be postulated, OLS results are not always reliable and are often biased. Secondly, the study applies the panel technique to confirm the causal links between the optimization mechanisms and the performance measures. However, although the panel effect model is used here to account for unobservable heterogeneity, it is unable to correct for endogenous causality problems. In line with the literature, the optimization mechanisms identified to assume managerial variables are exogenous factors to efficiency (Klein, 1998; Mehran, 1995). Also, several empirical researches Demsetz (1983); Demsetz and Lehn (1985); Himmelberg et al. (1999); Denis and Kruse (2000); Demsetz and Villalonga (2001); Dilling-Hansen (2005); Omar Farooque et al. (2007); Wintoki, Linck, and Netter (2010); Schultz et al. (2010) argue that the strategy optimization and management efficiency are endogenously determined: optimization can be one of the endogenous variables. The regressions are performed using Generalized Methods of Moments (GMM).

Where, financial efficiency is the dependent variable measured by return on assets (ROA), return on invested capital (ROIC), and Tobin's Q. The independent variables (the factors of optimized mechanisms control) include board size, independence of structure, decision forecasting, decision duality, independence of audit committee, audit frequency, autonomy of decision, equity, institutional input, public funding, foreign investment, management incentives and audit quality, and $\sum X$ is the vector of the other explanatory and control variables (firm size, Leverage, Growth prospects, risk, institutional age and LQ ratio), and μ is the error (see Table 9.1 for details).

TABLE 9.1 Variables Measures and Sources

Variables	Measurements	Symbols	Source
Dependent Variables			
Return on assets	It is calculated as the earnings before taxes to book value of the firm's total assets.	ROA	DataStream
Return on invested capital	It is calculated as (Net Income before preferred dividends + ((Interest expense on debt – Interest capitalized) × (1 – Tax rate)))/Average of last years and current year's (Total capital + Last year's short-term debt and current portion of long-term debt) × 100	ROIC	DataStream
Log Tobin's Q	It is calculated as the natural logarithm of the market value of total equity plus the book value of total debt relative to the book value of total assets.	TQ	DataStream

TABLE 9.1 *(Continued)*

Variables	Measurements	Symbols	Source
Independent Variables			
Structure size	It is the structure size for bank i in time t. It is calculated as the numbers of structure directors.	SSIZE	DataStream, Sirca, and Annual Reports
Independent of decision structure	It is the independent of structure of directors for bank i in time t. It is calculated as the proportion of outside independent directors.	SIND	DataStream, Sirca, and Annual Reports
Structure activities	It is the structure activities for bank i in time t. It is calculated as the numbers of structure director's connexion during the financial year.	SMEET	Annual Reports
Decision duality	It is a dummy variable taking the value 1 if the bank's decision is the chairman of the structure of directors, otherwise 0.	DUALITY	Annual Reports
Audit structure independence	It is the proportion of independent on the audit structure for bank i in time t.	ASIND	DataStream and Annual Reports
Audit structure activates	It is the audit structure activities for bank i in time t. It is calculated as the number of audit structure connexons during the financial year.	ASMEET	Annual Reports
Remuneration structure independence	It is the proportion of independent directors on the remuneration structure for bank i in time t.	RSIND	DataStream and Annual Reports
Managerial ownership	It is the total percentage of shares owned by the structure of directors for bank i in time t.	INSID	Annual Reports
Institutional ownership	It is the total percentage of shares owned by institutions investors for bank i in time t.	INS	Annual Reports
Government ownership	It is the total percentage of shares owned by the government for firm i in time t.	STATE	Annual Reports
Foreign ownership	It is the total percentage of shares owned by foreign individuals and institutional investors for bank i in time t.	FORGN	Annual Reports

TABLE 9.1 *(Continued)*

Variables	Measurements	Symbols	Source
Executive incentives	It is executive incentives for bank 'i' in time 't.' It is calculated as the natural logarithm of total salaries and benefits expenses.	LNSALARY	DataStream and Annual Reports
Audit quality	It is a dummy variable taking the value 1 if the ban's audited by audit firms, otherwise 0.	QAT	Annual Reports
Control Variables			
Log institution size	It is calculated as the natural logarithm of the total assets for bank i in time t.	ISIZE	DataStream
Leverage ratio	It is calculated by total liabilities over total assets for firm i in time t.	LR	DataStream
Log growth	It is calculated as the natural logarithm of the ratio of a firm's market value per share to its book value per share.	MBVE	DataStream
Bank and financial risk	It is the standard deviation of earnings (Beta). It is calculated as the historical beta local index for firm i in time t.	BFRISK	DataStream
Firm age	It is calculated as the number of years elapsed since the firm was incorporated.	IAGE	Website for each firm
Log liquidity	It is calculated as the natural logarithm of the current assets scaled by current liabilities	LQ	DataStream
Industry dummy	The industry classification is based on global industry classification standards (GICS) for Canadian Industries listed company.	IND-DUM	Website for ASX
Year dummy	7-years dummies	YEAR-DUM	–

Table 9.2 presents the set of descriptive statistics for the optimized mechanisms control, optimization, and control variables for Banking and financial institutions (i.e., Panel A for the financial performance, Panel B for planning and Panel C for control variables). In the case of the management efficiency variables, Panel A shows that Banking and financial institutions have an average (median) ROA of 5.3% (6.8%), the minimum reported

over the period is –172%, while the maximum is 135%. Also, the average (median) ROIC is 6% (8.5%), while the average (median) Tobin's Q is 2,530 (2,370): this suggests that the majority of the Banking sector has a higher return than other companies.

Panel B (Table 9.2) presents the descriptive statistics of the management optimization variables used in the study. The table shows that the average size of the management structure (SSIZE) and the threshold for neutral decision making (SIND) in Banking and financial institutions is between 7 and 8 decision-makers of which 59.7% are independent. The table states that the average number of times decision-making structures (SMEET) are run is 9 to 10 times per year. The minimum number of decision-making forecasts is 0, and the maximum number in one year is 37 for Canadian banks.

The table shows, among other things, that only 4.2% of the observations are characterized by dual decision-making. This means that companies in the banking and financial sector in Canada opt for variability in decision structures. With regard to the neutrality of audit structures (ASIND) and the frequency of decision making (ASMEET), Table 9.2 shows that the average (median) proportion is 85.7% (100%). The Canadian data shows a minimum value of 0 and a maximum of 1.

The average (median) number of audit structure consultations in Banking and financial institutions during the year is 3.74 (4) times per year (the data indicates a minimum number of 0, and the maximum number of 14 times in a year). Among other things, the table presents an average (median) of the proportion of structural bonus autonomy (RSIND) that is 82.2% (100%) for the Banking and financial sector.

The descriptive statistics for the investment variables in Table 9.2 show that the mean (median) value of the proportion of management capital (insid) reached 15.2% (12%). The Canadian data indicate that the minimum value of managerial (own) governance is 0, and the maximum value is 77%. The average (median) value of the proportion held by institutional investors (NSI) is 25.2% (23.2%). The average value of public funding (from STATE) is 0.5% (0). The median including foreign investment (FORGN) is 28.1% (27%). The percentage of companies in the banking and financial sector audited by QAT is 80.8%.

With respect to the control variables, Panel C determines that the median bank size (ISIZE) as measured by total assets is 13.101 ($13.140). The median leverage ratio (LR) is 23.5% (21.3%). The median book value (MBVE) is 75.8% (73.2%). The average value of enterprise risk (BFRISK) is 124.5% (113%). For the age of the institutions (IAGE), the descriptive statistics show that the average (median) age of the institutions is 21.82 (16) years. On average, the data shows that the minimum and maximum age of firms in the sector in Canada ranges from 0 to 187 years. Finally, the median LQ ratio is 130.4% (115.6%) of total assets.

Table 9.3 presents the correlation between the mechanisms for optimizing strategy and management efficiency in the Banking and financial sector. From this table, the following important relationships emerge:

- The size of the management structure (SSIZE) is positively significant and related to efficiency as measured by ROA, ROIC, and Tobin's Q ratio. Decision autonomy (SIND) and management conventions (SMEET) are also positively correlated with ROA and ROIC, but are not related to Tobin's Q. Decision duality is significantly negatively related to ROA and only to ROIC.
- The proportion of decision-making autonomy in relation to the audit structure (ASIND) is negatively correlated with the ROA as well as the return on capital employed (ROIC), and positively correlated with Tobin's Q. In addition, the working dynamics of the audit structure (ASMEET) has a positive association with ROA and ROIC.
- The number of shares held by management capital (insid) is positively correlated with ROA and ROI only. The share of institutional financing (INS) has no significant relationship with either ROA or ROI, but a significant positive relationship with Tobin's Q ratio. There is a significant positive correlation between foreign investment (FORGN) and outcomes measured by ROA and ROI. This management incentive (WAGE) has a significant positive association with all performance measures. The table also shows that QAT has a significant positive relationship with ROA and ROI.
- With regard to the control of variables, Table 9.3 shows that size (ISIZE) and the LR are significantly positive and consistent with the performance measures. They are negative and insignificant with the market measure. Growth (MBVE) is significantly positive with all the optimized mechanisms control measures. Age (IAGE) and LQ ratio are positively correlated with the accounting measures, but a negative correlation with Tobin's Q ratio is highlighted. Risk (BFRISK) has a significant negative correlation with all measures of OMCI.

TABLE 9.2 Summary Statistics of Firm Performance, Strategic Planning, and Control Variables (Canadian Bank and Financial Firms)

Variables	**Obs**	**Mean**	**Std**	**P25th Percentile**	**P50th Percentile (Median)**	**P75th Percentile**	**Min**	**Max**	**Skewness**	**Kurtosis**
Panel A: Performance										
Return on assets (ROA)	1,438	0.053	0.241	0.006	0.068	0.145	–1.727	1.351	–1.6207	13.0263
Return on invested capital (ROIC)	1,438	0.060	0.263	0.012	0.085	0.171	–1.958	1.353	–1.7257	12.3078
Log Tobin's Q ratio	1,438	2.530	1.113	1.660	2.370	3.230	0.360	8.920	0.8746	4.2970
Panel B: Strategic Planning Variables										
Structure size (SSIZE)	1,438	7.640	2.706	6.000	7.000	9.000	2.000	23.000	0.7328	4.1862
Structure independence (SIND)	1,438	0.597	0.196	0.500	0.625	0.750	0.000	1.000	–0.3483	2.3447
Structure connexions in year (SMEET)	1,438	9.579	4.840	6.000	9.000	12.000	0.000	37.000	0.9696	4.9990
Decision duality (DUALITY)	1,438	0.042	0.200	0.000	0.000	0.000	0.000	1.000	4.5837	22.0102
Audit structure independence (ASIND)	1,438	0.857	0.206	0.750	1.000	1.000	0.000	1.000	–1.2698	3.7178
Audit structure connection (ASMEET)	1,438	3.745	1.882	2.000	4.000	5.000	0.000	14.000	0.7569	4.4939
Remuneration structure independence (RSIND)	1,438	0.822	0.234	0.670	1.000	1.000	0.000	1.000	–1.0727	3.2002
Insider ownership (INSID)	1,438	0.152	0.112	0.070	0.120	0.210	0.000	0.770	1.2524	4.9040

TABLE 9.2 *(Continued)*

Variables	Obs	Mean	Std	P25th Percentile	P50th Percentile (Median)	P75th Percentile	Min	Max	Skewness	Kurtosis
Institutional ownership (INS)	1,438	0.252	0.126	0.159	0.232	0.332	0.029	0.872	1.0171	5.1582
Government ownership (STATE)	1,438	0.005	0.041	0.000	0.000	0.000	0.000	0.520	10.1012	112.9688
Foreign ownership (FORGN)	1,438	0.281	0.134	0.180	0.270	0.378	0.000	0.930	0.3716	2.9904
Log salary (LOG SALARY)	1,438	11.543	1.491	10.552	11.523	12.423	6.730	15.520	0.1251	3.3545
Audit quality (QAT)	1,438	0.808	0.394	1.000	1.000	1.000	0.000	1.000	–1.5645	3.4477
Panel C: Control Variables										
Log institution size (ISIZE)	1,438	13.101	2.257	11.500	13.140	14.800	4.750	18.930	–0.1042	2.7421
Leverage (LR)	1,438	0.235	0.204	0.089	0.213	0.332	–1.634	1.653	0.6768	14.6667
Log growth (MBVE)	1,438	0.758	0.786	0.223	0.732	1.255	–1.897	3.999	0.1674	3.4091
Bank and financial risk (BBFRISK)	1,438	1.245	0.801	0.720	1.130	1.650	–2.570	5.640	0.8669	6.0653
Institution age (IAGE)	1,438	43.258	42.106	13.000	26.000	56.000	0.000	187.00	1.3925	3.9935
Log liquidity (LQ)	1,438	1.209	0.929	0.581	0.873	1.595	–1.439	4.348	1.2786	4.1312

Variables definition: The sample covered a total of 206 banks and financial institutions over 2005–2011:

- ROA is the earnings before interest and taxes divided by total assets; ROIC is the net income before preferred dividends + ((interest expense on debt – interest capitalized) × (1-tax rate))/average of last year's and current year's (total capital + last year's short-term debt and current portion of long-term debt) × 100.
- Tobin's Q ratio is calculated as the natural logarithm of the market value of total equity plus book value of total debt relative to the book value of total assets.
- SSIZE is the number of members of the structure of directors.
- SIND is the proportion of non-executive directors on the decision structure.
- SMEET is the number of bank's structure connection during the year financial year.
- Decision duality is a dummy variable that takes a value of one if the decision structure is also the management structure.
- ASIND is the proportion of independent audit structure members to the total number of audit structure.
- ASMEET is the number of institution's audit structure connection during the year financial year.
- RSIND is the proportion of independent allocation structure members to the total number of the structure members.
- INSID is the percentage of institution shares owned by the structure of directors.
- INS is the percentage of shears held by institutions.
- STATE is the percentage of shares held by the government.
- FORGN is the percentage of shares held by foreign investors.
- Log SALARY is calculated as the natural logarithm of total salaries and benefits expenses.
- BI is a dummy variable, taking the value 1 if the institution was audited by a QAT auditor, otherwise 0.
- Log Institution Size Log ISIZE is calculated as the natural logarithm of total assets.
- LR is the total liabilities divided by total assets.
- Log growth MBVE is calculated as the natural logarithm of the end-of-year share price divided by book value per share; Bank and Financial Risk BFRISK is beta estimates.
- Age IAGE is the number of years elapsed since the institution was incorporated in ASX, and Log Liquidity Log LQ is calculated as the natural logarithm of the current assets scaled by current liabilities.

TABLE 9.3 Pearson Correlation for All Variables in the Canadian Bank and Financial Companies (N = 1,438)

	ROA	ROIC	LOG Tobin's Q	SSIZE	SIND	SMEET	Decision Duality	ASIND	ASMEET	RSIND
ROA	1.000	–	–	–	–	–	–	–	–	–
ROIC	0.949***	1.000	–	–	–	–	–	–	–	–
LOG Tobin's Q	0.039	0.036	1.000	–	–	–	–	–	–	–
SSIZE	0.120***	0.113***	0.124***	1.000	–	–	–	–	–	–
SIND	0.069***	0.085***	0.020	–0.012	1.000	–	–	–	–	–
SMEET	0.129***	0.131***	–0.028	0.191***	0.107***	1.000	–	–	–	–
DECISION DUALITY	–0.077**	–0.085***	–0.002	–0.008	–0.085***	–0.088***	1.000	–	–	–
ASIND	–0.009	–0.003	0.013	–0.053**	0.098***	0.007	0.006	1.000	–	–
ASMEET	0.105***	0.121***	–0.003	0.275***	0.142***	0.383***	–0.169***	0.092***	1.000	–
RSIND	0.007	0.012	0.026	–0.036	0.122***	0.020	–0.016	0.473***	0.103***	1.000
INSID	0.044*	0.056**	0.041	0.018	–0.114***	–0.018	0.193***	–0.001	–0.009	0.013
INS	0.001	0.003	0.308***	0.003	0.004	–0.080***	–0.023	–0.004	–0.018	–0.016
STATE	0.014	0.016	–0.001	–0.045*	0.032	0.072***	–0.027	–0.038	0.035	–0.011
FORGN	0.057**	0.058**	–0.037	0.020	–0.026	0.071***	–0.030	0.048**	0.080***	0.061*
LOG SALARY	0.101***	0.093***	0.053**	0.161***	0.22	0.177***	–0.113***	0.018	0.142***	0.040
QAT	0.178***	0.193***	–0.000	0.287***	0.153***	0.242***	–0.013	0.027	0.268***	0.054**
LOG ISIZE	0.282***	0.296***	–0.000	0.385***	0.205***	0.382***	–0.104***	0.078***	0.467***	0.038

TABLE 9.3 *(Continued)*

	ROA	ROIC	LOG Tobin's Q	SSIZE	SIND	SMEET	Decision Duality	ASIND	ASMEET	RSIND
LR	0.103***	0.109***	–0.018	0.065**	0.078***	0.175***	–0.072***	–0.016	0.153***	–0.039
LOG MBVE	0.105***	0.098***	0.205***	–0.012	–0.019	–0.111***	0.007	–0.032	0.044*	0.002
BFRISK	–0.107***	–0.123***	–0.071***	–0.072***	–0.109***	–0.091***	0.081***	–0.067**	–0.159***	–0.041
IAGE	0.098***	0.093***	–0.009	0.218***	0.190***	0.090***	–0.030	0.010	0.244***	0.056**
LOG LQ	0.057**	0.058**	–0.018	–0.006	–0.003	–0.003	–0.002	–0.004	–0.015	–0.004

* Denotes correlation is significant at the level 0.10 level (2-talied). All variables are as previously defined.

** Denotes correlation is significant at the 0.05 level (2-talied);

*** Denotes correlation is significant at the 0.01 level (2-talied);

TABLE 9.3 *(Continued)*

	INSID	INS	STATE	FORGN	LOG SALARY	QAT	LOG ISIZE	LR	LOG MBVE	BFRISK	IAGE	LOG LQ
INSID	1.000	–	–	–	–	–	–	–	–	–	–	–
INS	–0.008	1.000	–	–	–	–	–	–	–	–	–	–
STATE	–0.072***	–0.024	1.000	–	–	–	–	–	–	–	–	–
FORGN	0.033	–0.033	0.017	1.000	–	–	–	–	–	–	–	–
LOG SALARY	–0.065**	–0.039	0.014	0.084***	1.000	–	–	–	–	–	–	–
QAT	–0.012	0.002	0.035	0.121***	0.198***	1.000	–	–	–	–	–	–
LOG ISIZE	–0.063**	–0.078***	0.112***	0.085***	0.126***	0.443***	1.000	–	–	–	–	–
LR	0.067**	–0.016	0.011	–0.034	0.006	0.138***	0.300***	1.000	–	–	–	–
LOG MBVE	0.047*	0.127***	0.017	0.001	–0.009	–0.007	–0.129***	–0.046*	1.000	–	–	–
BFRISK	0.003	–0.066**	–0.004	–0.034	–0.001	–0.155***	–0.225***	–0.140***	0.029	1.000	–	–
IAGE	–0.139***	0.022	0.047*	–0.062**	0.112***	0.205***	0.396***	0.059**	–0.045*	0.210***	1.000	–
LOG LQ	–0.055**	0.060**	–0.056**	–0.073***	0.040	–0.023	0.028	–0.027	–0.049*	0.062***	0.018	1.000

* Denotes correlation is significant at the level 0.10 level (2-talied). All variables are as previously defined.

** Denotes correlation is significant at the 0.05 level (2-talied);

*** Denotes correlation is significant at the 0.01 level (2-talied);

Two key limitations are recognized in OLS regressions: biased estimation with spurious results. Therefore, given the nature of the data collected and in view of previous studies (Baysinger and Butler, 1985; Wintoki, et al., 2012), we will use the fixed effects (FEs) and RE models as regressions to control for possible unobserved heterogeneities in Banking and financial institutions. It must be said that panel data analysis under OLS regression can be inconsistent and meaningless under heterogeneity (Hsiao, 2003). In this regard, FEs, and RE models are used here to handle the heterogeneity of banking and financial institutions in order to promote the intercept of variables. The choice of the model is based on statistical tests (Lagrange multiplier tests among others). Thus, if the P-value is significant, the model is more suitable than the heterogeneity integrator model. Also, the Hausman test is applied to test the FEs model against the RE (Cameron and Trivedi, 2009). The null hypothesis opts for the RE model while the alternative hypothesis emphasizes the FEs. This means that if P-value is insignificant, the RE model should be validated, otherwise, the FEs model should be used. The Hausman test presents the regression models for each dependent variable of the banking and financial institutions under study.

The results of the regression using the FEs model are reported in column 3 of Table 9.4 along with the relationship between the optimized mechanisms control variables and efficiency as measured by the ROA. The results state that the Chi-square in the case of 434.57 has a significant statistic at the 1% level, which shows that the Panel models are more appropriate for the analysis of our data (better than the OLS regression which favors the Pooled model). The Hausman regression test is 33.61 (P-value significant at 1%). These results confirm the FEs model. For the optimized mechanisms control variables, the managerial parameter (insid) has a significant positive effect on the ROA. This indicates that the more integrative the structure, the more effective the optimized mechanisms control. This result is in good agreement with agency theory (equity capitalization cannot reduce agency costs but improve management efficiency). Furthermore, size (SSIZE) and decisional independence (SIND) have a positive but insignificant relationship with ROA. However, decision-making duality (DUALITY) and audit structure connections (ASMEET) have a significant negative effect on ROA. This result is consistent with Baysinger and Butler (1985). Fiduciary (QAT), institutional size (ISIZE) and book value (MBVE) have a significant positive effect on ROA. However, the LR and the age of institutions (IAGE) have a significant negative effect on ROA.

Table 9.5 provides the results of the FEs model regression of the relationship between optimization mechanisms and management efficiency as measured by the return on investment (ROIC) of capital of banking and financial institutions. The results indicate that the managerial decision (insid) has a significant positive relationship on ROIC. This finding is similar to the OLS regression. Also, audit structure (QAT) has a significant effect on ROI. On the other hand, dual decision making has a significant negative impact on ROI. It should be noted that no other management optimization mechanism is also significantly related to ROI. Book value (MBVE) has a significant positive effect on ROI: this finding is similar to the OLS estimate. However, the LR, risk (BFRISK) and age of institutions (IAGE) have a significant relationship with ROI.

Table 9.6 presents the results of the FEs regression. The Hausman test (regression) is 47.07. The P-value is significant at 1%. This result obtained from the Hausman test support indicates that the FEs model is more efficient than the RE model. Decisional independence (SIND), institutional financing (INS) and state capitalization have a significant positive effect on Tobin's Q. However, duality has a significant negative effect on Tobin's Q. In addition, the size of the management structure (SSIZE), the connections of the management structure (SMEET) and the autonomy of the audit structures (ASIND) have a positive, but relatively insignificant relationship with Tobin's Q ratio. As far as the control variables are concerned, the size of institutions (ISIZE) and book value (MBVE) have a significantly positive relationship with Tobin's Q, while structural risk (BFRISK) has a significant negative relationship with Tobin's Q.

The regression uses both OLS and panel models (FEs or RE models). However, OLS, and panel techniques can suffer from time-varying specific effects and endogenous and causality problems. The existence of endogeneity problems in OLS variables is examined here using the Durbin-Wu-Hausman test (Durbin, 1954; Hausman, 1978; Wu, 1973). The current tests, when using the performance variables as measured by Tobin's Q, ROA, and ROI, fail to accept the null hypothesis H0: the explanatory variables are thus exogenous. In other words, our results indicate that the hypothesis of no endogeneity is rejected. These results confirm the fact that OLS models and panel techniques are unreliable and biased. Therefore, this chapter concludes that GMM is the most appropriate approach. The results of the DWH tests point to endogeneity problems, which is another reason to opt for GMM regression.

Tables 9.4–9.6 present the results of the GMM regression on the relationship between the optimization and management efficiency variables measured by ROA, ROI, and Tobin's Q ratio, for banking and financial institutions. In these tables, the results of the OLS regression and the FEs model are presented in a way that facilitates a comparison with the GMM regression data. Several diagnostic tests are also reported in each table, such as the Sargan, AR (1) and AR (2) test. The results of these tests confirm the validity of the instruments used and reject possible serial/autocorrelation problems.

TABLE 9.4 System Generalized Method of Moments (GMM) Regression Results of the Relationship between Strategic Planning Mechanisms and Financial Performance as Measured by ROA for Canadian Bank and Financial Institutions

Dependent Variable ROA			
Independent Variables	**Pooled Model**	**Fixed Effects Model**	**Dynamic System GMM Model**
Const.	–0.393***	–	–0.064
	(–4.58)		(–0.35)
SSIZE	0.001	0.001	–0.001
	(0.09)	(0.50)	(–0.02)
SIND	0.013	0.008	0.139
	(0.38)	(0.23)	(1.17)
SMEET	0.002	–0.001	0.003
	(1.13)	(–0.80)	(1.33)
DUALITY	–0.055	–0.082**	–0.020
	(–1.23)	(–1.97)	(–0.37)
ASIND	–0.037	–0.027	–0.018
	(–0.97)	(–0.86)	(–0.51)
ASMEET	0.010**	–0.010**	–0.004
	(2.28)	(–2.03)	(–0.86)
RSIND	0.022	–0.001	0.036
	(0.83)	(–0.01)	(1.31)
INSID	0.121**	0.162**	0.170
	(2.11)	(2.06)	(0.79)
INS	–0.007	–0.009	–0.046
	(–0.15)	(–0.23)	(–1.61)

TABLE 9.4 *(Continued)*

Dependent Variable ROA			
Independent Variables	**Pooled Model**	**Fixed Effects Model**	**Dynamic System GMM Model**
STATE	–0.130**	–0.272	0.129
	(–2.11)	(–0.69)	(0.39)
FORGN	0.067	0.033	0.084**
	(1.61)	(0.63)	(1.99)
LOG SALARY	0.008*	0.001	–0.017
	(1.81)	(0.08)	(–1.34)
QAT	0.005	0.111***	0.088**
	(0.25)	(3.04)	(1.94)
LOG ISIZE	0.032***	0.070***	0.025*
	(6.40)	(7.42)	(1.79)
LR	–0.026	–0.138***	–0.169**
	(–0.61)	(–2.92)	(–2.04)
LOG MBVE	0.041***	0.026***	0.009
	(4.39)	(2.80)	(1.00)
BFRISK	–0.007	–0.014	–0.003
	(–0.79)	(–1.48)	(–0.23)
IAGE	–0.001	–0.010***	0.001
	(–0.49)	(–2.68)	(0.51)
LOG LQ	0.014***	0.013	0.026***
	(2.56)	(1.34)	(2.91)
IND-DUM	Yes	No	Yes
YEAR-DUM	Yes	Yes	Yes
SSIZE (t–1)	–	–	0.005
			(0.58)
SIND (t–1)	–	–	–0.073
			(0.74)
INSID (t–1)	–	–	–0.147
			(0.75)
ROA (t–1)	–	–	0.330***
			(11.06)
ROIC (t–1)	–	–	–
LOG Tobin's Q (t–1)	–	–	–

TABLE 9.4 *(Continued)*

Dependent Variable ROA			
Independent Variables	**Pooled Model**	**Fixed Effects Model**	**Dynamic System GMM Model**
Observations	1,438	1,438	1,232
No. Instruments	–	–	79
Hausman test	–	33.61**	–
Sargan test of over-identification	–	–	53.395
Sargan test (P-value)	–	–	0.3091
AR (1) test	–	–	–5.209***
AR (2) test	–	–	1.748
Lagrange multiplier test	–	434.57***	–

Note: ***, **, * represents statistical significance at 0.01, 0.05, and 0.10 levels, respectively. All variables are as previously defined. Sargan test of over-identification is under the null that all instruments are valid. AR (1) and AR (2) are tests for first-order and second-order serial correlation in the first-differenced residuals, under the null of no serial correlation.

TABLE 9.5 System Generalized Method of Moments (GMM) Regression Results of the Relationship between Strategic Planning Mechanisms and Financial Performance as Measured by ROIC for Canadian Bank and Financial Institutions

Dependent Variable ROIC			
Independent Variables	**Pooled Model**	**Fixed Effects Model**	**Dynamic System GMM Model**
Const.	–0.427***	–	0.023
	(–4.68)		(0.10)
SSIZE	–0.002	0.002	–0.001
	(–0.52)	(0.50)	(–0.01)
SIND	0.033	0.006	0.258**
	(0.88)	(0.15)	(2.02)
SMEET	0.001	–0.002	0.006***
	(0.44)	(–1.00)	(2.50)
DUALITY	–0.070	–0.097**	–0.018
	(–1.39)	(–2.18)	(–0.37)
ASIND	–0.044	–0.038	0.030
	(–1.07)	(–1.13)	(–0.76)

TABLE 9.5 *(Continued)*

Dependent Variable ROIC			
Independent Variables	**Pooled Model**	**Fixed Effects Model**	**Dynamic System GMM Model**
ASMEET	0.008*	–0.005	0.011
	(1.77)	(–1.00)	(0.27)
RSIND	0.024	0.004	0.055**
	(0.80)	(0.14)	(1.93)
INSID	0.164**	0.190**	0.068
	(2.66)	(2.28)	(0.28)
INS	–0.005	0.001	–0.021
	(–0.10)	(0.03)	(–0.62)
STATE	–0.131*	–0.071	0.384
	(–1.79)	(–0.17)	(0.67)
FORGN	0.067	0.030	–0.004
	(1.41)	(0.54)	(–0.08)
LOG SALARY	0.008*	–0.007	–0.021
	(1.67)	(–0.46)	(–1.45)
QAT	0.012	0.128***	0.105**
	(0.50)	(3.32)	(2.14)
LOG ISIZE	0.037***	0.065***	0.035***
	(7.37)	(6.48)	(2.44)
LR	–0.034	–0.120***	–0.217***
	(–0.91)	(–2.40)	(–2.69)
LOG MBVE	0.042***	0.019*	0.019**
	(3.95)	(1.92)	(2.12)
BFRISK	–0.009	–0.019*	0.010
	(–0.96)	(–1.90)	(0.72)
IAGE	–0.001	–0.008**	–0.001
	(–0.86)	(–2.12)	(–0.10)
LOG LQ	0.016***	0.013	0.021***
	(2.75)	(1.28)	(2.46)
IND-DUM	Yes	No	Yes
YEAR-DUM	Yes	Yes	Yes
SSIZE (t–1)	–	–	0.010
			(1.16)

TABLE 9.5 *(Continued)*

Dependent Variable ROIC			
Independent Variables	**Pooled Model**	**Fixed Effects Model**	**Dynamic System GMM Model**
SIND (t–1)	–	–	0.037
			(0.37)
INSID (t–1)	–	–	–0.139
			(–0.67)
ROIC (t–1)	–	–	0.297***
			(7.52)
Observations	1,438	1,438	1,232
No. Instruments	–	–	79
Hausman test	–	27.61*	–
Sargan test of over-identification	–	–	53.976
Sargan test (P-value)	–	–	0.2900
AR (1) test	–	–	–5.204***
AR (2) test	–	–	1.896
Lagrange multiplier test	–	516.03***	–

Note: ***, **, * represents statistical significance at 0.01, 0.05, and 0.10 levels, respectively. All variables are as previously defined. Sargan test of over-identification is under the null that all instruments are valid. AR (1) and AR (2) are tests for first-order and second-order serial correlation in the first-differenced residuals, under the null of no serial correlation.

TABLE 9.6 System Generalized Method of Moments (GMM) Regression Results of the Relationship between Strategic Planning Mechanisms and Financial Performance as Measured by Log Tobin's Q for Canadian Bank and Financial Institutions

Dependent Variable Log Tobin's Q			
Independent Variables	**Pooled Model**	**Fixed Effects Model**	**Dynamic System GMM Model**
Const.	–0.015	–	0.318
	(–0.11)		(0.68)
SSIZE	0.025***	0.008	0.082***
	(5.43)	(1.38)	(3.58)
SIND	0.087	0.112*	0.253
	(1.45)	(1.63)	(0.76)

TABLE 9.6 *(Continued)*

Dependent Variable Log Tobin's Q			
Independent Variables	**Pooled Model**	**Fixed Effects Model**	**Dynamic System GMM Model**
SMEET	–0.001	–0.04	–0.004
	(–0.66)	(–1.20)	(–0.87)
DUALITY	0.010	0.185**	0.139
	(0.17)	(2.31)	(1.07)
ASIND	0.024	0.072	0.079
	(0.41)	(1.19)	(1.06)
ASMEET	–0.014*	–0.005	–0.013
	(–1.83)	(–0.51)	(–1.02)
RSIND	0.070	–0.025	–0.039
	(1.23)	(–0.48)	(–0.56)
INSID	0.182*	–0.077	0.986*
	(1.80)	(–0.52)	(1.83)
INS	0.909***	0.742***	0.684***
	(11.79)	(9.69)	(7.51)
STATE	0.232	1.707**	1.425***
	(0.97)	(2.28)	(2.85)
FORGN	–0.087	–0.104	–0.022
	(–1.00)	(–1.04)	(–0.18)
LOG SALARY	0.023***	–0.078***	0.078**
	(2.91)	(2.59)	(1.98)
QAT	0.049	–0.208***	0.180
	(1.46)	(–2.98)	(1.57)
LOG ISIZE	0.009	0.062***	0.029
	(1.28)	(3.44)	(0.99)
LR	0.050	0.024	0.040
	(0.85)	(0.28)	(0.34)
LOG MBVE	0.106***	0.101***	0.063***
	(6.54)	(5.53)	(3.13)
BFRISK	–0.043***	–0.035**	–0.097***
	(–2.92)	(–1.94)	(–2.95)
IAGE	–0.001***	–0.009	0.002
	(–2.56)	(–1.35)	(0.69)

TABLE 9.6 *(Continued)*

Dependent Variable Log Tobin's Q			
Independent Variables	**Pooled Model**	**Fixed Effects Model**	**Dynamic System GMM Model**
LOG LQ	–0.012	–0.028	0.041*
	(–0.97)	(–1.50)	(1.69)
IND-DUM	Yes	No	Yes
YEAR-DUM	Yes	Yes	Yes
SSIZE (t–1)	–	–	–0.020
			(–0.98)
SIND (t–1)	–	–	0.164
			(0.69)
INSID (t–1)	–	–	–0.338
			(–0.65)
LOG Tobin's Q (t–1)	–	–	0.213***
			(6.22)
Observations	1,438	1,438	1,232
No. Instruments	–	–	79
Hausman test	–	47.07***	–
Sargan test of over-identification	–	–	43.127
Sargan test (P-value)	–	–	0.7090
AR (1) test	–	–	–6.504***
AR (2) test	–	–	–1.601
Lagrange multiplier test	–	308.75***	–

Note: ***, **, * represents statistical significance at 0.01, 0.05, and 0.10 levels, respectively. All variables are as previously defined. Sargan test of over-identification is under the null that all instruments are valid. AR (1) and AR (2) are tests for first-order and second-order serial correlation in the first-differenced residuals, under the null of no serial correlation.

The results of the GMM regression are reported in column 3 of Table 9.4. The GMM estimation (results reported in column 3 of Table 9.2) shows that even after adjustment for endogeneity, there is no significant relationship between the optimization variables and the selected ROA: thus, the assumptions on size (SSIZE) H1a, structure independence (SIND) H1B, audit structure connections (SMEET) H1c, decision duality (DUALITY) H2, audit autonomy (ASIND) H3A, audited institutions (ASMEET) H3b and optimized mechanisms control premiums (RSIND) H4 are rejected. In addition, past profitability (ROAt-1) and the size of the management structure (SSIZEt-1) have a significant impact on current performance (the presence of endogeneity problems). These results are consistent with the work of (Baysinger and Butler, 1985; Schultz et al., 2010; Wintoki et al., 2012).

With regard to the capital structuring variables, the results show that own funding (insid), institutional capitalization (INS), and public investment (STATE) are found to have no significant difference in direct and causal relationship with ROA. This is consistent with the results of Randall et al. (1988); Demsetz and Villalonga (2001); and Welch (2003). Therefore, the hypotheses on managerial decision (insid) H5a, institutional financing (INS) H5b and public investment (SATAE) H5C are rejected. On the other hand, only foreign funds (FORGN) have a significant positive effect, suggesting that foreign capitalization provides excellent opportunities for access to technology, managerial skills and more effective monitoring of applied management practices. This result is similar to the results reported by Wei et al. (2005); and Lee (2008) which provide a significant positive association between investment and performance in the context of optimization. Therefore, the H5D hypothesis is corroborated.

The Pooled OLS model indicates that bonuses have a significant positive relationship with ROA, but the GMM model shows that salary is not significantly related to ROA. Based on these results, the H6 hypothesis is rejected. On the other hand, the GMM system shows that QAT has a significant positive relationship with ROA, which implies that the quality of the optimized audit is one of the important factors for improving the effectiveness of management practices. Based on these results, the hypothesis on audit quality H7 is confirmed. For the control variables, the OLS results show that the coefficient of institution size (ISIZE) has a significant positive relationship with ROA. GMM also shows similar results, suggesting that banking and financial institutions that increase assets have better management efficiency. In addition, both OLS and GMM regressions show that log LQ has a significant positive relationship with ROA. However, the GMM results show that the LR is significantly negatively related to ROA. The negative relationship between optimized mechanisms control efficiency and the LR suggests that the increase in debt may lead to higher borrowing costs from banking and financial institutions, which absorbs the firm's profits.

With regard to the use of ROIC as a measure of management performance, the results of the GMM regression are reported in column 3 of Table 9.5. The results show that decision structure autonomy (SIND), connections

(SMEET) and compensation valuation (RSIND) have a significant positive relationship with ROIC. The significant positive effect of structure autonomy suggests that non-executive decision-makers in open economy banking and financial institutions contribute positively to management performance. Decision-making connections have a significant positive relationship with ROIC, indicating that banking and financial institutions with decision-making forecasts often tend to generate higher returns.

This finding is consistent with agency theory. The structural autonomy of payment is positively related to ROI, suggesting that it is composed only of independent decision-makers, and contributes to making recommendations for the optimized mechanisms control of the company. Thus, it mitigates conflicts of interest and improves management efficiency. However, the GMM approach does not find any significant relationship between other optimization variables such as (structural size (SSIZE), decision duality (DUALITY), autonomy of audit structures (ASIND), connections (ASMEET), optimized mechanisms control incentives (SALARY), QAT, and ROI. These results are similar to those found in Table 9.4. For the capital variables, the GMM regression results are similar to the results found in column 3 of Table 9.4. With regard to the control of variables, the size of institutions (ISIZE) and the LQ ratio are significant and positively related with ROI: these results corroborate those in Table 9.4. Finally, the GMM regression results reported in column 3 of Table 9.6 use Tobin's Q ratio as a measure of management performance. With respect to the management optimization mechanisms, the GMM results show that the size of the management structure (of SSIZE) has only a significant positive relationship with Tobin's Q ratio, indicating that the size of the management structure (SSIZE) seems to play an important role in mitigating agency conflicts. However, the results show that there is no evidence of a relationship between the other optimization variables (independence (SIND), connections (SMEET), duality (DUALITY), audit structure (ASMEET), remuneration (RSIND)) and Tobin's Q. These results confirm the previous findings reported in Table 9.4, except for structural size (SSIZE) which has a significant positive relationship with Tobin's Q ratio. The significant positive association with Tobin's Q indicates that the size of the management structure leads to better performance. This result is not consistent with the results obtained by Liption and Lorsch (1992); Yermack (1996); Hossain, Prevost, and Rao (2001); Bhagat and Black (2002); Cheng et al. (2008); and O'Connell and Cramer (2010) who found that the size of the structure has a significant negative effect on management effectiveness, but is consistent with the results of Dan Dalton, Daily, Johnson, and Ellstrand (1999); Larcker, Richardson, and Thon (2007); and Fauzi and Locke (2012).

For the capital structure variables (management (insid), institutional (INS) and public (STATE)) they have significant positive relationships with Tobin's Q ratio. These results are not consistent with the conclusions of Tables 9.4 and 9.5. A significant positive relationship between (insid) and Tobin's Q indicates that equity capital increases management performance. In addition, the GMM regression results regarding the inclusion of the decision capital structure as an endogenous variable, indicate that it has a significant positive relationship with Tobin's Q ratio. Although this result does not corroborate previous studies that failed to find a significant relationship between capital structure and managerial performance using the GMM model (Pham, Suchard, Zein, 2011; Schultz et al., 2010), this result is consistent with (Jensen and Meckling, 1976) who pointed out that the agency conflict between decision and managerial structures could be mitigated through managerial autonomy. This is because governance with a larger share of capital will be able to maximize managerial value to ensure the best managerial efficiency. This confirms the work of Randall et al. (1988); and McConnell and Servaes (1990), who found a significant positive relationship between these two parameters.

The institutional share (INS) has a significant positive relationship with Tobin's Q ratio, suggesting that the INS provides effective monitoring capacity to increase Tobin's Q ratio. This result is similar to the studies of Shleifer and Vishny (1997); and Henry (2010). Previous research is inconsistent with Tobin's Q ratio (Wei et al., 2005; Xu and Wang, 1999) Public funding (STATE) has a significant positive effect on Tobin's Q ratio. The results in Table 9.6 also show that optimized mechanisms control premiums have a significant positive relationship with Tobin's Q ratio. With regard to the control variables, the book value (MBVE) and the log LQ ratio have a significant positive impact on Tobin's Q ratio. This significant positive relationship remained stable in both OLS and GMM, except for the LQ ratio which is significant in GMM. However, the results in Table 9.4 show that the institutional risk (BFRISK) has a significant negative effect on Tobin's Q ratio. This negative sign also remained stable in both cases.

This chapter examines the effects of optimization on the management efficiency of banking and financial institutions in an open economy context. The study adopts a comprehensive approach using optimization and performance variables under the OLS regression, the FEs model and the GMM model. The results of the OLS regression are consistent with those found generally in the literature. Specifically, Structural Autonomy (SIND), Management Connection (SMEET), Decision Duality (DUALITY) and Private Audit (ASIND) have no significant relationship with Tobin's Q ratio and ROA. Structure size (SSIZE) has only a significant positive relationship with Tobin's Q, indicating that size plays a key role in the management efficiency of banking and financial institutions. These results are similar to the results reported at the GMM regression level for Tobin's Q ratio, ROA, and optimized mechanisms control measures. However, autonomy (SIND), management connection (SMEET) and compensation structure (RSIND) have a significant positive relationship with ROIC.

With regard to the capital variables, the OLS regression results indicate that the management structure (insid) has a significant positive relationship with ROA and ROI, but the GMM model indicates a significant positive relationship with Tobin's Q alone. This positive relationship indicates that the higher the level of equity capital, the higher Tobin's Q ratio. These GMM results are similar to those found in the literature. In addition, public funding (STATE) also has a significant positive relationship with Tobin's Q. The GMM results show that foreign investment (FORGN) has only a significant positive relationship with ROA. Institutional financing (INS) is the only one significantly positively related to Tobin's Q. The GMM results indicate that optimized mechanisms control premium (salary) has a significant positive relationship with Tobin's Q ratio. QAT also has a significant relationship with ROA and ROI, but no significant relationship with Tobin's Q ratio. These findings have important implications in an open economy context for the design of management optimization systems.

KEYWORDS

- **foreign investment**
- **leverage ratio**
- **liquidity ratio**
- **management optimization systems**
- **return on assets**
- **return on invested capital**

CHAPTER 10

The Standardization of the Measurement of the Effects of Optimization on Management Practices

After having analyzed the theoretical aspects of the integration of strategy optimization into the conceptual body of the theory of corporate efficiency, it is appropriate to question the empirical as well as the theoretical significance of these developments. In particular, a model should be sought to clarify the relationship between institutional competitive advantage and optimized mechanisms control in the Canadian context. In this respect, it is crucial to emphasize that the models invoked contain most of the claims that are reasonably susceptible to empirical testing. In particular, the measurement of efficiency and its impact on productivity; two parameters that are essential in addressing the interaction between optimized mechanisms control and competitive advantage in the banking and financial sector. The data used for this study comes from Statistics Canada. It consists of information on companies and financial institutions, grouped into categories of banks of activities over the given period. The four statistical analysis methods used are:

- Descriptive analysis (the evaluation of production outputs and the determination of the structure of company expenses);
- The non-parametric method of Performance Data Analysis Control (establishment of a synthetic indicator to measure the efficiency of banks in the case of input-oriented efficiency);
- AFM (categorizing banks according to all their reported characteristics); and
- Linear econometric modeling on panel data (determining the effects of an increase in management efficiency on output).

Value-Based Management in an Open Economy: Optimizing Strategy to Improve Business and Performance. Prof. Post-Dr Walter Amedzro St-Hilaire (Author)

It may be confusing to talk about empirical 'tests' of the models used in this chapter. Indeed, they are not models that can be immediately verified, assuming that such models exist in applied management. However, these models offer a unique framework for the simultaneous analysis of the effects of sectoral competitive advantage and optimization, in contrast to existing empirical studies which are based on very fragmentary, partial, and biased assumptions.

Firstly, we analyze the scope of the models selected in order to draw the main lessons from them. In particular, it is necessary to highlight:

- The role of competitive advantages in transmitting the effects of effective value-based management on the dynamics of industrial organization;
- The role of management flexibility in determining costs and competitive advantages for strategy optimization.

Then, we analyze whether the contribution of descriptive and statistical tools is a satisfactory framework for analysis or, on the contrary, whether we should accept the idea of "cross-sectional" models whose breakdown would obey criteria that are completely different from econometric considerations. In other words, at the statistical level, it is necessary to determine whether the industrial nomenclature in an open economy allows for the analysis of both market variables (production, imports, exports, and prices) and those of strategy optimization. Once this question has been resolved, it will be possible to determine which analysis scheme best corresponds to the reality of the Canadian case concerning this part of our study problem. As a reminder, we are looking for models that seem to contain the essence of the assumptions that are reasonably likely to provide a strong and relevant explanation of the complexity of the measurement of sectoral efficiency and its impact on the production of the open economy. It should be noted at the outset that in an open business context, firms need to be attractive not only to integrate into the market, but also to attract foreign investors. In the Canadian case, the business climate is favorable. The country occupies a prime position in terms of business facilities. This attractiveness has a natural impact on the efficiency of institutions and on economic growth is a variation in production, which itself is a sum of added values. These added values come in particular from the activities of companies.

Hence, the literature review on the relationship between the strategy optimization and the efficiency of companies, if it takes into account the process and the characteristics of the tool on performance, ignores the cost aspect as we have noted. In this respect, it is important to ask, in an open business context, what is the impact of the costs of optimizing strategies on efficiency and the profitability of companies on production? This question requires a number of further interrogations.

- How can we measure the performance of companies in an open economy and its impact on the strategy optimization?
- What is the best performing sectoral activities in the case of the banking?
- What is the effect of sectoral strategic costs on their optimized mechanisms control effectiveness and competitive advantage?
- What is the impact of the efficiency of the banking sector on the strategy optimization?

Specifically, this chapter will:

- assess the level of output, determine the structure of the banking sector's expenses and its impact on the strategy optimization in Canada;
- to construct a synthetic indicator for measuring company performance and its impact on optimized mechanisms control in the case of input-oriented efficiency;
- to combine banks into large groups according to their characteristics in terms of cost structure and strategy optimization;
- assess the impact of an increase in the optimization of Canadian companies and its impact on the strategy optimization.

The literature review clearly indicated that the competitive advantage of firms and institutions is the result of many factors, including strategy and management practices, but how much credit should be given to their ability to innovate, to investment in information technology and to the cost efficiency of optimized mechanisms control?

In addition to the agency theory, several studies on the effect of strategic management mechanisms on optimization improvement (Demsetz and Lehn, 1985; Morck et al., 1988) show that the objective of management strategies is the creation of the firm's wealth.

Thus, Louizi (2011) carried out on the causality between the strategic mechanisms of management (The practical mechanisms are apprehended by variables related to the functioning of the board of directors (size, meetings,

...); to the structure of ownership (shareholding, ...) and to the characteristics of the leaders (training, ...)) and the effectiveness of management. He manages to demonstrate, on the basis of an OLS on performance (Tobin's Q), that successful and failing firms are characterized by different management mechanisms. He argues that, in general, efficient, and competitive firms are better thought out and optimized than others. He also argues that growth opportunities are performance factors to be integrated. In relation to the place of optimized mechanisms control standards, Rosenstein and Wyatt (1990) highlight the fact that change affects the value and effectiveness of firms. Gilson (1990) in his study of financial distress, shows that firms increase the rotation of mechanisms in their optimization strategy when performance is poor. Similarly, Adams and Mehran (2005) analyze the relationship between decision optimization (measured by board size) and performance efficiency (measured by Tobin's Q) on a sample of banks with information collected over 10 years. They prove that, in contrast to the firm, banking institutions with large mechanisms are not the least efficient. Lahcen and Kharti (2011) looking for the determinants of financial efficiency of microfinance institutions, show that portfolio at risk, share of funds, total assets, and productivity influence institutional competitive advantage. Cauvin and Bescos (2006) determine those financial outcomes such as production cost reduction in strategy optimization, shareholder value creation and working capital requirements determine the competitive advantage of firms.

The phenomenon of innovation plays an important role in the success of products in the open market and thus in the efficiency of firms and institutions. In this respect, special attention has been paid to it in recent years by theorists of industrial economics. Naudhaus (1969) shows that the increase in production and the technological development achieved grows with the size of the market. Bellon (1991) states that a firm cannot be competitive if it cannot overcome its lack of technological advantage. Bismush and Oliveira (1986) state that, for a firm or an economy to be successful, it must be able to capture market share in both the domestic and foreign markets. Some studies have assessed the link between the degree of innovation and firm performance. Schumpeter's thesis that R&D-intensive innovations are the engine of economic development leads to the hypothesis that the most successful firms are those that are able to develop innovations with a high degree of novelty. While this hypothesis has long been accepted, recent analyzes on this issue do not allow for very explicit conclusions.

Thus, Freel (2000) instead highlights the non-linearity of the relationship between profitability and 'innovation, suggesting the significant 'influence of other unidentified factors. Moore (1995, cited in Freel, 2000) 'finds no systematic relationship between 'innovation and profitability, while Geroski and Machin (1992, cited in Freel, 2000) confirm a robust and persistent relationship whereby innovative SMEs have higher profit margins than non-innovative ones. Muller (2000) studying innovation strategy and new product performance, shows that the effect of innovation strategy on new product performance is enhanced when considering its intermediate influence on competitive response behavior. Hult et al. (2004) positively link innovation to different forms of performance (profitability, growth, market share, overall performance). Focusing on the innovation process, Das and Joshi (2012) show that innovation is positively related to the efficiency of technological services. This positive effect of firms' ability to innovate on their performance is not always demonstrated, however. Dibrell et al. (2008) point out that there is no direct relationship between innovation (product and process) and performance (measured by the profitability rate and the growth rate). Similarly, Jansen et al. (2006) find no direct effect of operational innovations on financial performance, although their study shows the moderating role of dynamism and contextual competitiveness, which generate significant cross effects on performance.

The use of strategic decision-making tools is growing in companies for an increasing number of tasks: communication, information retrieval, marketing of products and services, group work, company management, prospecting, etc. Their impact has long been questionable. Their impact has long been questionable. Indeed, following Solow's productivity paradox, some authors have been interested in the differences in productivity gains of decision-making tools in an open economy (Mairesse and Kocoglu, 2002). Beyond the macroeconomic issue, the use of decision-making tools in firms has been the occasion to look at the evolution of firms, their productivity (Benghozi, Flichy, and D'Iribarne, 2000) and the efficiency of firms (Greenan and Mairesse, 2006). The impacts of investment in tools and their use on the efficiency of companies is the main line of research. Thus, for (Leavitt and Whisler), the use of decision tools strengthens the efficiency of companies by eliminating middle managers. The central assumption of research on the impact of optimized strategy tools on productivity is that the implementation of a new decision-making system is not sufficient to generate positive effects on productivity.

Black and Lynch (2001) analyze the effects of optimal decision-making tools on sectoral productivity. They show that the strategy optimization via the decision tool has a positive and significant effect on productivity. Bresnahan et al. (2002) show that optimized strategy tools. Janod and St-Martin (2003); and Ben Aoun et al. (2010) demonstrate that optimization in the context of a reorganization has a positive effect on factor productivity even if it does not change the growth rate of the workforce and capital. They conclude that reorganizations based on decision tools are a source of performance improvement (more efficient use of strategy optimization factors without the need to increase them). Moumbe (2005), stresses that strategic tools can increase the performance of companies provided that they are accompanied by organizational change. Fadhel (2005), notes that at the level of the triangle formed by technological evolution, organizational change and the restructuring of competences, there is a relationship of co-evolution rather than causality.

Other non-financial factors unrelated to the practices of optimized mechanisms control strategies influence its level of performance. Lahcen (2011) shows that age is a determinant of optimization. Also, the outreach of MFIs' microfinance programs positively influences performance and the percentage of women clients negatively influences it. Jensen and Murphy (1990) demonstrate a relationship between the compensation system in firms and performance. Ferrary (2010) manages to show that when women reach a critical size in the company, it creates a set of interactions that facilitates the functioning of the company and leads to an improvement of its performance (growth, profitability, and higher productivity). Ittner et al. (2003); Banker et al. (2000); Gordon and Sohal (2001); Toni and Tonchia (2001) integrate other determinants of optimization (customer satisfaction, staff satisfaction, adaptation of products to market requirements, management mode, delivery times, strategic choices, size, advertising campaigns…).

The theory on strategic costs and performance relates the influence of costs on factor productivity and thus on the achievement of a level of performance by the company. Indeed, performance is here, as in several other studies, generally measured by factor productivity (single-factor, multi-factor, or synthetic index). However, costs can have an impact on other performance measurement indicators. On this basis, Kombou and Wanda (2006) show that the performance of companies obeys a logic of cost reduction. Razafindrakoto (1997), states that productivity growth trends are strongest for companies that have the possibility of circumventing difficulties such as costs and the supply of intermediate consumption in their strategic planning. Djimasra (2009) analyzing technical efficiency, productivity, and competitiveness shows that when the cost system is controlled and technical efficiency is gained, performance improves. Schreyer and Pilat (2001) show that the rate of change of multifactor productivity (strategic planning and value-added) is a function of strategic production costs. In this respect, the reduction of strategic costs has a positive effect on the productivity of companies, and in turn on performance and competitiveness.

Many authors have focused on measuring company performance. However, they have not all used the same measurement tools. Indeed, the consensus on the multidimensional aspect of optimization breaks down as soon as it comes to choosing the indicators for measuring it. As far as the company's accounting indicators are concerned, they are calculated on the basis of the company's operating account or balance sheet, and are associated with its economic and financial performance. They include turnover growth (Gauzente, 2011), strategy optimization, value-added, gross operating surplus, gross operating profit and net profit (Nalwango, 2009). As for financial ratios, some like Louizi (2011); Adams and Mehran (2005); Charreaux (1997) have used ratios such as: Profitability (the ratio of realized profit to sales), capital turnover (the ratio of sales to investment), ROI (the ratio of realized profit to investment), ROA (the ratio of net profit to total assets), Return on equity (the ratio of net profit to equity) and Tobin's Q. These accounting indicators and financial ratios are frequently used to understand performance. They provide an effective summary of the business, but they represent only a narrow view of optimization. They do not offer a multidimensional and comprehensive measure of efficiency as do the synthetic efficiency indicators.

Thus, it is now recognized that performance is a construct (Lenz, 1981). Composite indicators are therefore used, which can integrate several approaches to optimization. These are the productivity index (the ratio of outputs to inputs) and the Malmquist index (Färe et al., 1994).

The data used in this chapter come from our field survey conducted at the Ministry of Finance. However, the data obtained after the survey was supplemented by information collected from the companies concerned. It enabled a panel of data to be compiled (2006–2012). For reasons of confidentiality and relevance of the analysis, the companies were grouped by category of activity banks according to the national accounting nomenclature.

The evaluation of the optimization is a prerequisite for the factor analysis and the econometric regression. We use an efficiency frontier method which consists of estimating an efficiency function (Farrell, 1957). This efficiency frontier performance measure has the advantage of providing a multidimensional measure of optimization. It has several approaches (the non-parametric approach and the parametric approach). Given the deterministic interest and the lack of assumption on the functional form of the efficiency frontier of the non-parametric approach, we use it in this research under the name Performance Data Analysis Control. Building on the work of Farrell (1957), Performance Data Analysis Control requires the choice and definition of the efficiency under consideration (technical efficiency or allocative efficiency). Technical efficiency measures the ability of a production unit to produce the maximum output from a given level of input and a given technology (output orientation), or it measures the ability to use the minimum amount of input to produce a given level of output with a given technology (input orientation). Furthermore, taking into account the hypothesis of returns to scale, technical efficiency reflects the level of quality of resource management and organizational capacity of the production unit (pure technical efficiency) or it reflects the scale (optimal or not)1 at which the production unit is situated in a context of pure and perfect competition (scale efficiency). Allocative efficiency measures the ability of a production unit to choose the optimal combination of inputs, given their market prices and the budget allocated to acquire them.

Taking into account the absence of factor prices taken into account by allocative efficiency, we use technical efficiency for the measurement of firm performance. Under the assumption of constant returns to scale, if we consider a set of production units N each having inputs K and outputs M, for each production unity, the Performance Data Analysis Control is introduced in the form of an input-output ratio represented by the maximization program.

After the construction of the performance measurement index, a characterization of these Banking sector companies according to the relevant variables is carried out. For this purpose, Multiple Factor Analysis is used. This technique is applied in the case of multiple tables where the same group of individuals is described by several groups of variables. It is based on principal component analysis (PCA) when the variables in each group are quantitative and on multiple correspondence analysis (MCA) when the variables in each group are qualitative. Since in our case the variables are all quantitative, it is the PCA that is at the heart of the Multiple Factor Analysis. Thus, the data are presented in the form of multiple tables, i.e., in the form of an I observation table where the variables K are split into groups J. PCA is a method of factor analysis which consists of searching for reduced-dimensional subspaces which best fit the cloud of points-individuals (the rows of the initial data table representing the branches) and the cloud of points-variables (the columns of the initial data table representing the variables).

Following the construction of the performance measurement indicator and the characterization of the companies, we envisaged the realization of a linear regression model that would allow the impact of optimization on production to be detected. Opting for a longitudinal approach (dynamic aspect) which allows to observe over time the information relating to the companies, it is necessary to use a specific econometric modeling of linear regression (multiple linear regression on panel data). This panel data regression integrates the time series regression and the cross-sectional regression and is therefore the most suitable.

The following variables were therefore selected:

1. **Output (ptv) in Value:** This is the dependent variable to be estimated. In fact, the initial objective was to determine the impact of optimization on growth. However, given the difficulty of capturing growth for the categories of activity banks, we considered "output," available for all categories of banks, as a proxy for growth. Indeed, growth is a variation of GDP. GDP is the sum of the value-added of production units, the value-added being itself a difference between production and intermediate consumption. Economic growth is a macroeconomic characteristic of the economy, it is not determined at the microeconomic level;
2. **Performance (perf):** This is the synthetic performance measurement indicator obtained from the Data Analysis Control Optimization. This indicator is expressed here as a percentage and ranges from 0 to 100. It was used by Djimasra (2009) who demonstrated that the evolution of the growth rates of the factors was due to gains in technical efficiency, and therefore to optimization;
3. **Malmquist Productivity Index (malm):** This is a synthetic indicator for measuring overall productivity of production factors. This index is measured between two successive years. It is not calculated for the first year of operation. It was constructed using software. The link between growth and productivity is the work of many authors, notably Solow (1957);
4. **Technological Change (techn):** This is an indicator that quantifies the change in technology between two consecutive periods that a given category of business bank has had to face;
5. **Other Expenses (OTH):** This variable represents all expenses (expressed in millions) that do not fall under any of the expenditure items reported by enterprises. It represents a major component of the operating expenses of enterprises. Also, because of the diversity of elements that can enter into this variable, it has not been considered

as an input. It would therefore be interesting to see its influence on the production of the units considered.

Considering the variables presented above, the linear model is initially specified as follows:

$$ptvit = \alpha_i + \beta_{1i}\,perfit + \beta_{2i}\,malmit + \beta_{3i}\,technit + \beta_{4i}\,autreschit + \varepsilon_{it}$$

However, before launching the estimation using panel data, it is necessary to specify which type of model will be used. To do this, it is first necessary to check the homogeneity of the banks' behavior (individual effects) by means of a Fischer test:

- If the assumption of total homogeneity is validated, the appropriate model is an OLS estimation model.
- In the case of partial homogeneity, the panel structure is rejected and different data and models N have to be estimated per observation.
- In the case of complete heterogeneity, observations have characteristics that may affect the phenomenon under study. In this case, individual effects are said to exist, or the data are heterogeneous. In this case, a particular model of heterogeneity is used: a FEs model or a RE model.

In the case of total heterogeneity of behavior, this heterogeneity must be well modeled. Thus, to decide between the fixed-effects model and the random-effects model, a Hausman specification test is applied. If the null hypothesis of no correlation between the individual effects and the explanatory variables is validated, the RE model is used, otherwise the FEs model is used.

The turnover of the surveyed banks has an upward trend over the period 2006–2012. However, between 2008 and 2009, the total sales of these institutions declined considerably. This decline is partly explained by the 2008 crisis. The increase after this decline is explained by improved consumption following the recovery and the Bank of Canada's intervention on the policy rate. However, this improvement in turnover is not the same for the whole sector. Moreover, some institutions have higher turnover than others (Chart A1). Bancassurance leads all categories. Bancassurance is one of the most important categories of banks for the Canadian economy. The fact that petroleum products are easily sold on the market means that companies have many outlets.

Over the period 2006–2012, production, and value-added have an almost identical evolution with a decrease in value between 2008 and 2009 as well as turnover. The distribution of total production by category shows the importance of general banks with 81% of total production. The usefulness and diversity of the products of this category of bank gives it a wide range of customers and forces it to produce huge quantities. Thus, since GDP is a sum of VA, this banking category (general banks) has a positive influence on national output. The other main categories of banks that are the most productive are Bancassurance, Universal Banks, and Online Banks. They represent respectively 8.22%, 2.32% and 1.64% of the production. In addition, some categories of banks have negative VA. They thus have a negative effect on growth. These effects can be explained by the fall in prices, the insufficient production and the considerable intermediate consumption used.

As regards the general structure of the expenses of the banking institutions, the evolution is more or less the same as that of the turnover. This similarity may suggest a link between these two variables. This seems logical if we consider that the objective of maximizing profits implies a control of expenses and an improvement in sales. The costs incurred by banking institutions from 2006 to 2012 show an upward trend with a low point in 2009. In addition, between 2006 and 2012, operating expenses increased (41.53%). The banks with the highest strategy optimization costs spend more on strategy optimization costs than they sell their products and services. This explains the negative VA found.

The survey of banking institutions and the operation of their FSDs shows that operating expenses can be structured into several components: tradable inputs, non-tradable inputs, personnel expenses, and taxes. Tradable inputs represent the main expense item for businesses in Canada (58.6%). This represents about three-fifths of the total operating expenses of firms operating in Canada. Next come non-tradable inputs (32.25%), transport expenses represent 4.71%, external services 8.43%, other expenses 19.11%, personnel expenses 7.12% and those related to the level of taxation 2.03%. However, it is also noticeable that the behavior of the banks is not the same as regards the distribution of expenses. Thus, Bancassurance is the category for which purchase expenses represent more than half of the total operating expenses and which makes the least use of the other cost components of the strategy optimization.

Furthermore, the investment sector spends the largest share of its expenditure (28.14%) on its staff. This finding may suggest that the employees of investment banks are well supervised and that managers regard employees as essential factors of production. In general, the sector uses the least inputs. In contrast, other more material-intensive sectors spend just 0.78% of their total expenditure on staff. This can be explained by the fact that since the amount of trading and speculation is large, it represents a huge expense compared to wages and social contributions. Furthermore, it can be assumed that the more the Banks spend on inputs, the less they spend on personnel costs, optimization, and investments.

Given the lack of information on prevailing service prices, we apply the Performance Data Analysis Control method with an output orientation, using technical efficiency rather than allocative efficiency. In addition, we have chosen to perform a Performance Data Analysis Control with constant returns (to obtain total technical efficiency), and a Performance Data Analysis Control with variable returns (VRS; to obtain pure technical efficiency) in order to verify which category of bank operates at an optimal scale. For this purpose, efficiency scores were calculated on the basis of two outputs (turnover and VA) and four inputs (input purchases, transport expenditure, wages, and external services).

The results of the model present (for each year and for each category respectively) the technical efficiency, the pure technical efficiency, and the scale efficiency. Starting from constant output, a variable number of banks are technically efficient with given quantities of inputs, they maximize the output of their outputs. The years 2007, 2008, and 2012 recorded a minimum number of 4 categories of activities that are located on the constant return efficiency frontier (CRS). In 2009, a maximum of 6 banks are on this frontier. We also note that technical efficiency scores are generally quite high. The vast majority of banks are technically efficient at least once over the period 2006–2012. Only Deposit Banks and Direct Banks were never technically efficient at least once over the period. Corporate and investment banks were technically efficient four years.

Taking into account VRS, pure technical efficiency is measured for the categories of banking activities. Note that the CRS curve is fitted by a straight line while the pure technical efficiency frontier with VRS is fitted by a non-linear curve. As a result, the number of efficient banks in the latter case is higher than in the former. Thus, on an annual basis, the minimum number of purely and technically efficient banks is 06 in 2009 and the maximum is 11 in 2006. These extreme numbers are significantly higher than those seen in the CRS case. They are then characterized by good management and organizational practices resulting from rigorous optimization. Similarly, all the banks here were efficient at least once over the period.

Regarding scale efficiency, which measures whether the units are optimal, 4 to 6 are annually optimal. In other words, each year at least 4 banking categories have identical scores per CRS and per VRS, whether they are efficient or not. They then operate at optimal scales. That is, by proportionally increasing the level of all their factors, they reach the best possible and expected situation: the Deposit Banks have equal scores of 0.935 by CRS and VRS. They are operating at an optimal scale even though they are not efficient. Also, only the Direct Banks were never optimal over the period. These technical efficiencies considered as indicators of bank performance are presented in Table 10.1.

TABLE 10.1 Performance Measurement Indicator

Activity Bank Category	2006	2007	2008	2009	2010	2011	2012
Merchant banking	1,000	0,497	1,000	1,000	1,000	0,898	0,898
Investment banking	0,837	0,978	0,941	0,934	0,880	1,000	1,000
savings bank	1,000	0,916	0,985	0,895	0,750	0,803	0,934
Corporate banking	1,000	0,961	0,880	0,880	1,000	1,000	0,790
General bank	0,617	1,000	0,890	0,890	1,000	1,000	0,711
Corporate banking	0,571	0,973	0,974	1,000	0,883	0,929	0,785
Online banking	0,655	0,716	0,938	0,938	1,000	0,543	0,672
Bancassurance	1,000	0,996	0,910	0,906	0,931	0,835	0,777
Private banking	0,487	1,000	0,881	0,563	0,693	0,996	0,657
Depository bank	0,960	0,935	0,896	0,896	0,830	0,831	0,746
Universal bank	0,771	0,871	0,951	0,629	0,666	1,000	0,796
Electronic banking	1,000	0,925	0,924	1,000	0,701	0,701	1,000
Free bank	0,799	1,000	1,000	1,000	0,833	0,775	0,931
Cooperative bank	0,975	0,849	1,000	1,000	1,000	1,000	0,927
Direct bank	0,600	0,844	0,849	0,907	0,210	0,221	0,888
Network bank	0,844	1,000	1,000	0,734	0,903	0,814	1,000
General consulting bank	0,756	0,816	0,816	1,000	0,650	0,650	1,000

Source: Author based on data from the Bank of Canada.

The performance scores of the banks vary over the study period between 0.21 and 1. Specifically, they vary between 0.487 and 1 (in 2006); 0.497 and 1 (in 2007); 0.816 and 1 (in 2008); 0.563 and 1 (in 2009); 0.210 and 1 (in 2010); 0.221 and 1 (in 2011); and between 0.657 and 1 (in 2012).

Furthermore, it is noticeable that a certain category has for three years (2006, 2009, and 2012) obtained the lowest efficiency scores. Although performing well in 2007, these private banks are among those that negatively affect the economy. Indeed, they generally have very low scores compared to other categories of banks. This suggests that they do not have a significant impact on the Canadian market. The same is true of the Direct Banks, which obtain two consecutive years (2010 and 2011) of minimum performance scores and do not perform well over the period with generally very low scores compared to the others. This can be explained by the fact that these banks are generally public or Para public. They have a very important social role. They focus more on supporting the state in its regalian duties. They therefore do not prioritize the performance of their management systems and, in a way, are lax in dealing with certain internal problems such as corruption. Deposit money banks also never perform well over the period, but are optimal in 2007. This category of banks, when they increase the quantities of their factors, reach considerable levels of production. They therefore have increasing marginal effects.

Also, the four categories integrating optimization observed earlier, namely "General Bank," "Bancassurance," "Universal Bank," and "Online Bank," are performing well. They all performed at least once over the period. The General Banks even performed well three times (2007, 2010, and 2011) over the period. They are therefore the fourth-best performing category of banks in terms of the number of years of performance (their minimum performance being 0.617 obtained in 2006). Thus, this category can still increase its output by 38.3% with a given number of inputs. Bancassurance is one of those with very high-performance scores of around 80%. In 2012, because of its performance score of 0.777, it could still increase its output by 32.3% using the same number of inputs. However, it is the Corporate and Investment Banks that are the most frequent performers (number of years principle) over the period (04 years) and thus occupy the top three positions.

PCAs were performed on groups of annual variables. The optimization and performance variables used for each year are: inputs, wages, transport, taxes, external services, other expenses, outputs, turnover, value-added, and performance. Because of the length of their labels, the activity categories have been coded by numbers ranging from 1 to 17. The first factorial axis represents 48.28% of the total inertia and the second axis represents 20.53%. This factorial plane formed by the first and second factorial axes represents 68.81% of the total inertia and thus summarizes the positions of the banks over the entire period. With each of their contributions greater than 10, the banks that contributed most to the formation of this plane are, respectively, the Universal Banks, the Bancassurance, and the General Banks. On the other hand, those best represented (qualities higher than 0.8) are the General Banks. This factorial design (graph 5) shows a compact block formed by the largest number of categories (3, 4, 6, 9, 10, 12, 13, 14, 15, 16, 17) close to each other, characterized by similar behavior. This grouping means that the behavior of banks is homogeneous across years and also that the overall situation of the Canadian economy has not changed from 2006 to 2012. In contrast, banks 1, 2, 5, 7, 8, and 11 are each isolated, with distinct characteristics and behavior. This shows that the atypical categories of banks (isolated, with distinct characteristics and behaviors) are opposed to the homogeneous categories (homogeneous behaviors and close to each other).

The overall representation of the variables, taking into account the group membership of each variable, is shown in the data. Most of the variables are oriented along the first factorial plane. Thus, apart from the most influential variables identified above, a good number of other variables are also positively correlated with this axis, whether they are well or poorly represented. Removing the performance variables (which are poorly represented), all other variables with positive coordinates on each axis are positively correlated with each axis and are positively oriented along each axis. Consequently, it can be stated that this factorial plan does not show a structure of opposition between the variables. On the contrary, it reflects a side effect of all the variables. In other words, this design determines the variables with high values. This factorial design shows that the variables relating to production, VA, and other expenses generally have high values. Thus, in this design of the first two axes, the performance variables are poorly represented. They are not oriented along one of these axes. Also, they are represented by short segments and are far from the unit circle. The plan thus does not allow for any comment on the performance of banks over the period from 2006 to 2012.

However, the selected design provided very little information on categorization by including performance, which a priori has an ameliorative effect on bank characteristics. Hence the need to determine the impact of optimization on output. The results of the Fischer homogeneity tests show that the data are not homogeneous (p-value 1 = 8.337 × 10 – 11< 0.05). In other words, there are individual effects that differentiate the banks and are likely to significantly alter the phenomenon under study. Given the nature of this individual effects model, the Hausman test allows us to compare the estimates obtained in the case of the FEs model and the RE model. This test (5%

< p-value = 0.0906 with Chi-square = 8.03) shows an absence of correlation between the individual effects and the explanatory variables. Hence the use of the RE model whose estimator is the GCM.

The results of the random individual effects model present the likelihood ratio (Rbetween2) which indicates the share of the inter-individual variability of the interest phenomenon explained by the explanatory variables. In this case, 27.18% of the variability of the banks' output is due to all the explanatory variables. Moreover, the RE contributes only 5% to the model (R2 within) and the overall quality of the regression is only 19.76% (R 2overall). In addition, the Wald test (5.76 with 5% < P-value) shows that the explanatory variables are jointly insignificant in explaining output.

According to the results, the model is not globally significant in explaining output. This implies that a modeling problem has occurred. To solve the problems encountered, we will apply logarithmic transformations on two variables: the variable of interest "production" and the explanatory variable "performance." These transformations have a double importance. They allow the stabilization of the variance of the residuals in order to make the residuals normal and to improve the quality of the model. They also allow an interpretation in terms of elasticity of the impact of the optimization on production. The results show that the new model is globally significant in explaining output (p_value of the Wald test less than 0.05). Furthermore, 62.06% of the output variability in Canadian banks is explained by a set of phenomena including performance, productivity, technological change, and other operating expenses. Also, RE contributes more than 12% to the model, which has an overall regression quality of 37.17%.

Performance has a positive impact on production. Both variables move in the same direction. Indeed, increasing optimization by 1% leads to an increase in output of 0.054%. In other words, if planners can influence the organizational and managerial capabilities of firms to increase their performance by 1%, they would increase, for the same quantity of inputs, the quantities produced by 0.054%. Similarly, this positive relationship is also observed for the Malmquist index and output. The Breusch Pagan test confirms that the null hypothesis of non-significance of RE is rejected (p_value less than 5%). Thus, each category of bank has a significant random effect over the period which, all other things being equal, distinguishes it from the others.

Statistical analyzes show that the evolution of output and PV is similar, and that tradable inputs are the main expenditure item of banks and impact on output in Canada. This is followed by external services, wages, transportation, and taxes. Generally, banks behave differently depending on which category they fall into. However, four activities stand out in terms of their characteristics: "General Banking," "Bancassurance," "Universal Banking," and "Online Banking;" to which others can be added. However, the problem arises of being able to classify them according to their performance. Hence the need to construct a synthetic indicator to measure performance. The factor analysis enabled us to identify two groups of banks. The "atypical" group is characterized by high, heterogeneous, and distinct variable sizes. There is also the "homogeneous" group. These categories of banks are close to each other and have almost the same behavior. Regardless of the group to which they belong, these categories of banks have higher overall output, VA, and expenditure on other expenses.

At the end of this chapter, it is clear that the performance of banks and factor productivity have positive impacts on the growth proxy, i.e., output. Indeed, a positive fluctuation in performance of around 1% positively affects output by 0.054%. The notion of general equilibrium offers a unique framework for analyzing the effects of optimization on management tools. It requires some questionable simplifications, but allows us to highlight three essential elements for the analysis of this problem. The empirical results thus highlight the pattern of plurality of management practices and a clear homogeneity of these practices in relation to the categories of performance. It also appears that the Canadian case is best analyzed within the framework of the multi-faceted model, particularly with regard to recent developments. Finally, there is also evidence to suggest that Canadian exports are relatively more production intensive than imports.

KEYWORDS

- **constant return efficiency**
- **heterogeneity**
- **homogeneity**
- **multiple correspondence analysis**
- **principal component analysis**
- **variable returns**

CHAPTER 11

The Impact of Internal and External Strategy Optimization Mechanisms on the Productivity of Banking Institutions Using Firm-Level Data

This chapter analyzes the impact of internal and external strategy optimization mechanisms on the productivity of banking institutions using firm-level data. The internal strategy optimization mechanisms studied are investment decisions, dividend strategy, and optimization structure. External optimization, on the other hand, includes the competitiveness of the banking system and the legal and regulatory systems. Using cross-sectional series, we decided to use panel data for the empirical analysis in order to properly address the endogeneity issues. The analysis reveals that capital structure, dividend strategy, and optimization structure impact the productivity management of Canadian banks. The optimization systems of internal and external strategies have a significant influence on the ability of banks to manipulate earnings. The impact of strategy optimization systems on firm productivity, and in particular on productivity management.

The lack of effective monitoring of managerial productivity, the eradication of opportunistic behavior, and the application of incentives are real needs in complex performance management. Therefore, the main objective of this chapter is to analyze the impact of internal and external strategy optimization mechanisms on discretionary governance behavior. At the level of banking institutions, we study how capital structure and dividend decisions, as well as the optimization structure, determine accounting discretion. At the level of open economy variables, we focus on the banking system and the legal and regulatory systems as determinants of value-based management discretion.

Value-Based Management in an Open Economy: Optimizing Strategy to Improve Business and Performance. Prof. Post-Dr Walter Amedzro St-Hilaire (Author)

To do so, we conduct an empirical study over the period 1997–2015. This study takes into account not only the internal mechanisms of strategy optimization, but also the external mechanisms at the Canadian level: a first. It uses a novel statistical technique that effectively addresses endogeneity and heterogeneity problems not observed in the banking sector. In particular, panel data composed of a mix of cross-section and time series.

Agency theory is the theory that studies contractual conflicts caused by different incentives between contracting parties (Jensen, 1994; Jensen and Meckling, 1976). These problems are essentially asymmetries of interest between the parties. This conflict of interest requires mechanisms that encourage governance to behave in the best interest of the firms, not their own. This set of mechanisms is called the governance institutional optimization system. In a broad sense, strategic institutional governance optimization is the set of capital market rules related to equity investments in institutions (Claessens and Yurtoglu, 2013). This includes listing requirements, insider trading arrangements, disclosure, and accounting rules, and rights protection. These rules are essentially regulated by external mechanisms that are far from value-based management. This conceptualization of strategy optimization is close to that of Shleifer and Vishny (1997). Also, the interpretation of Zingales' (1998) approach to strategy optimization is defined as a complex set of constraints formed by the renegotiation of the rights of cash flows generated by institutions.

Institutional optimization systems are, by default, imperfect, and managers have an incentive to make discretionary decisions following their own interests rather than following wealth maximization. Therefore, techniques to manage returns could take the form of changes in accounting methods (Moses, 1987) such as changes in the inventory valuation system or in the accounting methods for recording and accounting for extraordinary services and expenses (Beattie et al., 1994). Performance is embedded in the productivity of the firm. Because governance has privileged information, they have the ability to manage the bottom line to maximize their returns. As a result, governance is more likely to engage in productivity management in the current period to ensure better future compensation conditions. Second, there is the constraint imposed by covenants. If the firm violates the covenants, it could face higher debt costs and reject potentially attractive investment projects (Angelo et al., 2004). Governance could overestimate profits in order to meet or exceed analysts' expectations. Whereas normal accruals are essentially means to improve the content of accounting information, discretionary (or abnormal) accruals are means to intentionally manipulate profits in favor of interests. Therefore, reporting can be meaningless if the accounts are seriously manipulated or distorted. Since the origin of discretionary accruals is based on the ability and motivation of managers to manipulate performance, on the one hand, and the productivity (or lack thereof) of strategy optimization mechanisms, on the other, the aim of the chapter is to study how institutional optimization systems impact on the discretionary use of accruals to manage income.

At the institutional level, three strategy optimization tools are studied: (i) capital structure decisions; (ii) dividend strategy planning; (iii) structure optimization. Capital structure optimization is a double-edged sword for productivity management. On the one hand, the level of debt could be used as a value-based mechanism for managerial behavior. The preference for consumption of benefits over wealth is mitigated through leveraged capital structures. High levels of debt reduce the power of governance (opportunists by reducing the cash flow available to spend on negative net present value projects (Jensen, 1986). An increase in leverage also increases the risk of bankruptcy which reduces value for money in non-value-adding activities. Therefore, governance avoids this opportunistic planning in favor of maximizing current interests. One should then expect that institutions with more structured and optimized leveraged capital would have lower productivity. Nevertheless, governance could make a good impression through performance in order to issue debt on more favorable terms and/or achieve the objectives imposed by the covenants (Mohrman, 1996). Given these arguments, the enabling or constraining roles of debt on productivity management seem to be a question justified by both positive and negative effects. Hence hypothesis 1 of our study:

- **H1:** A positive or negative relationship is expected between the productivity management optimization lever.

The decision of how much to disburse in dividends and what proportion or profits should be reinvested in the form of retained earnings is one of the most important issues facing governance. Lintner (1956) highlights the fact that optimization is reluctant to cut dividends, and targets long-term payout ratios when planning dividends. Brav et al. (2005) in examining payout practices find that maintaining dividend levels is the main variable in dividend optimization while payout ratios are of secondary importance. They conclude that since dividend cuts have negative consequences, it is quite possible that governance manipulates accounting information. Anglin et al. (2013) indicate that dividend optimization influences actual productivity management both upstream and downstream. They state that planners manipulate earnings upwards through actual activities to mitigate the shortfall (of pre-managed earnings over previous years' dividends) when these are lower than dividends paid for the previous year, suggesting that dividend levels are an important benchmark for earnings. They provide evidence of a strong relationship between changes in pre-generated earnings and actual productivity management by suggesting that dividend optimization impacts actual productivity management. Given the importance of dividend optimization for actual productivity management, traditional dividend planners manipulate earnings to a greater extent than those who do not plan. Therefore, hypothesis 2a is formulated:

- **H2a:** A positive or negative relationship between income and dividend optimization is expected.

Evidence of the interaction between dividend optimization and governance structures comes from analyzes that show that both the institutional and the legal effect payments. In fact, the dividend distribution strategy depends crucially on the ability of the system to compel dividend payments. Thus, we cannot dissociate the nature of the dividend mechanisms and the characteristics of the legal system. Therefore, our next hypothesis states that:

- **H2b:** In institutional contexts with weak strategy optimization systems, more productivity management would be expected to achieve certain objectives than in institutional contexts with relatively stronger governance systems.

The optimization structure plays a key role in the governance of institutions. In fact, Bebchuk and Hamdani (2009) argue that all the effort invested in creating a composite of strategy optimization is worthless unless the optimization structure is incorporated. As a result, governance decisions cannot be divorced from strategy optimization or from the opportunistic behavior of managers. There is a momentary overlap between governance and control that minimizes potential agency problems. Warfield et al. (1995) show that governance has a significant effect on discretionary decisions and that optimization is positively associated with the explanatory power of earnings and is inversely related to the extent of accrual adjustment. Governance is negatively associated with the size of accruals. Therefore, when institutional optimization is considered, a non-monotonic relationship could exist between institutional optimization and productivity management. Hence hypothesis 3a:

- **H3a:** Poor planning promotes alignment of interests and low probability of productivity management. Rigorous management optimization, on the other hand, favors a risk of entrenchment and consequently the likelihood of opportunistic management of accounting results. This means that a U-shaped relationship between institutional optimization and productivity is expected.

It has been widely argued that optimized governance structures solve some agency problems through direct supervision (Brown et al., 2011). This argument leads to a positive relationship between the optimization of governance and values as diffused by the supervision hypothesis. Therefore, productivity is reduced as a result of optimization structures. Thus, it can be deduced that vertical agency conflict and consequent management of accounting profits, could be effectively reduced by greater governance optimization (Shleifer and Vishny, 1986). Therefore, our hypothesis 3b states that:

- **H3b:** The more rigorous the optimization, the less opportunism through productivity management.

Positive accounting research provides evidence that choices of accounting optimizations are determined not only by regulatory systems, but also by factors that are specific to the institution, including its operating conditions and management preferences, all of which will result in a diversity of accounting treatments (Beattie et al., 1994). Accounting rules can limit a manager's ability to distort reported earnings, but the extent to which the rules influence reported earnings still depends on how the rules are applied (Wysocki, 2003). But such adoption of accounting standards is not associated with lower levels of discretionary accruals. This suggests that the choice of accounting method may be the result of opportunistic governance. What is expected from accounting standards is that the information published in the performance is more comparable and closer to potential discretionary adjustments. Therefore, it is assumed that institutions using accounting standards have greater incentives to publish transparently because they are subject to greater restrictions and exposed to a higher risk of litigation. Therefore, Hypothesis H4 regarding accounting standards adoption is:

- **H4:** Optimized mechanisms control of less discretionary productivity should be observed following the adoption of accounting standards.

Ball et al. (2003) argue that institutional arrangements are the most important factor in controlling interests, reducing opportunistic approaches reflected in productivity optimization, and improving the quality of accounting information. Therefore, one would expect that strategy optimization rules would limit the discretionary behavior of institutions in managing earnings. Ball et al. (2000) analyzes the effect of institutional legal systems such as the comparison between common law and code-law on the quality of accounting earnings. They find that the common law accounting result is indeed significantly more accurate than the code-law accounting result, but that this is entirely due to a greater sensitivity to economic losses (conservatism on income). Therefore, we can deduct from this that the legal origin of the country could also determine the way governance uses discretionary capacity to manage outcomes. Not only is the legal system relevant in the conduct of productivity management, but also the application of this legal entity. Naceur and Omran (2011) find that regulatory and institutional variables seem to have an impact on bank productivity. This allows us to propose the following hypothesis H5:

- **H5:** There is a negative association between legal enforcement, the effectiveness of the regulatory system and the optimization of productivity.

The role of banking institutions in open markets is to mediate between saving and borrowing units by reducing transaction costs. Banking efficiency improves capital allocation, LQ, firms' access to more sophisticated financial instruments, information flows, and reduces the cost of external financing, allowing firms to better exploit current growth opportunities. The banking sector influences the level and growth rate of per capita income (Rajan and Zingales, 1998) and in economic activities aggregates such as GDP growth, productivity, and investment growth rate (Ueda et al., 2008), impact on the evolution of accounting regulations (Rajan and Zingales, 1998), the availability and allocation of finance (Gonenc and Haan, 2014) and the optimization of productivity (Enomoto et al., 2014) among others. It is likely that banking efficiency improves market monitoring and control of accounting data, due to the strengthening of laws and regulations for investor protection and, by extension, due to the presence of more sophisticated market participants. Enomoto et al. (2014) show that quality accounting information is needed in open economies to discipline governance, mitigate its incentive to manage earnings and finds a correlation between banking efficiency and accounting institutions. Degeorge et al. (2013) highlight the role of the banking system on the efficiency of analytical mechanisms and productivity optimization. Therefore, banking efficiency asymmetrically impacts the extent of productivity optimization. We could thus assume that:

- **H6:** Productivity optimization is lower if the banking system is not efficient.

We consider cross-sectional and time-series information and use panel data for the empirical analysis. The statistical analysis is developed with a sample of 712 institutions. The data includes 4,550 observations for the period 1997–2013 with an average of 6.39 observations per institution. Recall that a minimum of 4 observations is required for effective use. The microdata are mainly financial reports and market information collected from the Bank of Canada and Finance Canada. Macroeconomic information was obtained from data updated by Statistics Canada. The governance indicators on accounting standards and the legal system were obtained from a literature search of the Chartered Professional Accountants and Industry Canada. Due to the structure of our data, we used the GMM in the econometric analysis. The panel data methodology allows us to control for two fundamental problems in this kind of study: unobservable heterogeneity and endogeneity problems (Arellano, 2002; Arellano and Bover, 1990). The relationships between the characteristics of institutions and their impact on the management of productivity optimization should be interpreted with caution because of the possibility of observing spurious relationships that would favor endogeneity problems. These models could also suffer from unobserved heterogeneity, where the identified relationships are symptoms of an unobservable factor driving the dependent and independent variables.

In such problems, because the independent variables are endogenous and correlated with the residuals of the regressions, OLS estimation is both biased and inconsistent (Brown et al., 2011). As a result, we address the problems of unobservable endogeneity and heterogeneity using the GMM estimator (Blundell and Bond, 1998; Bond, 2002) which provide us with efficiency gains. Several statistical tests are used. The Hansen/Sargan test assesses the specification of the validity of the model (Hansen, 1996). This test examines the lack of correlation between the instruments and the error term. The AR1 and AR2 statistics measure the correlation of the series. The Wald test is also used to assess the significance of all independent variables in the sample. The Lind-Mehlum contrast is used to study the statistical significance of the proposed non-monotonic relationships (the special case of the optimization structure). We use several alternative estimates for productivity optimization. The productivity management measures used in the analysis can be classified into two categories as suggested by Leuz et al. (2003); and Shen and Chih (2007).

With regard to the signs expected for β_1 and β_2 one could say that it is not trivial, except for β_2. In this case, it is expected that β_2 is expected to be negative, as depreciation has been included with a negative sign in the definition of total accruals (*ACC*). However, there is no clear prediction for the sign of β_1 because, on the one hand, the high level of sales could imply an increase in accounts, but, on the other hand, the increase in sales usually implies the increase in short-term debt. The net effect on working capital may not be determined a priori. Thus, the value of (*ACC*) in equation (2) is the level of total accruals as a function of banking activity and asset composition. Therefore, the error term in the regression, which is the difference between observed and estimated accruals as mentioned in equation (3), would become the part of the total accruals due to discretionary governance-planning behavior.

Among the independent variables, we have those that are directly related to our hypotheses (capital structure, dividend optimization, and optimization structure) and those that are usually called control variables, which are essentially variables that need to be introduced into the model in order to reduce sub-identification problems. The control variables are size, profitability, default risk, dummy variables, and time. The last two variables are necessary because of the panel data structure of the econometric models. We use two alternative measures for the capital structure: book value optimization leverage (BLOV) and market value optimization leverage (MOL). Nevertheless, Bowman's (1980) study provides empirical evidence on the comparability of book value and MOL measures in the systematic risk association tests. Parsons and Titman (2008) argue that the use of book value or market value leverage is a matter of debate. In fact, Bowman (1980) suggests that accounting measures of debt are statistically indistinguishable from market value measures. Welch (2011) points out that the conclusions of most studies are likely to be independent of the definition of optimization leverage. However, according to Lang et al. (1996), a market value-based measure risks giving too much weight to recent changes in equity. Therefore, we prefer to use both as robustness measures.

The variables used for dividend optimization correspond to two other measures of the payout ratio. The first one (Div1) is calculated on the basis of dividends and earnings per share and the second one (Div2) is calculated according to Pindado et al. (2006). For the optimization structure we also use three other measures for robustness purposes. The first one includes the governance held (PctShClHeld) which is represented by the assets. The second variable (Herf) corresponds to governance practices as measured by the Herfindahl-Hirschman index. The last measure relating to the optimization structure (Own1) is represented by the proportion of shares. Among the control variables, we included size, profitability (ROA), and default risk. For default risk, we chose two alternative measures (Z1 and Z2). The first (Z1) is the Z-score. We decided to also use the version of the score (Z2) for the open economy developed later (Altman, 2005). Productivity can often be the main reason why management engages in profit manipulation via aggressive revenue recognition techniques (Skinner et al., 1994). In fact, Richardson et al. (2002) suggest that productivity management is strongly affected by the size of the institution. Regarding the strategy optimization at the level of economies, we have the following institutional variables: accounting standards, the legal system and banking efficiency. For accounting standards, we use a dummy variable (IFRS) that takes the value 1 if the institution presents its performance statements in the IFRS framework and 0 otherwise.

Following Kaufmann et al. (2011), for the legal system we include variables resulting from the six dimensions of governance: (i) voice and accountability (VA), which is the process by which governments are chosen, controlled, and replaced; (Ii) political stability (PS), which measures perceptions of the likelihood that the government will be destabilized or overthrown by unconstitutional means; (iii) GE corresponds to the quality of public and civil services, and the degree of its independence from political pressures, the quality of policy formulation and implementation, and the credibility of the government's commitment to these policies; (iv) RQ, which measures perceptions of the government's ability to formulate and implement sound policies and regulations that enable and promote private sector development; (v) rule of law (RL), which reflects agents' confidence in and ability to abide by the rules of society, and in particular the quality of contract enforcement, governance rights, police, and courts; and finally (vi) control of corruption (CC), which measures perceptions of the extent to which public power is exercised for private ends, including grand and petty corruption and lobbying by elites and private interests. Although these six original indicators range from about –2.5 (low) to 2.5 (high) in governance productivity, the values are not at their extreme. We use several measures from the dataset to analyze banking efficiency. For example, we use six indicators for the efficiency of the banking system (three to measure the level of efficiency of the banking sector, and three others for the capital markets): (i) Central bank assets to GDP (CBAGDP); (ii) Other financial institutions assets to GDP (OIAGDP); Private credit by deposit money banks to GDP (PCGDP); Stock market capitalization to GDP (SMKGDP); International debt issues to GDP (IDGDP); Private bond market capitalization to GDP (PBGDP).

Table 11.1 describes the composition of the panel data used in the empirical analysis. Table 11.2 presents the alternative measures of discretionary accruals and the independent variables. The table shows the number of observations and the mean per variable. We can observe that the mean value for discretionary regularizations is always greater than 0. In Table 11.3 we have tested the hypothesis whether these values for the regularizations are statistically different from zero. We can observe in panel A that these average values are in fact statistically significant for zero. This does not allow us to reject the null hypothesis and therefore, in the same way as García-Meca and Sánchez-Ballesta (2009), it provides evidence that institutions manipulate their results either by increasing profits or by reducing them. The remaining data (Table 11.3) describes the average value for our three alternative measures for economic regulation.

TABLE 11.1 Composition of Panel Data

Country	Comments	Firms	Avg. Obs. per Country
Nunavut	231	35	6.60
British Columbia	1,686	270	6.24
Ontario	797	142	5.61
Quebec	186	33	5.64
Alberta	917	112	8.19
Manitoba	733	120	6.11
Total	4,550	712	6.39

TABLE 11.2 Descriptive Statistics by Variable

Variable	Acronym	Definition	Obs	Mean
Discretional accruals	DAcc1	Discretional accruals based on Model 1	4,545	0.0203
–	DAcc12	Discretional accruals based on Model 2	4,545	0.0697
Capital structure	LevB	Leverage at book values	4,550	0.4865
–	LevM	Leverage at market values	4,475	0.1742
Dividend policy	Div1	Payout ratio: DPS/EPS	4,550	1.0163
–	Div2	Payout ratio: Cash dividend/ NI(t–1)	4,381	0.0231
Ownership structure	Own1	Ownership structure: % shares in hands of the majority shareholder	4,546	0.2713
–	PctShClHeld	Ownership closely held: % shares in hands of managers and majority shareholder	2,556	0.5871
Firm size	Size	Ln(TA)	4,550	6.6236
Profitability	ROA	NI/TA	4,550	0.0688
Liquidity risk	Z1	Altman Z-Score for developed economies	4,550	4.7765
	Z2	Altman Z-Score for open economics	4,550	7.5574
Financial development	CBAGDP	Central bank assets/GDP	4,549	6.5495
–	OIAGDP	Other financial institutions assets/ GDP	4,549	10.7839
–	PCGDP	Private credit by deposit money banks/GDP	4,549	27.3127
–	SMKGDP	Stock market capitalization/GDP	4,549	43.7889
–	IDGDP	International debt issues/GDP	4,549	9.2647
–	PBGDP	Private bond market capitalization/ GDP	4,549	11.1535
Legal system	IFRS	0 if local and 1 if IFRS	4,550	0.3514

TABLE 11.2 *(Continued)*

Variable	Acronym	Definition	Obs	Mean
–	GOES	Voice and accountability	4,549	0.3852
–	PS	Political stability	4,548	–0.2409
–	GE	Effective governance	4,549	0.1678
–	RQ	Regulatory quality	4,549	0.4132
–	RL	Rule of law	4,549	–0.1236
–	CC	Control of corruption	4,549	0.0990

TABLE 11.3 Descriptive Statistics of Alternative Productivity Management Measures for the Whole Sample and by Country

Panel A: Productivity Management for the Whole Sample					
Variable	**Obs.**	**Mean**	**Std. Error**	**Std. Dev.**	**p-Value**
DAcc1	4,545	0.0203	0.0003	0.0229	(0.0000)
DAcc2	4,545	0.0697	0.0013	0.0882	(0.0000)
Panel B: Productivity Management (Model 1) by Country					
DAcc11	Obs.	Mean	Std. Error	Std. Dev.	p-Value
Nunavut	231	0.0204	0.0012	0.0188	(0.0000)
British Columbia	1,788	0.0177	0.0004	0.0184	(0.0000)
Ontario	789	0.0238	0.0012	0.0323	(0.0000)
Quebec	87	0.0216	0.0042	0.0394	(0.0000)
Alberta	917	0.0203	0.0006	0.0196	(0.0000)
Manitoba	733	0.0227	0.0008	0.0225	(0.0000)
Panel C: Productivity Management (Model 2) by Country					
DAcc2	Obs.	Mean	Std. Error	Std. Dev.	p-Value
Nunavut	231	0.0670	0.0051	0.0770	(0.0000)
British Columbia	1,788	0.0730	0.0021	0.0905	(0.0000)
Ontario	789	0.0708	0.0034	0.0962	(0.0000)
Quebec	87	0.0595	0.0071	0.0658	(0.0000)
Alberta	917	0.0616	0.0027	0.0829	(0.0000)
Manitoba	733	0.0726	0.0031	0.0852	(0.0000)

In terms of the leverage position (LevB) we see that institutions have a debt level of about 48.65% of their total assets. For market values, this ratio (LevM) is 17.2%. The market value of this ratio is relatively lower than the book value due to the fact that capitalization is overvalued (Lefort and Walker, 2007; Saona and San Martín, 2015). In addition to this, we observe that institutions pay a large proportion of their earnings in the form of dividends (Div1) in open economies. This is consistent with the argument that in open economies with weak investor rights protection, shareholders demand higher dividend yields, and therefore firms pay larger cash dividends. This can be seen as a short-term approach to investment, which is looking for a short payback period in order to recoup it fairly quickly. This weak protection is also reflected in the structure of the company's optimization (Espinosa, 2009; Khanna and Palepu, 2000; Lefort, 2005), which is characterized by the presence of large pyramid structures, institutional investors, and highly qualified governance.

Table 11.4 presents the correlation matrix results for the productivity management measures and the independent variables. For leverage, we observe a negative correlation with two of our three measures for productivity optimization. The payout ratio, as well as decision power, are both negatively correlated with productivity management measures. For most of the external strategy optimization indicators (those measuring the efficiency of the banking system as well as those measuring the adequacy of the legal and regulatory systems), we observe a negative correlation between them and discretionary accruals.

Table 11.5 presents the regressions between the independent variables measuring the internal strategy optimization systems and the management of productivitics measured according to model 1 (DAcc1). In all regressions, we observe that the independent variables are jointly statistically significant according to the Wald test. There is no second-order autocorrelation between the variables, although in some regressions we find only first-order autocorrelation. This fact does not invalidate our results. Since the establishment of panel data, a certain level of first-order autocorrelation is expected for a 5% confidence interval according to Arellano and Bond (1991). As far as the conditions are concerned, both Sargan and Hansen's overidentification tests did not reject the overidentification constraints, which means that the set of instruments is orthogonal to the estimated residuals. Thus, the results presented in Table 11.5 are robust to the standard diagnostic tests for panel data.

TABLE 11.4 Multivariate Analysis: The Dependent Variable is DAcc1 and the Independent Variables are the Internal Strategic Drivers of Optimization

Variables	**(1)**	**(2)**	**(3)**	**(4)**	**(5)**	**(6)**
Constant	0.0223***	0.0278***	0.0290***	0.0468***	0.0355***	0.0256***
–	(0.00166)	(0.000309)	(0.000258)	(0.000490)	(1.42e-06)	(5.61e-06)
LevB	–0.00602*	–0.0168***	–0.0146***	–0.0301***	–0.0277***	–0.0256***
–	(0.00322)	(0.000257)	(0.000135)	(0.000433)	(2.43e-06)	(3.29e-06)
Div1	–	0.000165***	1.64e-07	2.15e-05***	2.21e-05***	2.63e-05***
–	–	(2.19e-06)	(5.88e-07)	(7.24e-07)	(3.42e-09)	(5.71e-09)
Own1	–	–	–0.00854***	–0.00864***	–	–
–	–	–	(0.0001)	(5.85e-05)	–	–
PctShClHeld	–	–	–	–	0.0148	–0.0632***
–	–	–	–	–	(5.28e-07)	(1.47e-05)
$PctShClHeld^2$	–	–	–	–	–	0.0486***
–	–	–	–	–	–	(1.64e-05)
Critical value	–	–	–	–	–	0.6502
SIZE	–	–	–	0.000296***	0.000324***	0.000335***
–	–	–	–	(6.07e-05)	(4.83e-07)	(9.24e-07)
ROA	–	–	–	–0.00815***	–0.0228***	–0.0195***
–	–	–	–	(0.000550)	(5.07e-06)	(6.43e-06)
Z1	–	–	–	–0.00113***	–0.00150***	–0.00143***
–	–	–	–	(1.97e-05)	(2.19e-07)	(1.65e-07)
Comments	11,072	6,449	4,541	4,541	2,940	2,940

TABLE 11.4 *(Continued)*

Variables	(1)	(2)	(3)	(4)	(5)	(6)
Number of iden	949	841	710	710	624	624
Industry effect	Yes	Yes	Yes	Yes	Yes	Yes
Temporal effect	Yes	Yes	Yes	Yes	Yes	Yes
Country FE	Yes	Yes	Yes	Yes	Yes	Yes
No instruments	–	275	379	379	426	426
Wald-test	–	47.05	22518	29186	4.110e+10	4.110e+10
p-value	0.0000	6.06e-11	0	0	0.736	0
AR(1)	–3.5640	–6.013	–3.418	–3.418	–2.295	–2.295
p-value	0.00037	1.83e-09	0.000630	0.000631	0	0.0217
AR(2)	0.68400	–1.844	1.162	0.866	1.232	1.232
p-value	0.49400	0.0652	0.245	0.386	0	0.218
Sargan-test	683.10000	717.3	772.6	749.8	679.7	679.7
p-value	0.00000	0	0	0	0.0217	0
Hansen-test	378.50000	435.4	408.6	403.1	400.3	400.3
p-value	0.30300	1.12e-09	0.112	0.128	0.218	0.736
Lind Mehlum-test	–	–	–	–	–	1876.73
p-value	–	–	–	–	–	0.000

The table shows the coefficient estimates. Standard errors in brackets. *** $p < 0.01$, ** $p < 0.05$, * $p < 0.1$.

TABLE 11.5 Analysis of the Factoring of the Main Component*

Variables	Factor	Eigenvalue	Difference	Proportion	Cumulative
Panel A: Financial Development Variables					
CBAGDP	Factor1	3.5559	2.44791	0.5927	0.5927
OIAGDP	Factor2	1.1080	0.29466	0.1847	0.7773
PCGDP	Factor3	0.8133	0.4773	0.1356	0.9129
SMKGDP	Factor4	0.3360	0.21741	0.056	0.9689
IDGDP	Factor5	0.1186	0.05051	0.0198	0.9886
PBGDP	Factor6	0.0681	0	0.0114	1
Panel B: Legal and Regulatory Systems					
GOES	Factor1	5.0000	4.3130	0.8333	0.8333
PS	Factor2	0.6870	0.5310	0.1145	0.9478
GE	Factor3	0.1561	0.0701	0.0260	0.9738
RQ	Factor4	0.0859	0.0346	0.0143	0.9882
RL	Factor5	0.0514	0.0318	0.0086	0.9967
CC	Factor6	0.0196	0	0.0033	1.0000

*Banking competitiveness and legal and regulatory systems.

In the first regression, we observe that optimizing leverage (LevB) has a negative impact on managing productivities (DAcc1). This finding indicates that the resources available for benefit consumption are minimized as higher levels of debt imply large cash flow disbursements that could otherwise be used opportunistically by governance. In addition to this, it appears that productivity optimization is seen as a consequence of the covenants contained in debt contracts. Creditors and institutions will set more restrictive clauses in contracts and eventually demand more information on productivity. As a consequence of better monitoring, governance will have less room for opportunistic manipulation of performance statements. This finding allows us to accept our first hypothesis, which suggests that higher optimization leverage has a negative influence on productivity management.

The results of the second regression suggest that there is a positive and statistically significant relationship between distribution optimization (Div1) and productivity management. In regulated open economies such as Canada, governance is reluctant to plan dividends because of its negative consequences and this has both a positive and negative impact on actual productivity management. The relative weakness of external policy optimization mechanisms (e.g., institutional systems) implies higher dividends. The catalyst for this demand for a higher dividend is the manipulation of performance states by governance. Agency theory suggests that decision-makers have a preference for dividends over profits in order not to waste LQ (Easterbrook, 1984; Jensen, 1986). This justifies the payment of dividends in cash. These results allow us to accept hypothesis H2a.

The results of the third and fourth regressions show that governance (Own1) is negatively related to productivity. This finding allows us to accept hypothesis H3b which states that the more rigorous the optimization, the lower the opportunism through productivity management. Therefore, we can observe that optimized governance structures effectively solve agency problems by directly supervising performance. The optimization structure performs well. The last regression (Table 11.5) is used to test hypothesis H3a. In this hypothesis, we proposed a non-linear relationship between institutional optimization and productivity (PctShClHeld). In this case, we used the variable PctShClHeld and its square calculation (PctShClHeld2) to test this hypothesis. As can be seen, there is in fact a quadratic relationship between this variable and the overestimation of financial reports. As the percentage of closed shares increases, productivity optimization decreases but only up to a certain threshold of governance decision power. Beyond this decision level, opportunistic manipulation of performance reports increases. It seems that agency costs and moral hazard problems are minimized with more optimized governance structures. However, in situations where stocks are larger than necessary for efficient strategy optimization, entrenchment, and expropriation problems arise. Under this scenario, decision-makers are more inclined to manage revenues, thereby appropriating parts of the wealth of minority shareholders. Such a threshold or critical point can be determined essentially by optimizing, for example, the sixth regression in Table 11.5. When less than 65.02% of the shares are held by the planner, productivity management is minimized, ceteris paribus. Otherwise, the problems of entrenchment and expropriation appear and are eventually materialized in the opportunistic manipulation of performance states. This corroborates hypothesis H3a even if the probability of managing gains increases as a consequence of the high risk of entrenchment. Therefore, a U-shaped relationship exists between PctShClHeld and DAcc1. To ensure that this relationship exists, we used the Lind-Mehlum test (2010). As can be seen at the bottom of the table in the sixth regression, the test accepts the U-shaped relationship between these two variables as a hypothesis.

In order to avoid under specification problems, in the last three regressions the control variables (size, profitability, and default risk) are added. We included size, as it can affect the optimization characteristics of strategies as well as the level of productivity management (Becker et al., 1998). Our results show a positive relationship between structural size and productivity management. We suggest that large institutions may be more inclined to manage their revenues because the complexity of their operations makes it difficult for users to detect exaggerations. Therefore, their planning may apply more aggressive accounting methods (Richardson et al., 2002). In the same regression, we can see that profitability (ROA) is negatively related to productivity optimization. In other words, we observe that profitable institutions do not find it necessary to manipulate performance in their strategic planning.

Finally, in Table 11.5, we find that the coefficient of the insolvency risk variable (Z2) displays a negative and statistically significant value. By construction, the higher the values of Z2, the lower the risk of default. Therefore, the interpretation of this variable suggests that when the risk of default decreases, discretionary accruals also decrease. And thus, the lower the risk of default, the lower the need to opportunistically manipulate income in either direction.

In addition to the variables of optimization of internal strategies studied in the previous section, the set of dimensions of external optimization coming from institutional systems are presented in Table 11.6. Recall that these variables are determined exogenously (Aerts et al., 2013). Among these variables, we include a set of measures for banking system efficiency (first six regressions) and another set of variables that measure the efficiency of legal and regulatory systems as indicators of strategy optimization (last six regressions). First, we observe that accounting standards are negatively related to productivity management. There is less optimized productivity value-based management following the adoption of accounting standards, which proves hypothesis H4. Institutions that use harmonized accounting standards have greater incentives for transparency because they are subject to higher restrictions. For the time period considered in the analysis, we can state that 35.14% of our observations include institutions that have published their accounts according to harmonized standards (see Table 11.2). And according to the results presented in Table 11.6, institutions reporting under the harmonized standards reduce the extent of their productivity management by about 1.1% compared to those reporting under provincial accounting standards.

The first six regressions in Table 11.6 include indicators of external optimizations as measures of the evolution of the banking system. As stated in our hypothesis, it is likely that the level of efficiency of banking optimization strengthens the monitoring and control of accounting data, due to laws and regulations in open economies. Our results reject the H6 hypothesis. In fact, productivity optimization seems to be higher when the banking system is efficient (see the first three regressions in Table 11.6). According to this result, we can suggest that governance incurs moral hazard problems in exaggerating performance to meet the requirements of, for example, more sophisticated banking instruments. This result could be corroborated with the bank development index which, although not reported in this study, shows a negative and statistically significant relationship with the discretionary accruals measure (DAcc1). Therefore, it seems that, at least with regard to the efficiency of the banking system, productivity optimization is a reaction to more sophisticated financial instruments and needs. The variables used to measure this impact are i) the size of the assets of the central bank (CBAGDP) and ii) other financial institutions (OIAGDP) relative to national GDP, and iii) the size of private credit by deposit banks (PCGDP) relative to GDP.

For the capital market, we have included regressions 4 to 6 in Table 11.6. We observe that neither the stock market capitalization nor the private bond market capitalization is statistically significant. Therefore, these two indicators of capital market competitiveness do not determine institutional optimization in terms of productivity. The issue of international debt as a fraction of GDP is, however, the only important variable concerning capital market competitiveness. Once again, we can observe that discretionary earnings management is a response to a highly regulated system. Therefore, we suggest that the widely tested hypothesis in open economies of an inverse relationship between banking system competitiveness and productivity optimization is adapted to the Canadian regulatory context. Despite the above results on banking competitiveness, the results on legal and regulatory systems are overwhelming. For all six indicators (VA, PS, GE, RQ, RL, and CC presented in regressions 7 to 12 in Table 11.6) on legal and regulatory systems, we observe a positive and statistically significant relationship with the variable DAcc1. Therefore, we can argue that the better the regulations, the smaller the room for discretionary decision making. This allows us to accept hypothesis H5.

Since we are working with a number of variables used as measures for the external indicators, and because all these variables are highly correlated (see Table 11.4) we cannot include them completely in the regression and have chosen to introduce them individually as presented in Table 11.6. In order to overcome this drawback in modeling discretionary regularizations, we decided to apply the principal component factorization technique to enter all these variables in the same regression and take advantage of their information content. The main advantages of this technique are that the factor(s) created are uncorrelated, on the one hand, and capture to a large extent the variability of the individual variables used in the factor estimation. Table 11.7 presents the number of factors generated for the variables used to measure banking competitiveness and the variables used for the legal and regulatory system. In Panel A, we can observe that there are two factors with Eigenvalues higher than 1 that measure the financial performance of Canada. These first two factors account for about 77.73% of the variability of the six alternative variables used to measure banking competitiveness by province. Panel B, however, shows that a single factor accounts for more than 80.00% of the variability of the variables used to measure legal and regulatory systems. These three factors were introduced in the regression analysis presented in Table 11.8. All regressions show that the factors are statistically significant. Therefore, we can assume that to a large extent, banking competitiveness as well as law enforcement framed by legal and institutional systems, is a determinant of discretionary regularizations (either increasing or decreasing revenues). Furthermore, we found that the direction of this relationship is consistent with the results presented: positive for banking competitiveness and negative for legal and regulatory systems. These results could be considered as robustness checks of our main results.

The last three regressions in Table 11.8 show the estimated models using the PctShClHeld variable. These regressions calculate the critical values of the optimization in which the discretionary manipulation of performance states is minimized. The U-shaped relationship between PctShClHeld and DAcc1 is tested empirically by the Lind-Mehlum contrast. According to this test, we can accept that there is a statistically significant U-shaped relationship between PctShClHeld and DAcc1.

The consistency of our results is tested by using as estimated dependent variable the one based on model 2 (see equation 6). This variable is slightly different from DAcc1 as it is adjusted by the change in accounts receivable to take into account all the discretion of the accounting when collecting credit offers. Table 11.9 includes six regressions representing the internal strategy optimization variables (those measuring capital structure, dividend plans and optimization structure), in addition to the external strategy optimization systems (the bank performance variables and the legal and regulatory systems variables). We find that optimization leverage (LevB) is negatively correlated with discretionary accruals. This confirms what was found above: higher levels of leverage reduce the opportunistic use of optimization tools through covenants and lower levels of cash flow available for discretionary use. The management of the distribution as well as the optimization structure, shows the same relationships as in our previous results. Therefore, we can observe that our hypotheses regarding internal strategy optimization systems are accepted.

TABLE 11.6 Regression Results Including Factors for Banking Competitiveness and Legal and Regulatory Systems*

Variables	(1)	(2)	(3)	(4)	(5)	(6)
Constant	0.0371***	0.0382***	0.0368***	0.00305***	0.00290***	0.00169***
–	(0.000637)	(0.000609)	(0.000642)	(6.12e-05)	(1.97e-05)	(3.07e-05)
LevB	–0.0330***	–0.0323***	–0.0322***	–0.00999***	–0.00757***	–0.00858***
–	(0.000452)	(0.000504)	(0.000480)	(5.63e-06)	(3.36e-06)	(6.59e-06)
Own1	–0.00570***	–0.00395***	–0.00453***	–	–	–
–	(5.87e-05)	(5.45e-05)	(6.12e-05)	–	–	–
PctShClHeld	–	–	–	–0.0485***	–0.0479***	–0.0479***
–	–	–	–	(2.51e-05)	(5.35e-06)	(9.23e-06)
$PctShClHeld^2$	–	–	–	0.0381***	0.0380***	0.0377***
–	–	–	–	(2.52e-05)	(6.73e-06)	(9.68e-06)
Critical Value	–	–	–	0.63648	0.63026	0.63528
Div1	4.94e-05***	5.51e-05***	5.53e-05***	7.47e-06*	2.34e-05	1.71e-05*
–	(4.89e-07)	(5.74e-07)	(6.45e-07)	(1.47e-08)	(1.72e-08)	(1.06e-08)
SIZE	0.00187***	0.00158***	0.00188***	0.00223***	0.00204***	0.00226***
–	(7.07e-05)	(6.91e-05)	(7.41e-05)	(1.71e-06)	(1.67e-06)	(1.86e-06)
ROA	–0.0101***	–0.0126***	–0.0131***	–0.0411***	–0.0361***	–0.0369***
–	(0.000736)	(0.000632)	(0.000820)	(4.11e-06)	(7.63e-06)	(5.16e-06)
Z1	–0.000942***	–0.000927***	–0.000912***	–0.00136***	–0.00139***	–0.00145*
–	(2.33e-05)	(2.56e-05)	(2.65e-05)	(8.47e-07)	(3.90e-07)	(9.91e-07)
IFRS	–0.0111***	–0.0111***	–0.0120***	–0.00851***	–0.00691***	–0.00711***
–	(0.000259)	(0.000310)	(0.000335)	(4.59e-06)	(2.11e-06)	(2.87e-06)

TABLE 11.6 *(Continued)*

Variables	(1)	(2)	(3)	(4)	(5)	(6)
Fac1LegEnv	–0.00197**	–	–0.00277***	–0.00228**	–	–0.000876***
–	(8.79e-05)	–	(0.000100)	(8.12e-07)	–	(5.93e-07)
Fac1FinDev	–	0.000437***	0.000963***	–	0.00146***	0.00126***
–	–	(0.000146)	(0.000144)	–	(1.18e-05)	(1.56e-05)
Fac2FinDev	–	0.000653***	0.00131***	–	0.00194***	0.00149***
–	–	(9.02e-05)	(0.000108)	–	(3.90e-06)	(4.43e-06)
Comments	4,539	4,540	4,539	2,892	2,939	2,892
Number of iden	710	710	710	622	624	622
Industry effect	Yes	Yes	Yes	Yes	Yes	Yes
Temporal effect	Yes	Yes	Yes	Yes	Yes	Yes
Country FE	Yes	Yes	Yes	Yes	Yes	Yes
No instruments	407	407	407	407	407	407
Wald-test	3.480e+10	3.480e+10	3.480e+10	3.480e+10	3.480e+10	3.480e+10
p-value	0	0.215	0.215	0.215	0.0289	0
AR(1)	–2.185	–2.185	–2.185	–2.185	–2.185	–2.185
p-value	0.0289	0.0289	0	0.727	0.727	0
AR(2)	1.241	1.241	1.241	1.241	1.241	1.241
p-value	0.215	0.727	0.727	0	0.215	0.215
Sargan-test	670.1	670.1	670.1	670.1	670.1	670.1
p-value	0	0	0.0289	0	0	0.0289
Hansen-test	378.6	378.6	378.6	378.6	378.6	378.6

TABLE 11.6 *(Continued)*

Variables	**(1)**	**(2)**	**(3)**	**(4)**	**(5)**	**(6)**
p-value	0.727	0	0	0.0289	0	0.727
Lind Mehlum-test	–	–	–	885.67	943.98	704.56
p-value	–	–	–	0.000	0.000	0.000

*DAcc1 dependent variable.

TABLE 11.7 Robustness Analysis Including Factors for Banking Competitiveness and Legal and Regulatory Systems*

Variables	**(1)**	**(2)**	**(3)**	**(4)**	**(5)**	**(6)**
Constant	0.00851***	0.0261***	0.0251***	–0.00666***	0.0209***	0.0231***
–	(0.00240)	(0.00257)	(0.00276)	(4.12e-05)	(0.000128)	(0.000120)
LevB	–0.0471***	–0.0495***	–0.0491***	–0.0555**	–0.0536**	–0.0539**
–	(0.00232)	(0.00232)	(0.00237)	(4.61e-05)	(2.41e-05)	(4.84e-05)
Div1	9.07e-05	6.40e-05**	6.81e-05*	6.46e-05***	4.19e-05***	1.51e-05***
–	(1.06e-05)	(9.03e-06)	(1.01e-05)	(1.23e-07)	(1.33e-07)	(1.44e-07)
Own1	–0.00602***	–0.00119*	–0.00125**	–	–	–
–	(0.000642)	(0.000625)	(0.000633)	–	–	–
PctShClHeld	–	–	–	–0.05078***	–0.04169***	–0.04321***
–	–	–	–	(0.000115)	(0.000113)	(0.000142)
$PctShClHeld^2$	–	–	–	0.03700***	0.02993***	0.03064***
–	–	–	–	(0.000115)	(0.000112)	(0.000146)
Critical value	–	–	–	0.06862	0.06965	0.070495
IFRS	–0.0264***	–0.0369***	–0.0372***	–0.0293***	–0.0516***	–0.0511***

TABLE 11.7 *(Continued)*

Variables	(1)	(2)	(3)	(4)	(5)	(6)
–	(0.00121)	(0.00128)	(0.00131)	(2.57e-05)	(2.44e-05)	(2.73e-05)
SIZE	0.00152***	–0.000359	–0.000151	0.00349***	0.00133***	0.000993***
–	(0.000478)	(0.000494)	(0.000534)	(7.38e-06)	(5.80e-06)	(1.27e-05)
ROA	–0.279***	–0.261***	–0.260***	–0.263	–0.250	–0.244*
–	(0.00370)	(0.00383)	(0.00405)	(2.99e-05)	(4.74e-05)	(4.45e-05)
Z1	–0.00183	–0.00184*	–0.00184*	–0.00248*	–0.00221	–0.00219*
–	(5.69e-05)	(6.53e-05)	(6.57e-05)	(1.64e-06)	(1.37e-06)	(7.80e-07)
Fac1LegEnv	–0.00198***	–	0.000660	–0.00774***	–	–0.00434***
–	(0.000494)	–	(0.000506)	(9.80e-06)	–	(1.04e-05)
Fac1FinDev	–	0.0114***	0.0116***	–	0.0177***	0.0168***
–	–	(0.000761)	(0.000768)	–	(8.33e-05)	(0.000120)
Fac2FinDev	–	0.000607	0.000770	–	0.000157***	0.00166***
–	–	(0.000593)	(0.000603)	–	(1.41e-05)	(1.59e-05)
Comments	4,539	4,540	4,539	2,892	2,939	2,892
Number of iden	710	710	710	622	624	622
Industry effect	Yes	Yes	Yes	Yes	Yes	Yes
Temporal effect	Yes	Yes	Yes	Yes	Yes	Yes
Country FE	Yes	Yes	Yes	Yes	Yes	Yes
No instruments	407	407	407	407	407	407
Wald-test	4.880e+09	4.880e+09	4.880e+09	4.880e+09	4.880e+09	4.880e+09
p-value	0	0	0	0	0.872	0

TABLE 11.7 *(Continued)*

Variables	**(1)**	**(2)**	**(3)**	**(4)**	**(5)**	**(6)**
AR(1)	–7.182	–7.182	–7.182	–7.182	–7.182	–7.182
p-value	0	0	0.634	0.872	0	0
AR(2)	–0.477	–0.477	–0.477	–0.477	–0.477	–0.477
p-value	0.634	0.634	0.872	0.634	0.634	0.634
Sargan-test	373.2	373.2	373.2	373.2	373.2	373.2
p-value	0.778	0.872	0	0.778	0.778	0.872
Hansen-test	363.2	363.2	363.2	363.2	363.2	363.2
p-value	0.872	0.778	0.778	0	0	0.778
Lind Mehlum-test	–	–	–	432.36	607.98	350.83
p-value	–	–	–	0	0	0

*DAcc2 dependent variable.

TABLE 11.8 Comparative Analysis by Institutional Context*

Variables	**(1)**	**(2)**	**(3)**	**(4)**
	Ontario + British Columbia	**Other Countries**	**Ontario + British Columbia**	**Other Countries**
Constant	0.0298***	0.0423***	0.00834***	0.0448***
–	(6.13e-06)	(4.67e-06)	(3.62e-05)	(0.000630)
LevB	–0.0141***	–0.0301***	–0.00611**	–0.0311*
–	(2.65e-06)	(1.90e-06)	(2.30e-05)	(0.000352)
Div1	0.000026***	0.000043***	–0.000100***	0.000019**
–	(1.39e-08)	(9.98e-09)	(7.18e-08)	(6.97e-07)

TABLE 11.8 *(Continued)*

Variables	(1) Ontario + British Columbia	(2) Other Countries	(3) Ontario + British Columbia	(4) Other Countries
Own1	–0.00620***	0.00167***	–	–
–	(3.09e-07)	(1.25e-06)	–	–
PctShClHeld	–	–	–0.0385***	0.0197***
–	–	–	(5.65e-05)	(0.000319)
$PctShClHeld^2$	–	–	0.0303***	–0.0102***
–	–	–	(4.33e-05)	(0.000327)
Critical Value			0.6353	0.9656
IFRS	–0.0120***	–0.00388***	–0.00848*	–0.000287**
–	(6.28e-06)	(7.32e-07)	(1.70e-05)	(5.68e-05)
SIZE	0.00155***	–0.000344***	0.00199***	–0.00220***
–	(3.10e-06)	(8.17e-07)	(2.53e-06)	(7.43e-05)
ROA	–0.0314***	0.0175***	0.0241	0.0199*
–	(3.66e-06)	(3.08e-06)	(2.39e-05)	(0.000456)
Z1	–0.00114***	–0.000914***	–0.00127***	–0.00122***
–	(4.96e-07)	(4.07e-08)	(8.48e-07)	(8.43e-06)
Fac1LegEnv	0.00575***	–0.00115***	0.00283***	0.00587***
–	(3.21e-06)	(4.45e-07)	(6.02e-06)	(6.02e-05)
Fac1FinDev	0.000663***	–0.00230***	0.000643*	0.0199**
–	(1.86e-06)	(1.27e-06)	(8.96e-06)	(0.000169)

TABLE 11.8 *(Continued)*

Variables	(1)	(2)	(3)	(4)
	Ontario + British Columbia	**Other Countries**	**Ontario + British Columbia**	**Other Countries**
Fac2FinDev	0.00287***	0.00122***	0.00127*	–0.00147**
–	(2.67e-06)	(5.85e-07)	(6.22e-06)	(3.55e-05)
Comments	2,573	1,966	1,992	900
Number of iden	411	299	376	246
Industry effect	Yes	Yes	Yes	Yes
Temporal effect	Yes	Yes	Yes	Yes
Country FE	Yes	Yes	Yes	Yes
No instruments	377	377	324	324
Wald-test	1.770e+10	1.770e+10	1.550e+08	1.550e+08
p-value	0	0	0	1
AR(1)	–2.625	–2.625	–3.793	–3.793
p-value	0.00867	0.223	0.000149	0.000149
AR(2)	1.217	1.217	0.578	0.578
p-value	0.223	0.858	0.563	0.563
Sargan-test	688.8	688.8	703.9	703.9
p-value	0	0	0	0
Hansen-test	337.1	337.1	170.8	170.8
p-value	0.858	0.00867	1	0
Lind Mehlum-test	–	–	390.16	8.31
p-value	–	–	0	0.744

*DAcc1 dependent variable.

External strategy optimization systems have been introduced into the regressions through the different factors which are essentially a composite of the financial and legal systems. These three factors (Fac1LegEnv, Fac1FinDev and Fac2FinDev) are statistically significant although some significance is lost in regressions 3 and 4. Cautiously, we could still suggest that, apart from this lack of significance, the results in general are still very consistent using the alternative variable DAcc2.

The last part of the empirical analysis offers a comparison by institutional context. In this case, we decided to divide the sample into two large groups according to the relative effectiveness of their legal and regulatory systems. To do this, we calculated the average of the IL, SP, EG, QR, LR, and CC by province. Only Ontario and British Columbia have a positive value. The other provinces have a negative average. Ontario and British Columbia have a relatively better institutional environment than Quebec, Alberta, and Manitoba. Therefore, we re-estimate the regressions with these two groups of provinces. When we look at the dividend strategy in detail, we observe that the coefficient of Div1 is higher for provinces with relatively strong strategy optimization (Ontario and British Columbia in regression 1) and lower for provinces with relatively weaker systems (Quebec, Alberta, and Manitoba in regression 2). We rejected the hypothesis that these coefficients are statistically equal at the 5% confidence interval. Therefore, we can accept our hypothesis H2b, which suggests higher management productivities in institutional settings with weak strategy optimization systems to achieve certain cash dividend targets than in provinces with strong systems. This result could be reinforced by the findings of regressions 3 and 4 in Table 11.10. Between these pairs of regressions, we observe that, in fact, dividend strategy negatively influences accounting optimization for the group composed of Ontario and British Columbia, but is positive for all other provinces.

Regarding the governance structure measured by the Own variable, we observe that it is an effective strategy optimization tool in provinces with better protection of investors' rights (Ontario and British Columbia), negatively impacting the discretionary capacity of governance. However, under the institutional contexts, more optimized governance structures lead to higher managerial discretion. Similar results are presented when PctShClHeld is considered (see regressions 3 and 4 in Table 11.10). In the third regression we still observe that the U-shaped relationship between PctShClHeld and DAcc1 occurs in institutional settings with relatively better strategy optimization systems. However, such a relationship is the opposite in an institutional environment with relative investor rights protection (see regression 4). In fact, it can be observed that for most of the PctShClHeld variables (96.56%), strategic management of productivity increases as PctShClHeld also increases. Moreover, as stated in regression 4, according to the Lind-Mehlum test, there cannot be a U-shaped relationship between PctShClHeld and DAcc1. Finally, we observe that, irrespective of how well or poorly policy optimization is done under the institutions, the adoption of the Harmonized Standards reduces discretionary management accounting.

KEYWORDS

- **book value optimization leverage**
- **control of corruption**
- **government effectiveness**
- **international debt issues to GDP**
- **market value optimization leverage**
- **political stability**
- **regulatory quality**

CHAPTER 12

The Linearity of the Links Between Financial Tools and Operational Efficiency

The chapter examines the performance of strategy optimization tools in the banking sector. Many researchers have preferred to use stock price-based measures rather than accounting measures when evaluating strategy optimization. Overall, our results are consistent with those of previous authors who suggest that the productivity of management tools declines in the post-planning period of strategies. This deterioration starts in the year when firms introduce strategy optimization tools and intensifies in magnitude in subsequent years. One year after the introduction of strategy optimization, banking institutions experienced a 43% decline in their ROA. On average, the ROA is 47% lower between the years defined before and after the introduction of the optimization mechanisms. After that, we focus on the factors that may have affected productivity. Thus, contrary to Jain and Kini (1994), we find that the lower productivity of strategic management tools is actually related to the absence of opportunity assumptions. Although the introduction of strategy optimization into management mechanisms in the banking sector has maintained similar levels of service growth and capital expenditure (CEG) to other open markets, we find that the growth rate in the pre-planning period of strategy is better and stronger than the growth rate in the post-planning period of strategy. This finding supports the opportunity gap theory.

In addition, the relationship between changes in structural productivity and governance is examined. The results show that the negative relationship between productivity and value-based management exists in the context of strategy optimization. For every increase in retention by the original owners,

Value-Based Management in an Open Economy: Optimizing Strategy to Improve Business and Performance. Prof. Post-Dr Walter Amedzro St-Hilaire (Author)

the productivity of strategy optimization decreases due to the increase in agency costs between owners and shareholders. Furthermore, a significant positive relationship exists between strategy optimization style and productivity. Mikkelson et al. (1997) find that complex planning tends to perform better than traditional optimization. Finally, we do not find a relationship between the age of the firm and productivity.

In the literature on productivity optimization, ROA is the most widely used index to assess the productivity of banking and financial institutions. This chapter assesses the impact of business efficiency on productivity optimization tools by using assets to generate revenues. A higher ratio is, in this respect, considered an indication of better productivity, and vice versa. The ROA ratio can be calculated using the following equation:

$$\text{ROA} = (\text{Profit after depreciation, interest, and taxes/Total assets}) \times 100 \quad (1)$$

The second most commonly used ratio to measure the productivity of optimization tools in the banking sector is the return on sales (ROS), or profit margin. ROS is more appropriate for measuring profitability, as optimization measures show a strong increase in assets but no immediate increase in revenues. Thus, the profitability of management tools should be assessed in relation to net services. We use both ratios to assess the productivity of the banking sector. The ROS is calculated using the following equation:

$$\text{ROS} = (\text{Profit after depreciation, interest, and taxes/Total sales}) \times 100 \quad (2)$$

In addition, the asset turnover, AT (sales to asset ratio) is used to measure the effectiveness of the decision-making tools in Canadian banks. Asset turnover is an important index used to analyze how assets are used to generate revenue, and is indicative of the speed at which banks are increasing their services relative to the increase in their assets. Asset turnover is calculated using the following equation:

$$\text{TA} = (\text{Net sales/total assets}) \times 100 \quad (3)$$

We thus implement a matched-pairs approach by comparing the productivity of the adoption of optimization tools before and after the decision is made. This approach thus helps us to compare the change in productivity of Canadian banks between two periods, before, and after the adoption of strategy optimization, to draw a conclusion about the change in productivity. If productivity in the post-strategy planning period is better, then there is reason to conclude that management tools have been improved. But if the post-planning period of the strategies is not profitable, then it can be inferred that productivity has a negative effect on the banks' optimized mechanisms control tools. The study period (2003–2010) will be divided into three segments for each strategic decision.

The first concerns the pre-planning period of the strategies, labeled as "-years," and grouping the years before the adoption of the optimization of the strategies. Thus, in the case of the adoption of the strategy optimization in 2006, the pre-planning years of the strategies will be 2005, 2004, and 2003, which correspond to –1, –2, and –3, respectively. The second tranche will cover the year of adoption of strategy optimization (Y0). From there, if 2006 is the year the bank adopted strategy optimization (2006 = Y0). Finally, the third segment is the post-planning period of the strategy (+ years = the years following strategy optimization). For a bank that adopts strategy optimization in 2006, 2007, 2008, 2009, 2010, and 2011 will be considered as post strategy optimization and correspond to +1, +2, +3, +4 and +5, respectively. In this research we use different objective criteria. We compare Y–1 (the base year) to each of the five post-planning years of the strategies, including Y0. We divide the time interval differently by comparing Y–1 to Y+1, and the average of (–3, –2, –1) to the average of (+1, +2, +3). As a result, we derive a performance line that reflects the variation in performance from before to after strategy optimization.

Year 2010 is chosen as the cut-off date to allow a minimum of one-year post-strategy optimizing by the end of 2011 for banks that adopt optimization mechanisms in 2010. Banks whose management tools are optimized before 2010 will have included additional years in their post-strategy planning period. The average number of years before strategy optimization is still three because all banks provided data on their management tools to the Bank of Canada over this three-year period. The year of the adoption of strategy optimization or Y0 will be excluded from the comparison, as it has a mixed property. The change in each variable between two periods and for each bank will be examined using the median (Jain and Kini, 1994) since the estimation of operational productivity can be biased and averages are particularly sensitive to outliers. The analysis is based on raw data due to the limited number of banks participating in the study. The Wilcoxon signed-rank test is used to test whether the difference between the pre- and post-planning periods of the strategies is significantly different from zero.

Almost all previous studies report a significant decrease in post-planning productivity of strategies. This decline in operating performance is found in open economies (Jain and Kini, 1994; Mikkelson et al., 1997; Cai and Wei, 1997; Kutsuna et al., 2002, Pagano et al., 1998, Chun et al., 2000; Kim et al., 2004). On this basis, the results on the decrease in productivity under the adoption of strategy optimization should be extended to the Banking sector. Hence the following hypothesis:

- **Hypothesis 1:** The operational performance of banks in Canada deteriorates after the adoption of strategy optimization mechanisms (pre-planning strategy productivity is better than post-planning strategy productivity).

The loss of productivity can be explained by a decisional failure to generate positive net present values (NPV) (pre-planning of strategies) or by not maintaining the same level of CEG (Jain and Kini, 1994). Productivity could fall during the course of investment. This is not consistent with the no opportunity hypothesis because productivity falls despite strong growth in services and CEG. It is important to note that the growth in services was measured from year –1 to year +1. Therefore, the growth rate between these years is considered instead of measuring the difference in growth between the two periods (before and after policy optimization) separately. The rates of the two growth periods are compared separately, in addition to measuring the growth between Y–1 and Y+1. To test for any association between lack of opportunity and lower productivity when adopting strategy optimization, the following hypothesis is invoked:

- **Hypothesis 2:** The absence of opportunity does not explain the variation in productivities in the context of the adoption of strategy optimization in the Banking sector.

Services growth (SG) and CEG are measured using the following equations:

$$SG = [(\text{services in year } 1 - \text{services in year } 0)/\text{Services in year } 0] \times 100 \quad (4)$$

$$CEG = [(\text{CE in year } 1 - \text{CE in year } 0)/\text{CE in year } 0] \times 100 \quad (5)$$

The total debt ratio (TDR) is added to the growth in services and CEG to assess whether the value of services is being used to maximize debt-reducing projects. This ratio is essential to assess the ability of banks to meet long-term obligations. The level of debt and equity is fundamental to shareholders and debt holders. For shareholders, the level of debt is an indication of the risk of non-payment of dividends, while for lenders, the level of debt can be used to assess the ease of obtaining loans. A strategy optimization greatly impacts the debt structure in the sense that companies can use the proceeds to repay the debt (Pagano et al., 1998). The TDR will be used as a measure of the debt level of banks after the adoption of strategy optimization. The TDR indicates the proportion of assets financed by debt. A lower ratio suggests a better financial situation. The TDR can be calculated using the following equation:

$$\text{Total Debt Ratio (TDR)} = (\text{Total Liabilities/Total Assets}) \times 100 \qquad (6)$$

Another explanation for the decline in productivity after the adoption of strategy optimization is 'timing.' Firms may time their decision to coincide with higher productivity that may be difficult to maintain in the future. This action could explain the productivity spike in strategy optimization adoption in the last year. Cai and Wei (1997) note that the adoption of strategy optimization in the financial sector tends to increase productivity in the year before restructuring. Governance could use the adoption of strategy optimization to manipulate accounting data to make firms' services and offerings more attractive to potential customers and investors. In both cases, Y–1 is a good indication to test the time lag in cases of unusual increases in accounting data in the year before the adoption of strategy optimization.

For Laughran and Ritter (1995) windows of opportunity are an explanation for the decline in productivity. These "windows of opportunity" suggest that banks are opportunistic in exploiting upward trends. When the market rises, investors tend to be more optimistic and have high expectations about stock returns. This excessive optimism creates a unique opportunity for banks, and insiders, to achieve a higher price for stocks. Brau and Fawcett (2006) find that insiders are opportunists, and this finding could explain the decline in operational productivity in the context of adopting strategy optimization.

Jain and Kini (1994) link the decline in productivity with the adoption of strategy optimization to a number of reasons, one of which is the decline in incentives, due to the change in the management structure. The adoption of strategy optimization usually leads to a significant change in the management structure. Mikkelson et al. (1997), find no link between the change in management structure and the decline in productivity in the situation of the adoption of strategy optimization. Unlike and Kini (1994) who examine the change in management structure only in the year of adoption of strategy optimization, Mikkelson et al. (1997) critically evaluate the change in structure by measuring it over time after adoption. Cai and Wei (1997) argue that there is no link between change in management structure and changes in productivity. One of the main reasons for these conflicting results regarding the impact of change in management structure on productivity change is methodological bias. Nofsinger et al. (2004) argue that the relationship between productivity change and change in management structure is not linear. They examine three different models: one linear and two non-linear relationships. Alignment of interests occurs at various levels and entrenchment occurs at the 'intermediate' level. Therefore, the following hypothesis on the impact of the change in management structure on bank productivity is formulated:

- **Hypothesis 3:** Change in management structure is associated with variation in productivity between contexts of adoption of strategy optimization.

In addition, it has been suggested that age and size of the firm may have an impact on the productivity of management practices. Mikkelson et al. (1997) find that large, well-established firms are more successful than small firms adopting a change in management structure (Pagano et al., 1998). Of the many control variables, the length of a firm's operating history is the only robust predictor of change in productivity. Strongly established firms have better long-run operational productivity than nascent firms. Based on these empirical results, the following hypothesis on firm age and size is to be tested:

- **Hypothesis 4:** Age and size of banking institutions are associated with variations in the productivity of management tools so that rigorous and hard planning is more successful than other types of strategy optimization.

Age is measured by the length of operational history and size is measured by the natural logarithm of total assets.

The first model tests the association between the adoption of strategy optimization and ROA over time. The model is as follows:

$$\text{Firm productivity} = \beta + \text{it } 0\ \beta\text{Strategic planning}1 + \beta\text{GFC}2 + \beta\text{log}3\ (\text{Sales})\ \text{it} + \text{BAge}4\ \text{it} + \beta\text{TDR}5\ \text{it} + \beta\text{log}6\ (\text{CE})\ \text{it} + \varepsilon\text{it} \quad (7)$$

In this model, all data on optimization adoption across all available years for each structure is pooled. Thus, the analysis consists of unbalanced panel data and includes 393 observations. The dependent variable is productivity, as measured by gross ROA. The first independent variable is the adoption of strategy optimization, which is a dummy variable that takes the value of 1 for the years of adoption and subsequent years and 0 for the pre-planning period of strategy. This variable captures the effects of productivity on the adoption of strategy optimization. This variable is expected to have a negative impact on the productivity of the firm. The second variable is the GFC, which is a dummy variable that takes 1 for the years during the banking crisis in 2008 and 2009 and 0 for the other years. This variable is included to control for the negative effect of the banking crisis on productivity. The natural logarithm of sales (banking services) is included to capture the effect of revenue on productivity, and this variable is expected to have a positive impact. Sales also represents the impact of size. The use of total assets is avoided to minimize the multicollinearity effect. Age is found to have a positive impact on productivity because older firms have higher productivity. Finally, the total annual debt ratio and the natural logarithm of CEGs are included to determine whether these factors could explain the variation in productivity. Although the banking crisis occurred in mid-2007, the effect on company performance began to emerge in 2008, and most companies started to recover in 2010.

Hypotheses 3 and 4 are tested by looking for an association between the change in productivity of strategy optimization and the change in management structure, age, and size. Based on the work of Kim et al. (2004); and Wang (2005), the following model is proposed:

$$\Delta \text{ Productivity} + 1 \text{ à} - 1 = \beta + 0 \text{ }\beta\text{Structure1} + \beta\text{Age2} + \beta\text{Size3} + \text{BSG4} + \beta\text{CEG5} + \beta\text{TDRC6} + \varepsilon \quad (8)$$

The dependent variable in this model is either the change in ROA or ROS between Y+1 and Y–1. The independent variable of structure represents the shareholding (in percentage) at adoption. It is expected that the decision structure will have a negative impact on the change in productivity due to the increase in layout costs. A higher rate of the procedure results in higher agency conflict and lower productivity. Age is the difference between the year of implementation and the year of adoption. Size is the natural logarithm of total assets in the year of adoption of strategy optimization. SG represents the growth in sales (banking services) from Y–1 to Y+1. The age and size variables are positively related to the change in productivity. Thus, they are included in the analysis. Also included is the growth in CEG to control for any increase in assets. The total change in the debt ratio between Y+1 and Y−1 (TDRC) has been included to control for optimization leverage.

Data is limited by the information that is available. This implies that one of the problems of the research is the use of a highly limited and specific dataset in a way that allows the hypotheses to be tested, the results obtained and the conclusions reached in the pre-existing literature. The initial database consists of banking institutions that adopted strategy optimization in the years from 2003 to 2010. To study the impact of adopting strategy optimization mechanisms on productivity, the following filters were applied: institutions adopting strategy optimization as a nascent enterprise were excluded because pre-planning strategy data were not available; institutions without pre-planning strategy data were excluded because of the lack of comparability between the pre- and post-planning strategy periods. Institutions that adopted strategy optimization mechanisms in 2011 were excluded because 2010 is the cut-off date for comparison in this study. The 2010 cut-off was chosen to allow for at least one year of post-strategy planning productivity at the time this study was conducted. These filters reduced the final sample to 52 adoptions of strategy optimization, and the majority of these adoptions are in Quebec (30). 9 adoptions in Ontario, 7 adoptions in British Columbia, 4 adoptions in Saskatchewan, and Nunavut and Newfoundland and Labrador have 1 strategy optimization adoption each.

Two sources of data were used to implement this study, the first being prospectuses. Industry Canada requires that every banking institution provide three years of audited accounting information, which includes the balance sheet, income statement and cash flow. These prospectuses also include general information about the offering, such as prices, share units, etc. These data are kept on the website of the Capital Market Authority which is publicly available for collection and evaluation. The second data source consists of annual reports and a questionnaire. The annual management reports, a questionnaire, and in case these annual reports are not available, the official websites of the banks were consulted.

The descriptive statistics is for 52 banking institutions. The Canadian dollars is used to control for exchange rate variations. The hourly investment price at the mechanism level varies between a minimum of 0.28 and a maximum of 136.5, with an average value of 12.7. The table also reveals a substantial undervaluation of 82% and 48% as shown by the mean and median, respectively. The maximum undervaluation comes from the Abu

Dhabi National Energy Company, which shows a price explosion on the first day and a return of 684%. The total investment of the 52 amounted to 23.1 billion. The total funding and total assets variables indicate wide variations in the size of strategy optimization, as indicated by the 25th and 75th percentiles. Medians and other statistical measures show similar trends. In addition, the length of the operating history of strategic planning is longer than those described in the previous literature. The mean and median ages of strategy optimization are 20 and 18 years, respectively, and the oldest institution has an operating history of over 50 years. Finally, the shareholding is inspected. Here is the percentage of shares that are held by the banks at the time of the adoption of strategy optimization. All statistical measures indicate that the Banks at the time of adoption of the optimization of the strategies retain a significant shareholding. The average and median rates are 65.5% and 70%, respectively, while 25 adoptions retain 70% and 6 others retain more than 70% of the stake.

Descriptive statistics for 52 adoptions from banks in Canada between 2003 and 2010 are presented. The adoption price is the standardized hourly price for all adoptions. Initial returns (percentage undervaluation) are calculated by taking the difference between the strategic closing price on the day of listing and the issue price. The yields are calculated by multiplying the number of options offered by the hourly price. Total assets are the total value of assets in the strategy planning year. Age is the length of the operating history of the strategy optimization calculated by taking the difference between the strategy optimization year and the implementation year. Governance is the percentage of shares held by the original owners at the time of the IPO.

The measures of accounting profitability are presented in Table 12.2 for the entire sample of 52 organizations that adopted the strategy optimization mechanisms. Group A shows a sharp decline in profitability in the period from pre-strategy planning to the post-strategy planning period. The mean (median) deteriorates in all profitability ratios. The average (median) ROA drops significantly from 14% to 8%, which is a sharp decline of about 43%. Furthermore, all profitability measures (ROS) and (A/S) show the same observed trends of severe decline. Section B of Table 12.2 indicates that the deterioration in productivity has intensified in magnitude. The number of adoptions that are included in the survey has decreased to 46 because few adoptions in the sample have no more than 1 year of post-planning data on strategies. The mean ROA, ROS, and S/A decreased in the second year of post-adoption by 43%, 24%, and 20%, respectively, and the decrease in median change for ROA and ROS in the second year increased by –56% and –18%, respectively. Group C presents a comparison between the average profitability for the years before and after adoption. The average ROA for all banks before adoption is 13%, while this value is 7% after (a decrease of about 46%). The ROS and S/A show a similar severe deterioration. The average (median) measures of change for the two proxies are –11% (–3%) and –15% (–8%), respectively. Overall, it is concluded that the operational productivity of strategy optimization is worse in the post strategy optimization period than in the pre strategy optimization period.

The profitability ratios are ROA (ROA = net income/total assets), return on sales (ROS = net income/total turnover) and sales of assets (S/A = total turnover/total assets). For each variable, the usable observations are reported, the mean and median values, the change in these values from before to after the adoption of strategy optimization. The Wilcoxon test (with its z-statistic) is used as a significance test for the change in median values. Group A shows a comparison between the year before the adoption of strategy optimization (Y–1) and the year after the adoption (Y+1). Group B compares Y–1 with the second year after the IPO (Y+2). Finally, Group C compares the average for all available years before the adoption of optimization (Y) with the average for all available years after adoption (Y+).

Table 12.3 shows the estimates of the change in ROA of the adoption of strategy optimization over time and in all years before adoption and all years after adoption. The univariate estimates were inspected, and it was observed that the adoption of optimization and the dummy variables are significant and provide the expected negative signs. In this case, the adoption of strategy optimization presents a major change in the management structure. As data on the change in structure overtime is not available, the impact of the change on an annual basis is not possible. However, the use of the adoption of strategy optimization as a dummy variable in this model captures the change in structure that occurs upon adoption. Furthermore, age is found to be positively associated with better productivity over time, in line with the findings of Mikkelson et al. (1997); and Balatbat et al. (2004) whose studies have shown a positive impact of institutional age on productivity. This finding also suggests that as the age of institutions increases, productivity improves from year to year. In addition, it is found that optimizing strategies that increase the level of debt, as measured by the TDR, shows lower productivities over time.

Models 1 to 3 show the multivariate association between productivity of the adoption of optimization and other variables. The strategy planning variable remains negatively significant in all models, confirming earlier findings that strategy optimization has a negative impact on productivity due to the change in management structure. The GFC remains negative, as one would expect, but this value is negligible in all models. As expected, sales, and the TDR are inversely associated with the productivity of strategy optimization. Therefore, an optimization adoption with more services and sales has higher productivities, and adoptions with higher debt levels have lower productivities. No significant relationship is found between the change in CEG over time and the productivity of strategy optimization.

Examination of panel data on the productivity of strategy optimization over time shows the relationship between the productivity of optimization mechanisms and several variables over time. The data are based on 393 observations pooled from 52 optimized institutions between 2000 and 2011. The dependent variable is the productivity of strategy optimization mechanisms as measured by ROA = net income/total assets. The independent variables are: Strategy optimization, the dummy variable takes a value of 1 from the year of adoption until the following years and 0 before, GFC the dummy variable which takes a value of 1 for the years during the banking crisis of 2008 and 2009, and 0 otherwise, Sales is the natural logarithm of profitability in the context of strategy optimization over time, Age is the length of the operating history of institutions with strategy planning mechanisms, Debt ratio is calculated as total liabilities/total assets, CEG is the natural logarithm of optimized expenditure on acquisition or upgrading of physical assets.

Contrary to what was assumed in this study, Group A shows a significant decrease in sales growth in the post-strategy planning period compared to the growth rate in the pre-strategy planning period. The median profitability growth dropped from 25% to 8% (a drop of about 68%). In addition, the growth in CEG reveals a similar result of severe deterioration in the post-planning period of the strategies. The average growth has decreased by about 76%. Both variables indicate that the growth rates in the pre-planning period of the strategies are better than the growth rates in the post-planning period of the strategies. This result proves the absence of opportunity. In addition, the TDR (insignificant) suggests that strategic plans rely more on debt after adoption than in the pre-planning period of the strategies. In groups B and C of Table 12.4, growth rates and CEG are measured between Y–1 and Y0 and between Y–1 and Y+1, respectively.

This approach allows for direct comparisons. The results indicate that adoption-maintained growth in both variables, and that sales increased significantly from Y–1 to Y0 and Y–1 to Y+1 by 17% and 47%, respectively.

In addition, adoption shows a significant increase in CEG by 11% and 102%, respectively, but no significant change is observed in the TDR. These results suggest that, although the optimization mechanisms maintain a certain level of growth in services and CEG, this growth is not really satisfactory compared to the pre-planning period of the strategies. More importantly, the results indicate that there is a methodological problem in examining the lack of opportunities and support the explanation of the time lag. Overall, the decline in productivity is a function of several factors, and no single theory can provide a satisfactory explanation.

Empirical results for 36 institutions that have adopted strategy optimization are available. Group A compares the average growth for the pre-planning years of the strategies to the average of the post-planning years of the strategies. Groups B and C compare the pre-IPO year Y−1 to the year of the advent of strategy planning Y0, and the year following strategy optimization Y + 1, respectively. Growth in profitability, growth in CEG and TDR are increasing between the two time periods calculated as in Eq. 4, 5, and 6, respectively.

It is confirmed that the decrease in productivity is associated with the adoption of strategy optimization mechanisms. Several techniques were used to match pairs before and after adoption and to pool all data. We analyze the impact of governance on productivity and the type of association that exists between these factors. Table 12.5 shows the estimates of equation 8. In Group A, it is found that a linear association between a change in productivity (ROA) and a change in the optimization structure exists. As hypothesized, a change in the optimization structure has a significant negative impact on productivity in the context of adopting strategy optimization. This result suggests that a high retention rate leads to a targeted drop in productivity. The univariate analysis provides no new explanation for the variation in productivity. However, the multivariate analysis of Group A suggests that the size of the institutions adopting strategy optimization has a significant positive impact on the variability of productivity. Formal planning shows superior results to non-formal optimization. The variation in CEG is also significant and negative due to the impact that CEG has on assets and increases the denominator when calculating ROA.

Group B, shows that a change in the optimization structure has a negative impact on the ROS. The univariate analysis also suggests that strategic planning with higher CEG had a better change in SWR. Surprisingly, size has a negative impact on SWR, but a positive one when multivariate analysis is used. In model 2 of group B, the 2adjusted R is very high, 81% of the explanatory power, and most variables have the expected sign at significant levels. Governance remains negative, suggesting that an increase in governance retention leads to a larger decrease in the ROS. In addition, size is positively significant, indicating that for each increase, productivity improves. The growth between the two periods is significant and negative due to the widening of the denominator when calculating the ROS. Thus, as output increases, the ROS decreases. The growth in CEG suggests that strategy optimization and large investments have a better variation in the ROS. Similarly, strategic planning that increases its optimization leverage experiences a better change in the ROS.

The sample consists of 30 adoptions completed in Quebec between 2003 and 2010. The others were excluded to keep the sample homogeneous. The association between the change in productivity and the change in the optimization structure is presented. The dependent variable in Group A is the change in ROA between Y+1 and Y–1, while in Group B, the dependent variable is the change in ROS. The independent variables are: Governance, which is retention; Age is the length of operating history; Size is measured by the natural logarithm of total assets; SG is the growth in return from Y–1 to Y+1; CEG is the growth in CEG from Y–1 to Y+1; TDRC is the change in TDR from Y–1 to Y+1.

In this chapter, the operational productivity of 52 banks that adopted strategy optimization mechanisms between 2003 and 2010 was examined. The results indicate that productivity deteriorates in the post-strategy period. The average ROA and ROS in the post-strategy period decreased by 47% and 25%, respectively. Based on 393 observations, it is indicated that the mechanisms of strategy optimization are significantly associated with a decline in productivity. Several reasons for the decline in productivity were investigated and it was found that several factors are instrumental in this dynamic. Firstly, the results indicate that although strategy planning maintains strong growth in output and CEG, this growth is not comparable to the strength of growth in the pre-planning period of strategies. This suggests that the adoption of optimized mechanisms have either not achieved the necessary level of growth in the post-strategy planning period or that they are involved in “window dressing” behavior, which makes the accounting figures for the pre-strategy planning period better than they actually are. It is found that the change in governance structure that results from the adoption of strategy optimization has a significant negative impact on productivity trends. The relationship is linear and shows that the cost of agency increases as governance increases retention, leading to a decrease in productivity due to increased agency conflict. Size has a significant positive impact on productivity, in the sense that large optimized banks perform better than small ones, while age does not seem to be as important. In sum, the explanation for the decline in operational productivity under adoption is a complex function of several factors, and no single hypothesis can explain the decline.

KEYWORDS

- **capital expenditure**
- **net present values**
- **return on assets**
- **return on sales**
- **services growth**
- **total debt ratio**

CHAPTER 13

Rationalization Factors of Research Approach for Business Improvement Process Deployment

Understanding the relationship between strategy optimization and company performance is essential for understanding company strategy, analyzing company performance, and characterizing company management systems. The last decades have been marked by the rise of the globalization of trade in goods and services, allowing companies to have access to new growth opportunities. Thus, globalization and internationalization have become strategic management tools that can increase the overall performance of institutions, but their actual impact on financial performance is still somewhat unclear. Companies and institutions in an open economy must therefore deploy strategies and resources to meet the new challenges posed by a new competitive context, to stand out thanks to the competitive advantage in order to achieve a higher level of performance than the competition. The effect of the strategy optimization on the performance of companies in an open economy is thus exercised through institutional restructurings subject to the new conditions of international competition. Also, the internationalization of competition requires a global presence on the market and a dynamization of exchanges. This confrontation pushes actors to reorganize, in particular by improving the management of strategies in order to be more efficient on the market. The question of the optimization of companies and institutions remains an important subject in the strategic research of the last decades and leads to several questions.

The aim of the analysis of the interactions between the strategy optimization and the performance of firms in an open economy thus raises a double question: the first concerns the impact of the strategy optimization on the

Value-Based Management in an Open Economy: Optimizing Strategy to Improve Business and Performance. Prof. Post-Dr Walter Amedzro St-Hilaire (Author)

performance of firms, while the second concerns the role of performance in the strategy optimization. The aim of this question on the interactions between strategies, performance, and forms of operational management is to understand how the efficiency of companies guides strategies and how they are modified according to the evaluation of the optimization process and the conduct of changes in an open economy. The second perspective is more interested in highlighting the role of performance in determining management practices and tools. The question will mainly be asked from the point of view of the role of economic productivity in the determination of optimal strategies.

This chapter aims to present the results of a body of research that links business efficiency and strategy optimization in an open context in the Banking sector. The organizational approach on which this study is based on the triangulation approach as much in terms of data collection, methodology, and results presented. This approach makes it possible to solve a central problem by using several methods in order to give a more appreciable scientific status to the conclusions. The first chapter presents the methodology used. The second chapter presents the empirical and descriptive results of the study. Finally, the last chapter will discuss the results in the light of the theory and empirical literature.

In order to facilitate the understanding of the approach followed throughout this empirical section, the objective of this section is to present the framework followed to carry out the field research. To do so, it will begin with the presentation of the theoretical model that led to the statement of the hypotheses of this study (Section 1). Subsequently, we will present the characteristics of the sample and the variables selected for the study (Section 2). Finally, the presentation of the research methodology (Section 3) will conclude this part.

In order to explain the performance of companies through the use of management practices and the role of this performance in the development of optimization, it is necessary to first define the hypotheses to be tested (I), which relate the variables under study; then to simplify the phenomenon that we are seeking to study by constructing a model (II), understood as a schematic and partial representation of a naturally more complex reality.

The evaluation of strategy optimization is a major issue for management and industrial economics professionals. With this in mind, many studies have been conducted with the aim of assessing the effects of strategy optimization on the performance of companies. Indeed, the synthesis of theoretical contributions elaborated in the first part of this study led us to note that, although it is difficult to demonstrate the existence of direct relationships between strategy optimization and firm performance, efforts to examine the inter-correlation between the two have been made using both quantitative and qualitative approaches. In particular, these have demonstrated the impact of management practices on the building of competitive advantage and the innovativeness of firms, which are presented in the literature as key components of firm optimization. Empirically, a number of studies help us to understand to some extent the link between these two concepts. Theoretically, however, the LCAG model has strongly inspired our research. Historically one of the first models to address the link between strategy optimization and firm performance, it aims to diagnose the internal and external situation of the firm and to confront it with the context in which it evolves (open economy in this case).

The aim of our research is then to consider the interconnection between strategy optimization and business performance around the following four specific questions:

- what factors favor the role of performance in the strategy optimization;
- what links exist between the dimensions of strategy optimization on the efficiency of companies;
- what economic and managerial factors influence the interconnection between strategy optimization and firm productivity in an open economy;
- how, at the organizational level, does the efficiency of firms drive innovative strategies and how do they change as a result of evaluating the optimization process and driving change in an open economy

Based on these four objectives, we were able to study the impact of strategy optimization on the performance of companies and to understand the role of performance in the development of management practices based on the following hypotheses:

1. **Optimization Inefficiency:** It refers to certain flaws that may occur in strategic management mechanisms. These flaws in turn refer to logistical failures and breakdowns. Logistical breakdown refers to difficulties arising from the process and practices indiscriminately established in strategic matters and refers directly to the concept of practices enacted in the sector. Dysfunction, on the other hand, generally represents the prevailing norm in Banking and financial institutions. The search for a link between the quality of the strategy

optimization process and the competitive advantage of the banking and financial sector in an open economy is an important concern for policy makers. It is right then that the fundamental question in industrial economics and value-based management is whether better operational optimization results in and contributes to better industrial competitiveness.

2. **The Value-Based Management Structure:** This refers to the type of optimized mechanisms control in use within the company. We attempt to identify the structural effects of the practical mechanisms of optimized mechanisms control on sectoral performance in Canada. Since the financial crisis, there has been a particular interest in the importance of this issue in the banking and financial sector in open economies. It is well recognized that optimal decision making plays an effective role in the management of firms, the production of reliable financial information and the enhancement of investor confidence. In addition, this form of strategy optimization would play a key role in enhancing transparency while mitigating conflicts of interest. In this study, we are interested in the relationship that decision structure, optimized remuneration, audit quality, and capital structure have on management efficiency.

In form and substance, decision structure refers to both structural size and the independence of decision-making mechanisms. A number of models have addressed the structural effect of strategic decision-making mechanisms on firm effectiveness. Several authors agree that size has a significant relationship with the effectiveness of firm structural optimization (Jensen, 1993; Cheng, Evans, and Nagarajan, 2008; Eisenberg, Sundgren, and Wells, 1998; Yermack, 1996). However, Elsayed (2011) concludes that there is a significant positive relationship between size, management effectiveness and planner quality in the context of non-duality. Thus, a negative relationship is found between size and managerial quality regarding the effectiveness of firms. For Bhagat and Black (2002), on the other hand, the size of the organization has no significant impact on the effectiveness of optimized mechanisms control. This result differs from that reported by O'Connell and Cramer (2010) showing a significant negative relationship between firm size and effectiveness. However, they also found a significant positive relationship between the ratio of external strategists and practice effectiveness.

As well as structural size, the independence of the mechanisms plays an important role in the process of reducing conflicts of interest between the majority investor (the capitalist) and governance (strategists). The fit between the mechanisms and the firm's best practices suggests that the management structure must be independent of decision making in order to reduce the risk of moral hazard. According to Jensen (1993); and Fama and Jensen (1983) the management structure and the decision-making structure must be dissociable in order to avoid the evaluation capacity of management practices is affected (Jackling and Johl, 2009). While for Jensen (1993) decision-making concentration can lead to partisan decision-making to the detriment of stakeholders, Bradbury (1990); and Gendron and Bédard (2006) defend decision-making duality as something positive for the effectiveness of optimized mechanisms control because if the overall power of the firm is concentrated within the same structure, there will be fewer conflicts of interest and management is facilitated, thus allowing for greater efficiency of practices. To this effect Bhagat and Black (2002); and Bonn et al. (2004) find a negative relationship between the independence of mechanisms and the efficiency of practices.

There is a relationship between audit quality and improved business efficiency (Keung, Robin, and Tessoni, 2007; Collier and Zaman, 2005). Optimized auditing allows for the monitoring of the integrity of performance statements and reports (Zoort et al., 2002), the supervision of external controls and activities (Krishnan, 2005). Audit quality as an external management optimization mechanism could reduce agency costs and information asymmetry (Adeyemi and Fagbemi, 2010; Willenborg, 1999). To get the most out of optimized audit practices, it is imperative that the audit mechanisms are effective and active. Previous empirical studies argue that companies that have a high external audit requirement, demonstrate more industry expertise and de facto are more likely to discover performance irregularities. These conclusions are supported by numerous studies (Francis and Krishnan, 1999; Willenborg, 1999; Lennox, 1999; Jagan Krishnan and Schauer, 2000; DeFond, Francis, and Wong, 2000).

Regarding optimized compensation, the literature is more in favor of a positive and significant relationship between optimized bonuses and firm efficiency than the opposite (Laing and Weir, 1999; Klapper and Love, 2004). Optimized mechanisms control bonuses play an important role in mitigating agency problems. Moreover, optimized compensation mechanisms are positively related to financial performance (Jensen and Murphy, 1990; Kren and Kerr, 1997; Christensen, Kent, and Stewart, 2010; Benson, Hutchinson, and Sriram, 2011). However, Jensen and Murphy (1990) show a negative relationship between optimized premiums and managerial performance. Similar results are reported by Core, Holthausen, and Larcker (1999); and Chen and Jermias (2012). These results differ from those described by Leonard (1990); Gregg, Machin, and Szymanski (1993); and Banghøj, Gabrielsen, Petersen, and Plenborg (2010) as they find a weak and insignificant relationship between managerial efficiency and optimized bonuses although bonus optimization policies provide incentives for better management practices (Fama and Jensen, 1983). Thus, a positive relationship between monetary incentives and firm efficiency is found by Jensen and Meckling (1976); Coughlan and Schmidt (1985); Abowd (1990); Sun, Cahan, and Emanuel (2009); Ozkan (2011); and Farmer, Archbold, and Alexandrou (2013).

Finally, regarding capital structure and the efficiency of best practices, studies show that there is either a positive or a negative or endogeneity relationship. Demsetz (1983); Demsetz and Lehn (1985); and Demstez and Villalonga (2001) argue that the composition of capital is endogenously determined and that there should be no systematic relationship between capital structure and changes in firm practices. For Farooque et al. (2007), capital composition is negatively related to the efficiency of management practices. Hyeon Cho (1998) showed that there was no relationship between capital structure and management efficiency. Rogers, Dami, Ribeiro, and Sousa (2007) report the lack of influence of capital composition on the effectiveness of optimized mechanisms control. Chiang (2005) investigates the relationship between management optimization variables and operating performance.

The results indicate a significant negative correlation between the proportion of capital held by governance and practice efficiency. However, the relationship between various forms of capital and management efficiency is significantly positive. Also, Wei, Xie, and Zhang (2005) found that the share of equity and institutional capital has a significant negative impact on management efficiency. In addition, foreign capital has a strong and positive impact on the efficiency of best practices. Although, Wu and Cui (2002) found that firms with high capital concentration have better accounting profits, but this realized performance is poorly represented by the ratio of market to book value and costs to profit. In other words, they show a non-linear relationship between capital structure and firm contribution. Although Shleifer and Vishny (1986); Alonso-Bonis, and Andrés-Alonso (2007) find a positive relationship between capital concentration and managerial efficiency, Shleifer and Vishny (1986) and Morck, Shleifer, and Vishny (1988) after examining the relationship between internal capital structure and market valuation measured by Tobin's Q, also find a non-linear relationship between capital concentration and managerial efficiency. Other authors like Morck et al. (1988) show similar non-linear or curvilinear relationships (McConnell and Servaes, 1990; Hermalin and Weisbach, 1991; Hyeon Cho, 1998; Himmelberg, Hubbard, and Palia, 1999). Jensen and Meckling (1976) argue that capital diversity has different effects on firm efficiency.

The adoption of strategy optimization: we want to know whether the adoption of management practices by a firm is beneficial for it or not. A number of studies have found a significant decrease in productivity after strategy optimization in open economies (Jain and Kini, 1994; Mikkelson et al., 1997; Cai and Wei, 1997; Kutsuna et al., 2002, Pagano et al., 1998, Chun et al., 2000; Kim et al., 2004). However, it is important to study this relationship from the point of view of duration in order to determine whether this negative effect is only short-term or whether it continues in the medium to long term. We simultaneously assess the effects of the adoption of optimization on the competitive advantage and performance indicators of banks. As we are aware that several factors may influence the post-optimization results, we are also interested in both organizational and market factors when estimating our model. Several authors have attempted to explain the loss of productivity following the adoption of management practices. For Jain and Kini (1994), the absence of arbitrage opportunities could be an explanation. Indeed, firms find it difficult to generate positive NPV (pre-strategy optimization) or do not maintain the same level of CEG. Another explanation for the decline in productivity after the adoption of strategy optimization is 'timing.' Firms may time their decision to coincide with higher productivity that may be difficult to maintain in the future. This action could explain the productivity peak in strategy optimization adoptions in the last year before restructuring (Cai and Wei, 1997). Governance could use the adoption of strategy optimization to manipulate accounting data in order to make firms' services and offerings more attractive to potential customers and investors. For Laughran and Ritter (1995) windows of opportunity are an explanation for the decline in productivity.

- These "windows of opportunity" suggest that banks are opportunistic in exploiting upward trends. When the market rises, investors tend to be more optimistic and have high expectations of stock returns. This excessive optimism creates a unique opportunity for banks, and insiders who are opportunistic, to achieve a higher price for stocks (Brau and Fawcett, 2006). Finally, Jain and Kini (1994) link the decline in productivity after the adoption of strategy optimization to a number of reasons, one of which is the reduction in incentives, due to the change in the management structure. The adoption of strategy optimization usually leads to a significant change in the management structure. Mikkelson et al. (1997), find no link between the change in management structure and the decline in productivity in the situation of adopting strategy optimization. Cai and Wei (1997) argue that there is no link between change in management structure and changes in productivity. One of the main reasons for these conflicting results regarding the impact of change in management structure on productivity change is due to methodological bias.

Finally, it has been suggested that the age and size of the firm may have an impact on the productivity of management practices. Mikkelson et al. (1997) find that large, well-established firms are more successful than small firms adopting a change in management structure (Pagano et al., 1998). Age is measured by the length of operating history and size is measured by the natural logarithm of total assets.

- Internal Optimization Mechanisms: These mechanisms represent the company's own characteristics that allow it to distinguish itself from others and to build up a competitive advantage. With regard to internal optimization mechanisms, we study three optimization tools, namely capital structure decisions, dividend policy and optimization structure. Capital structure optimization is a double-edged sword for productivity management. On the one hand, the level of debt could be used as a value-based mechanism for managerial behavior. Managers' preference for consumption of benefits over shareholder wealth is mitigated through more leveraged capital structures. High levels of debt reduce the power of opportunistic managers by reducing the cash flow available to spend on non-discretionary assets, such as benefits or negative net present value projects (Frank and Goyal, 2009; Harris and Raviv, 1991; Jensen, 1986). Nevertheless, capital structure decisions could work against the interest of shareholders and in favor of opportunistic behavior by managers. Managers could make a good impression through financial statements in order to issue debt under more favorable conditions and/or achieve the objectives imposed by the covenants (Mohrman, 1996).

In addition, the decision of how much to pay out in dividends to current shareholders and what proportion or earnings should be reinvested in the form of retained earnings is one of the most important issues facing managers. Brav et al. (2005) in examining payout practices find that maintaining dividend levels is the main variable in dividend optimization while payout ratios are of secondary importance. They conclude that since dividend cuts have negative consequences, it is quite possible that governance manipulates accounting information. Anglin et al. (2013) indicate that dividend optimization influences actual productivity management. As a result, planners who follow traditional dividend policies manipulate earnings to a greater extent than those who do not follow this policy. However, the dividend strategy depends critically on the ability of the legal and institutional system to compel dividend payments (Porta et al., 2000; Mitton, 2004).

Finally, the ownership structure plays a key role in value-based management. Indeed, when the owners are also the managers of the firms, there is an overlap between ownership and control, and therefore potential agency problems are minimized. Warfield et al. (1995) show that governance has a significant effect on discretionary decisions and that optimization is positively associated with the explanatory power of earnings and is inversely related to the size of the accrual's adjustment. Therefore, when institutional optimization is considered, a non-monotonic relationship could exist between it and productivity management. Furthermore, it has been widely argued that optimized governance structures solve some agency problems through direct supervision (Brown et al., 2011). Thus, it can be inferred that vertical agency conflict and consequent management of accounting profits, could be effectively reduced by further optimization of governance (Shleifer and Vishny, 1986).

- The Mechanisms for Optimizing External Strategies: These mechanisms are derived from the environment in which the company operates and which may or may not favor management practices. They concern the general context, i.e., the demographic, economic, institutional, natural, technological, and cultural environment. According to Wysocki (2003), accounting rules can limit a manager's ability to distort earnings, but the extent to which the rules influence reported earnings still depends on how the rules are applied. Therefore, it is assumed that institutions using accounting standards have greater incentives to publish transparently because they are subject to greater restrictions and a higher risk of litigation. However, the choice of accounting method could be the result of opportunistic governance. Ball et al. (2003) argue that institutional arrangements are the most important factor in controlling interests, reducing opportunistic approaches reflected in productivity optimization, and improving the quality of accounting information. Therefore, one would expect that strategy optimization rules would limit the discretionary behavior of institutions in managing earnings. Naceur and Omran (2011) find that regulatory and institutional variables seem to have an impact on bank productivity.

The impact of the banking sector is found at several levels of the economy (Rajan and Zingales, 1998; Ueda et al., 2008; Gonenc and Haan, 2014). It impacts, among others, the evolution of accounting regulations (Rajan and Zingales, 1998), the efficiency of analytical mechanisms (Degeorge et al., 2013) and the optimization of productivity (Enomoto et al., 2014). Bank efficiency reduces discretionary management (Enomoto et al., 2014), due to the strengthening of laws and regulations for investor protection and, by extension, due to the presence of more sophisticated market participants.

- Management Efficiency: It can be measured by both financial and non-financial indicators. Louizi (2011) in a study linking strategic management mechanisms and management efficiency, manages to demonstrate that successful and struggling firms are characterized by different management mechanisms. In his study on financial distress, Gilson (1990) shows that firms accentuate the rotation of mechanisms in their optimization strategy when performance is weak. Industrial economics is interested in the role of innovation in the efficiency of firms and institutions. Based on Schumpeter's theses, it concludes that the most successful companies are those that manage to design innovations with a high degree of novelty. Thus, Bellon (1991) easily states that a company cannot be competitive if it cannot overcome its lack of technological advantage. In contrast, Freel (2000) highlights the non-linearity of the relationship between profitability and innovation and Moore (1995) does not observe any systematic relationship between innovation and profitability.

The above discussion has led to the formulation of five main hypotheses for this study:

- **H1:** Ineffective policy optimization has a negative effect on the performance of Canadian banks and financial institutions.

- **H2:** The value-based management structure has a positive or negative effect on the performance of companies.
- **H3:** The adoption of strategy optimization has a negative effect on company performance in the short term and a positive effect in the medium term.
- **H4:** Mechanisms for optimizing internal strategies have a positive effect on company performance.
- **H5:** Mechanisms for optimizing external strategies have a positive effect on company performance.
- **H6:** The level of performance of companies determines the management mechanisms in place.

The first five hypotheses relate six variables grouped into five explanatory variables (or independent variables), namely: inefficiency of optimization, management structure, adoption of strategy optimization, internal mechanisms of optimization and external mechanisms of optimization, and one explained variable (or dependent variable), namely: performance. We have introduced intermediate variables at this level. The first intermediate variable is "competitive advantage." It links the first three hypotheses to performance. According to Ma (1999), competitive advantage is at the origin of any superior performance. Also, for Fahy (2000), the achievement of a sustainable position of competitive advantage can lead to superior performance, as measured by financial indicators. The second intermediate variable is 'governance.' Managers do not always make decisions that put the interests of the company first. According to postmodernism, the reduction of opportunistic strategies through a number of practices such as the decentralization of executive power within the company should increase the company's performance. We therefore relate this variable to the internal and external mechanisms of optimization. When the competitive advantage increases, this is likely to increase the productivity of projects. Conversely, when the discretionary management of governance decreases, this has a positive effect on the level of performance of the company because its interests are put forward. The last hypothesis links the explanatory variable, "level of performance" to the explained variable "management mechanisms."

Once the hypotheses and the conceptual model have been clearly defined, it is appropriate to proceed with the presentation of the sample on which these hypotheses have been tested, as well as the items used to measure the variables defined above. The second section aims to answer the following questions: Who are the subjects surveyed? How were these subjects chosen?

By what means were they interviewed? And finally, how were the variables measured?

In order to validate our assumptions, the mixed approach to data collection was used. Indeed, the quantitative or qualitative approach alone could not allow us to account for the reality studied, as biases would have remained. The subjective and objective nature of our database was given a longitudinal and cross-sectional aspect to the data because it was collected over a certain period of time and simultaneously from different entities. The data covers over 700 North American banks and financial institutions based in Canada. The most relevant data sources, which collects information to quantify the relationship between strategy optimization and business performance, are the Multilateral Investment Guarantee Agency and the Bank of Canada. Both institutions have multiple objectives and the data is structured around a number of modules corresponding to the areas of investigation according to the objectives. However, we also solicited data from Finance Canada, Statistics Canada, Industry Canada, DataStream, SIRCA, the Ministry of Finance and the websites of the institutions studied. Four types of data were used to study the relationship between our two key concepts.

The first is the questionnaire survey that was conducted in 2015. This was useful on two levels. Firstly, it enabled us to determine the management breakdowns and dysfunctions in force and was conducted with the Canadian Bankers Association. More than 30 stakeholders at all levels were asked to comment on and validate the different versions of the questionnaire, especially with regard to its content. The dysfunctions surveyed met two criteria: they corresponded to reality (validity), and they were noticed in a fairly common way, without this implying that all of them had noticed them (variance). The respondent, i.e., the head of the optimization and control department, had to note the presence (yes) or absence (no) of malfunctions. The survey also aimed to collect some indicators of logistical breakdown not available at the Bank of Canada. It should be noted that all variables used here are measured at the design level and refer to the year 2014.

The survey was conducted in the summer of 2015. Of the 68 questionnaires mailed out, 50 were returned. The return rate was 74.4%, which is higher than what is usually observed in studies of this kind (Becker and Huselid, 1998). The second type of questionnaire was conducted at the Ministry of Finance. However, the data obtained after the survey was supplemented by the collection of information from the companies concerned. The field survey and information gathering provide so-called primary data, which can be quantitative, qualitative or both, depending on the researcher's objectives. The collection period was from 1 January to 30 July 2015. It allowed us to build up a panel of data over a period of 6 years (2006–2012). For the purposes of confidentiality and relevance of the analysis, the companies were grouped by category of activity banks according to the national accounting nomenclature. These are: investment banking, savings banks, corporate banking, online banking, bancassurance, private banking, deposit banking, universal banking, electronic banking, free banking, cooperative banking, direct banking, network banking and general consultancy.

The second is made up of various databases and constitutes the majority of the secondary sources. On the one hand, the Bank of Canada provided a lot of information on the competitive advantage of companies in the Banking and financial sector. On the other hand, it, and Finance Canada provided financial reports and market information. This second database was supplemented by data updated by Statistics Canada.

The third type of data is desk research. This was used to collect governance optimization indicators on accounting standards and the legal system from the Association of Professional Accountants and Industry Canada. The fourth type of data is the document review. To collect data on management optimization and financial data for the period 2005–2011. This method was used to complete the information on the age of the institutions. This was best done by consulting the websites of the financial institutions studied. As mentioned, the problem lies at two levels. First, we want to determine the impact of strategy optimization on the performance of banks in Canada. We have six variables, five of which are independent or explanatory variables and one dependent or explained variable. Next, we determine the role of performance in strategy optimization. To do this, we use one independent or explanatory variable and one dependent or explained variable.

Our problematic pursues two goals as mentioned above. It sometimes involves the performance variable as an explained variable and sometimes as an explanatory variable. Thus, performance will be called "management efficiency" when it is an explanatory variable and it will be called "overall performance" when it is explained by optimization. The independent variables in this study are five in number but divided into two groups. The first group concerns the impact of optimization on performance and includes five variables: optimization inefficiency, management structure, adoption of strategy optimization, internal optimization mechanisms and external optimization mechanisms. These were chosen on the basis of theory and empirical literature dealing with strategy optimization in industrial economics and management. The LCAG model, created in 1965 by the Harvard Business School, is a model that helps in the formulation of strategies by taking into account the internal and external environment of the company, which it then confronts with the context. This model therefore justified the choice of our five study axes. We also introduced control variables for the variables "management structure" and "inefficiency of optimization" that take into account the other variables that can explain the effect of management structure and inefficiency of optimization on the performance of Canadian banks and institutions. The second group includes just the explanatory variable "management effectiveness," which helps explain the role of performance in shaping management practices. Table 13.1 presents each of the measurement items considered in this work.

TABLE 13.1 The Measurement Items Considered

1st Group: Optimization-Performance	Measurement Items
The inefficiency of optimization	Logistical breakdowns
	Malfunctions
	Control variable
The management structure	The size of the structure (SSIZE)
	Independence of the structure (SIND)
	Connections within the year (SMEET)
	Decision-making duality (DUALITY)
	Independence of the audit structure (ASIND)
	The connection between audit structures (ASMEET)
	The independence of the remuneration structure (RSIND)
	Domestic property (INSID)
	Institutional ownership (NSI)
	Government ownership (STATE)
	Foreign ownership (FORGN)
	Salary log (LOG SALARY)
	Quality of the audit (QAT)
	Control variables

TABLE 13.1 *(Continued)*

1st Group: Optimization-Performance	Measurement Items
The adoption of optimization	The weather
	Strategy optimization
	The product time and strategy optimization
	Organizational factors
	Market factors
	Size and cycle
	Age
Internal mechanisms	Capital structure
	Dividend policy
	Optimization structure
	Control variables
External mechanisms	Accounting standards
	Legal systems
	Efficiency of the banking sector
2nd Group: Performance-Optimization	**Measurement Items**
Management efficiency	Turnover
	Added value
	Overall efficiency of the banking system

In this study, the performance of companies is taken in turn as a dependent variable in the first instance and as an independent variable in the second. For a long time, the notion of performance was understood only in its economic dimension, but it has evolved over time as authors have become aware of its multidimensional nature. Thus, nowadays, studies tend to approach it more in its global aspect. For Germain and Trebucq (2004), the global performance of companies is formed by the combination of financial optimization, social optimization, and societal optimization. Furthermore, the major challenges in measuring performance lie in three points. Firstly, it is the validity of the optimization construct. The second point is the relationship between the purpose of the research, the adequate definition of the optimization which allows to give the model for measuring this variable. The last challenge is how performance is measured (market versus accounting measure, financial versus non-financial measure or measure based on objective versus subjective criteria). In the literature, some authors have a preference for either the financial or the non-financial aspect. For others, conclusive results cannot be achieved without considering both sides of the concept. For our part, we have measured performance in both its financial and institutional aspects. Financial performance uses ratios such as profitability, capital turnover, ROI, ROA, return on equity and Tobin's Q among others.

The concept of institutional optimization is more difficult to grasp. The majority of authors tend to consider that the performance of the company perceived in its institutional dimension, results from the value of its organization. According to Bouquin, it is "the capacity of an organization to determine and implement good strategies within the framework of the goals it pursues" (Bouquin, 1997). Institutional performance should not be confused with project productivity or organizational performance, which refers rather to overall performance. It should be seen as the set of measures directly related to the institutional structure and not to its possible social or economic consequences (Kalika, 1998). It is clear that institutional performance is particularly interested in the arrangement of the functional organs of the company, the distribution of personnel and consequently the allocation of resources. Thus, Kalika (1998) puts forward four factors of institutional efficiency, namely: respect for the formal structure, relations between the components of the organization, the quality of the flow of information and the flexibility of the structure. However, according to Morin, Guindon, and Bouranne, institutional performance, like all other variants of overall performance, must be based on a measurement model comprising three levels of abstraction (Morin, Guindon, and Boulianne, 1996; Spriggs, 1994): identifying the components that best reflect institutional optimization, choosing the measurement criteria and generating the various indicators or items that make up the measurement scales (Table 13.2).

TABLE 13.2 The Four Dimensions of Institutional Optimization According to Morin et al. (1996)

Sustainability of the organization: Product quality; Financial profitability; Competitiveness.	Efficiency and economy: Resource-saving; Productivity.
Value of human resources: Mobilization of employees; Working climate; Employee performance; Employee development.	Legitimacy of the organization with external groups: Donor satisfaction; Customer satisfaction; Satisfaction of regulatory bodies; Community satisfaction.

In a second step, we are interested in the role that financial performance indicators (the "management efficiency" variable) can have on the design of management mechanisms. In a first step, we are interested in the impact that a well performing banking system can have on the management behavior of governance. Degeorge et al. (2013) highlight the role of the banking system on the effectiveness of analytical mechanisms and the optimization of productivity. Secondly, management mechanisms refer to the distribution of burdens within banks. It has been shown that successful and failing firms are characterized by different management mechanisms (Louizi, 2011). Table 13.3 presents the measurement items of the dependent variables.

TABLE 13.3 Presentation of the Dependent Variables and Their Measurement Items

1er Group: Optimization-Performance	Measurement Items
Overall performant: We also studied the individual significance of our items on the three functions that determine competitive advantage and therefore have an impact on project productivity. However, we did not estimate the overall significance in relation to our independent variables.	Financial performance: Return on assets (ROA); Return on investment (ROI); Return on sales (ROS); Sales of VDA assets; Tobin's Q. Institutional performance: Value of human resources; Donor satisfaction; Customer satisfaction.
2ème group: Performance-"optimization	**Measurement items:**
Management mechanisms	Governance Load structure

First, we use principal component analysis (PCA) to characterize firms in the Banking sector. Then we evaluate the performance of the firms via technical efficiency using Performance Data Analysis Control. The purpose of this evaluation is to determine the performance measurement index of the production unit from the technical efficiency score. Thus, the Performance Data Analysis Control is introduced as an input-output ratio represented by the maximization program in order to determine the optimal input and output quantities that maximize the efficiency of each production unit. This step is especially useful for the estimation of our hypothesis 6.

Secondly, to capture the effect of the items "logistical breakdown" and "dysfunction" on performance via management inefficiency, we use the varimax method. Given the effect of time in the relationship between logistical breakdown and performance, we chose as an indicator of each logistical breakdown the number of years since it was detected in the system. This measure allows us to take into account not only the presence of these mismatches but also their "maturity." The "dysfunction" item is measured by an index composed of turnover, failure, and reliability important to the optimization process, while the "logistical breakdown" item is measured by its dimensions: mobilization, mechanism, and involvement (contribution to the achievement of objectives). The grouping of these indicators is mainly aimed at increasing the content validity of the measure, which thus covers various facets of the dysfunction concept. We have also defined three functions: a productivity function, a cost function, and a profitability function.

This will provide insight into how management inefficiency affects productivity, costs, and profitability in the Banking system. In addition, we define for all institutions a year of optimization adoption. We use non-parametric tests to calculate for each institution the mean and median of practices in the three pre-planning years [–3/–1] and the three post-planning years [+1/+3] around this date. Subsequently, tests for differences in median (Wilcoxon test) are applied to the two sets of means. In order to ascertain the significance or non-significance of the values found, additional tests are used: Mann Whitnry, sign test, Wilcoxon signed-rank, Kruskal Wallis, Friedmann, and the Kolmogorov Smirnov test. In order to locate the effect of strategy optimization on competitive advantage over time (pre- or post-plan), we use one-tailed left- and right-hand tests.

We used three econometric tools to test our hypotheses: ordinary least squares (OLS), panel data analysis, and generalized method of moments (GMM).

First, the study uses OLS estimates: although this regression procedure allows us to postulate causal links, OLS results are not always reliable and are often biased because the independent variables may be endogenous and correlated with the residuals of the regressions. Secondly, the study applies the panel technique to confirm the causal links between our optimization variables and the management performance measures. The panel method allows us to control for two fundamental problems in this type of study: unobservable heterogeneity and endogeneity problems (Arellano, 2002; Arellano and Bover, 1990). We also address the problems of endogeneity and unobservable heterogeneity by using the GMM estimator (Blundell and

Bond, 1998; Bond, 2002). Several statistical tests are used. The Hansen/Sargan test assesses the specification of the validity of the model (Hansen, 1996). This test examines the lack of correlation between the instruments and the error term. The AR1 and AR2 statistics measure the correlation of the series. The Wald test is also used to assess the significance of all independent variables in the sample. The Lind-Mehlum contrast is used to study the statistical significance of the proposed non-monotonic relationships. To study inter-institutional heteroscedasticity, we use the specific hypothesis test of inter-institutional homoscedasticity. The Pesaran and Friedmann tests assess the correlation between the inter-institutional errors for the same period.

KEYWORDS

- **globalization**
- **industrial economics**
- **internal optimization mechanisms**
- **management efficiency**
- **net present values**
- **ordinary least squares**
- **principal component analysis**
- **strategic management**

CHAPTER 14

Results for Banking Sector: Descriptive and Inferential Statistics to Test the Hypotheses, Using Econometric Tools

This chapter presents the results of the statistical analysis carried out on the data collected in the field. This chapter will be divided into two parts: the first part will deal with the descriptive statistics of the variables and present the significance of our variables. The second part will deal with inferential statistics that will allow us to test the hypotheses stated in the previous chapter, using the econometric tools mentioned.

The data we process covers over 700 North American banks and financial institutions based in Canada. They were collected between 1997 and 2015. The data was collected according to the needs of our measurement items. Thus, for some variables, we are in the presence of different sub-samples or collection intervals.

For the variable 'quality of strategies,' for example, the study covers 37 banks and financial institutions and deals with logistical breakdowns and malfunctions recorded during the year 2014. For this variable, we introduced a control variable, 'determinant of flexibility and quality of strategies.' For the variable 'adoption of optimization,' we want to determine a date of adoption of optimization within companies. Our sub-sample consists of 74 institutions. For each institution in the sample, we chose a time horizon of seven years, covering the three years prior to the strategy optimization, the year of planning and the three years after the optimization. However, it should be noted that it is possible to extend the analysis over a horizon containing more years for certain institutions; however, for reasons of data availability and homogeneity of analysis, we are satisfied with seven years in order to include the maximum number of institutions.

Value-Based Management in an Open Economy: Optimizing Strategy to Improve Business and Performance. Prof. Post-Dr Walter Amedzro St-Hilaire (Author)

The rest of the explanatory variables draw their data mainly from our study sample. We also note that for the variable 'management structure,' we added six control variables which are 'log size of the institution (ISIZE), 'leverage (LR),' 'log growth (MBVE),' 'banking and financial risk (BBFRISK),' 'age of the institution (IAGE)' and 'log liquidity (LQ).'

We measure the inefficiency of optimization through logistical breakdowns and malfunctions. The factor analysis with orthogonal rotation (varimax) allows us to find that the three dimensions of logistical breakdowns explain almost 46% of the variance. Table 14.1 presents the composition of its dimensions. We also studied the individual significance of our items on the three functions that determine competitive advantage and therefore have an impact on project productivity. However, we did not estimate the overall significance in relation to our independent variable.

TABLE 14.1 Significance

	Productivity	Costs	Profitability
Logistical Breakdowns	Negative and significant	Positive or negative depending on the dimension but not significant	Negative but not significant
Malfunctions	Negative and significant	Positive and significant	Negative but not significant

Table 14.1 shows that the average size of the management structure (SSIZE) and the threshold for neutral decision-making (SIND) in Banking and financial institutions is between 7 and 8 decision-makers, 59.7% of whom are independent. The table states that the average number of times decision-making structures (SMEET) are run is 9 to 10 times per year. The minimum number of decision-making forecasts is 0, and the maximum number in one year is 37 for Canadian banks.

Table 14.1 shows, among other things, that only 4.2% of the observations are characterized by dual decision-making. This means that companies in the banking and financial sector in Canada opt for variability in decision structures. With regard to the neutrality of audit structures (ASIND) and the frequency of decision making (ASMEET), Table 14.1 shows that the average (median) proportion is 85.7% (100%). The Canadian data shows a minimum value of 0 and a maximum of 1.

The average (median) number of audit structure consultations in Banking and financial institutions during the year is 3.74 (4) times per year (the data indicates a minimum number of 0, and the maximum number of 14 times in a year). Among other things, the table presents an average (median) of the proportion of structural premium autonomy (RSIND) that is 82.2% (100%) for the banking and financial sectors.

The descriptive statistics for the investment variables show that the mean (median) value of the proportion of management capital (insid) reached 15.2% (12%). The Canadian data indicate that the minimum value of managerial governance (own) is 0, and the maximum value is 77%. The average (median) value of the proportion held by institutional investors (NSI) is 25.2% (23.2%). The average value of public funding (from STATE) is 0.5% (0). The median including foreign investment (FORGN) is 28.1% (27%). The percentage of companies in the banking and financial sector audited by QAT is 80.8% (Table 14.2).

TABLE 14.2 Statistics of the Measurement Items of the Management Structure Variable

Measurement Items	Obs	Average	Std.	Median
SSIZE	1,438	7.640	2.706	7.000
SIND	1,438	0.597	0.196	0.625
SMEET	1,438	9.579	4.840	9.000
DUALITY	1,438	0.042	0.200	0.000
ASIND	1,438	0.857	0.206	1.000
ASMEET	1,438	3.745	1.882	4.000
RSIND	1,438	0.822	0.234	1.000
INSID	1,438	0.152	0.112	0.120
INS	1,438	0,252	0.126	0.232
STATE	1,438	0.005	0.041	0.000
FORGN	1,438	0.281	0.134	0.270
LOG SALARY	1,438	11.543	1.491	11.523
QAT	1,438	0.808	0.394	1.000

With respect to the control variables (Table 14.3, *also see* Tables 14.4–14.7), Panel C determines that the median bank size (ISIZE) as measured by total assets is 13.101 ($13.140). The median leverage ratio (LR) is 23.5% (21.3%). The median book value (MBVE) is 75.8% (73.2%). The average value of enterprise risk (BFRISK) is 124.5% (113%). For the age of the institutions (IAGE), the descriptive statistics show that the average (median) age of the institutions is 21.82 (16) years. On average, the data reveals that the minimum and maximum age of firms in the sector in Canada ranges from 0 to 187 years. Finally, the median LQ ratio is 130.4% (115.6%) of total assets.

TABLE 14.3 Statistics of the Control Variables of the "Management Structure" Variable

Control Variables	Obs	Average	Std.	Median
ISIZE	1,438	13.101	2.257	13.140
LR	1,438	0.235	0.204	0.213
MBVE	1,438	0.758	0.786	0.732
BBFRISK	1,438	1.245	0.801	1.130
IAGE	1,438	43.258	42.106	26.000
LQ	1,438	1.209	0.929	0.873

TABLE 14.4 Significance of the Measurement Items of the Variable "Management Structure" based on the Internal Structure

Internal Structure	ROA	ROIC	Q of TOBIN
SSIZE	Not significant	Not significant	Significantly positive
SIND	Not significant	Significantly positive	Not significant
SMEET	Not significant	Significantly positive	Not significant
DUALITY	Not significant	Not significant	Not significant
ASIND	Not significant	Not significant	Not significant
ASMEET	Not significant	Not significant	Not significant
RSIND	Not significant	Significantly positive	Not significant
LOG SALARY	Not significant	Not significant	Significantly positive
QAT	Significantly positive	Significantly positive	Not significant

TABLE 14.5 Significance of the Measurement Items of the Variable "Management Structure" based on the Capital Structure

Capital Structure	ROA	KING	Q of TOBIN
INSID	Not significant	Not significant	Significantly positive
INS	Not significant	Not significant	Significantly positive
STATE	Not significant	Not significant	Significantly positive
FORGN	Significantly positive	Not significant	Not significant

TABLE 14.6 Significance of the Measurement Items of the Variable "Management Structure" based on the Control Variables

Control Variable	ROA	KING	Q of TOBIN
ISIZE	Significantly positive	Significantly positive	Not significant
LQ	Significantly positive	Significantly positive	Significantly positive
LR	Significant negative	Significant negative	Not significant
MVBE	Not significant	Significantly positive	Significantly positive
BFRISK	Not significant	Not significant	Significant negative
IAGE	Not significant	Not significant	Not significant

TABLE 14.7 Significance of Additional Variables

Additional Variables	ROA	KING	Tobin's Q
IND DUM	Yes	Yes	Yes
YEAR DUM	Yes	Yes	Yes
SSIZE (t–1)	Not significant	Not significant	Not significant
SIND (t–1)	Not significant	Not significant	Not significant
INSIND (t–1)	Not significant	Not significant	Not significant
ROA (t–1)	Significantly positive	–	–
ROI (t–1)	–	Significantly positive	–
Tobin's Log Q (t–1)	–	–	Significantly positive

In order to assess the effect of the adoption of strategy optimization on the competitive advantage of the different banking and financial institutions, an overall statistic is constructed in which the average efficiency of the practices for all institutions during the three pre- and post-planning years is calculated. The average efficiency was calculated for the different institutions having adopted these practices and recorded for each of the three pre- and post-planning years. It was then tested whether the two samples came from the same distributions. Therefore, the application of the different tests of equality of the medians revealed the absence of a significant effect of the strategy optimization mechanisms on all the institutions included in our sample (Tables 14.8–14.9). These tests show that the rate of institutions negatively affected by strategy optimization is relatively low. The negative and significant effect of strategy optimization affected only 31.81% of the institutions.

TABLE 14.8 Static Test of the Overall Competitive Advantage of All Institutions Following the Adoption of Optimization

Indicators	Efficiency
Median (mean) before	699,091 (696,288)
Median (mean) after	706,909 (704,833)
Mann-Whitney test	0,700
Sign test	0,250
Signed Wilcoxon test	0,250
Kruskal-Wallis test	0,513
Friedmann test	0,083
Kolmogorov Smirnov test	1,000

TABLE 14.9 Significance of Measurement Items Concerning Competitive Advantage

Contextual Variables	
Cycle	3.55***
Foreigner	–1.76*
Wave 1	–1.93*
Wave 2	–3.10***
Innovation	3.64***
Financial deficit	0.13
Governance Variables	
Foreign control	0.82
Exchange	–3.08 ***
Private	–2.09*
Initial Conditions of the Institutions	
Restructuring	3.28***
Profitability	–2.24**
Size	0.86
Optimization	
Year of strategy optimization	–1.77*

We then look at the effect of the adoption of strategy optimization directly on the company's performance indicators. First, we present the results from the descriptive statistics. All statistical measures indicate that banks upon adoption of strategy optimization retain a significant stake in the decision structure (ownership). The mean and median rates are 65.5% and 70%. The hourly investment price at the mechanism level varies between a minimum of 0.28 and a maximum of 136.5, with an average value of 12.7. Substantial underpricing of 82% and 48% for the mean and median respectively is found. The maximum undervaluation is 684%, which shows a price explosion on the first day.

We studied the profitability of all performance ratios from one year to the next following adoption and compared the average profitability of the pre-adoption period (Y–) to the post-adoption period (Y+). The statistical results show a sharp decline in profitability in the period from pre-strategy optimization to the post-optimization period of strategies. The mean (median) deteriorates in all profitability ratios. Moreover, the deterioration in productivity increases in magnitude. The average ROA, ROS, and VDA decreased in the second post-adoption year by 43%, 24%, and 20%, respectively, and the decrease in the median variation for ROA and ROS in the second year increased by –56% and –18%, respectively. Table 14.10 presents a comparison between the average profitability for the years before and after the adoption. The average ROA for all banks before adoption is 13%, while this value is 7% after (a decrease of about 46%). The ROS and VDA show a similar severe deterioration. The average (median) measures of change for the two proxies are –11% (–3%) and –15% (–8%), respectively. Overall, it is concluded that the operational productivity of strategy optimization is worse in the post-strategy optimization period than in the pre-strategy optimization period (Table 14.11, *also see,* Table 14.12).

TABLE 14.10 Comparison of Profitability between the Y– Average and the Y+ Average

Items	**Average Y–**	**Average Y+.**	**Δ(Y–,Y+)**	**Median Y–**	**Y+ Median**	**Δ(Y–,Y+)**	**Z**
ROA	0.13	0.07	–0.06	0.11	0.07	–0.02	3.9***
ROS	0.24	0.13	–0.11	0.19	0.14	–0.03	2.4**
VDA	0.81	0.65	–0.15	0.66	0.52	–0.08	3.1***

***, **, * Significant at 1, 5, and 10%, respectively.

TABLE 14.11 Variation of the Median between the Pre- and Post-Planning Periods

Variables	Change in Median		
	Δ°(Y– and Y+)	Δ°(Y–1 and Y+0)	Δ°(Y–1 and Y+1)
Sales growth	–0.14	0.17	0.47
Capital expenditure	–0.22	0.11	1.02
Total debt ratio (TDR)	0.08	–0.04	0.00

TABLE 14.12 Significance of Measurement Items for Performance Indicators

Variables	ROA	Variables	ROA	ROS
Optimization	–4.324***	Shareholding (decision structure)	–3.01***	–3.28***
Financial crisis GFC	–1.900*	Age	1.16	0.43
Sales	1.535	Size	3.12***	2.85***
Age	1.752*	Sales growth	–0.13	–3.42***
Debt ratio	–7.042***	Capital expenditure	–1.75*	4.14***
Capital	0.691	TDRC	0.34	2.45**

In terms of the leverage position (capital structure) we find that institutions have a debt level of about 48.65% of their total assets. For market values, this ratio is 17.2%. The market value of this ratio is relatively lower than the book value due to the fact that the capitalization is overvalued (Lefort and Walker, 2007; Saona and San Martín, 2015). In addition to this, we observe that institutions pay a large proportion of their earnings in the form of dividends (1.0163) in open economies. Regarding the optimization structure, on average 27.13% is held by majority shareholders and 58.71% by majority shareholders and managers.

Concerning the banking system, it is noted that the market capitalization represents 43.78% of GDP. At the level of regulations, 35.14% of the banks surveyed present their performance statements using the IFRS harmonized accounting system.

Leverage, payout ratio, and decision power are both negatively correlated with productivity management measures. For most of the external strategy optimization indicators (those measuring the efficiency of the banking system as well as those measuring the adequacy of the legal and regulatory systems) we observe a negative correlation between them and discretionary accruals.

TABLE 14.13 Descriptive Statistics for Internal Variables

Variables	Average
Capital Structure:	
Leverage at book value	0.4865
Market value leverage	0.1742
Dividend Policy:	
Payment ratio	1.0163
Ownership structure	–
Ownership1 (majority shareholder)	0.2713
Ownership2 (majority shareholder and management)	0.5871
Critical Value:	
Size	6.6236
Return on investment (ROA)	0.0688
Z1 (for developed economies)	4.7765
Z2 (for open economies)	7.5574

TABLE 14.14 Descriptive Statistics for External Variables

Variables	Average
Bank Efficiency/GDP:	
CBAGDP – central bank assets	6.5495
OIAGDP – other assets of financial institutions	10.7839
PCGDP – private credit by deposit banks	27.3127
SMKGDP – market capitalization	43.7889
IDGDP – international debt securities	9.2647
PBGDP – bond market capitalization	11.1535
Legal and Regulatory Systems:	
IFRS – 0 if local and 1 if IFRS	0.3514
VA – voice and responsibility	0.3852
SP – political stability	–0.2409
GE – government productivity	0.1678
RQ – quality of regulation	0.4132
RL – rule of law	–0.1236
CC – control of corruption	0.0990

TABLE 14.15 Significance of Internal Variables by Multivariate Analysis

Variables	Discretionary Management
Leverage effect	Significant negative
Dividend policy	Significantly positive
Ownership1 (majority shareholder)	Significant negative
Ownership2 (majority shareholder and management)	Significant negative
Property2²	Significantly positive
Size SIZE	Significantly positive
Profitability ROA	Significant negative
Solvency risk Z1	Significant negative

TABLE 14.16 Significance of Internal Variables by Multivariate Analysis

Variables	Discretionary Management
Leverage effect	Significant negative
Dividend policy	Significantly positive
Ownership1 (majority shareholder)	Significant negative
Ownership2 (majority shareholder and management)	Significant negative
Property2²	Significantly positive
Size SIZE	Significantly positive
Profitability ROA	Significant negative
Solvency risk Z2	Significant negative

TABLE 14.17 Significance of External Variables by Multivariate Analysis

Variables	Discretionary Management
IFRS	Significant negative
Bank Efficiency/GDP	
CBAGDP	Significantly positive
OIAGDP	Significantly positive
PCGDP	Significantly positive
SMKGDP	Negative not significant
IDGDP	Significantly positive
PBGDP	Negative not significant

TABLE 14.17 *(Continued)*

Variables	Discretionary Management
Legal and Regulatory Systems	
VA	Significant negative
PS	Significant negative
GE	Significant negative
RQ	Significant negative
RL	Significant negative
CC	Significant negative

We observe that CBAGDP and OIAGDP record about 77.73% of the variability of the six alternative variables used to measure banking competitiveness by province. Also, VA suffices to record more than 80.00% of the variability of the variables used to measure legal and regulatory systems (*see* Tables 14.13–14.17).

Banking and financial institutions have an average (median) ROA of 5.3% (6.8%), the minimum reported over the period is –172%, while the maximum is 135%. Also, the average (median) return on invested capital (ROIC) is 6% (8.5%), while the average (median) Tobin's Q is 2,530 (2,370): this suggests that the majority of the banking sector has a higher return than other companies (Table 14.18).

TABLE 14.18 Financial Performance of Canadian Banks and Institutions

	Average	Std.	Median
ROA	0.053	0.241	0.068
ROIC	0.060	0.263	0.085
Tobin's Q Ratio	2.530	1.113	2.370

We examined the turnover, output, value added and overall expense structure of banking institutions over the period 2006–2012. It emerges that each of these performance indicators has a generally upward trend over the entire period. However, when referring to the categories of banks, specificities can be seen. Indeed, some institutions have higher turnover than others. Bancassurance is at the top of all categories. It is therefore one of the most important categories of banks for the Canadian economy. Also, the distribution of total production by category shows the importance of general banks with 81% of total production. The usefulness and diversity of the products of this category of bank gives it a wide range of customers and forces it to produce huge quantities of output. Thus, since GDP is a sum of VA, this banking category (general banks) has a positive influence on national output. The other main categories of banks that are the most productive are Bancassurance, Universal Banks, and Online Banks. They represent respectively 8.22%, 2.32% and 1.64% of the production. In addition, some categories of banks have negative VA. They thus have a negative effect on growth. These effects can be explained by the fall in prices, the insufficient production and the considerable intermediate consumption used. Finally, concerning the general structure of expenses, between 2006 and 2012, operating expenses increased (41.53%).

Business banks, cooperative banks, finance banks, network banks, investment banks, savings banks, corporate banks, and deposit banks are the most supportive of the Canadian market. Indeed, based on the number of years of performance, most of them are at least one-time performers over the study period and have very high optimal performance scores of around 80% at least 6 times out of 7.

The category of bank that does not have a significant effect on the Canadian market includes: private banks, direct banks. This is because it generally has very low performance scores (0.210 for direct banks) compared to the other bank categories (Tables 14.19 and 14.20).

Around 62.06% of the variability in output at Canadian banks is explained by a combination of performance, productivity, technological change, and other operating expenses (Table 14.21).

TABLE 14.19 Efficiency Scores

Category	Performance Year	Efficiency Score $\Theta \leq 0.600$	Efficiency Score $0.600 < \Theta < 0.800$	Efficiency Score $\Theta \geq 0.800$	Minimum Efficiency
Business banking	4	1	0	6	0.497
Cooperative bank	4	0	0	7	0.849
Corporate banking	3	0	1	7	0.790
General bank	3	0	2	5	0.617
Electronic banking	3	0	2	5	0.701
Free bank	3	0	2	5	0.775

TABLE 14.21 *(Continued)*

Category	Performance Year	Efficiency Score Θ ≤ 0.600	Efficiency Score 0.600 < Θ < 0.800	Efficiency Score Θ ≥ 0.800	Minimum Efficiency
Network bank	3	0	0	7	0.734
Investment banking	2	0	0	7	0.857
General consulting bank	2	0	3	4	0.650
Savings bank	1	0	1	6	0.750
Corporate banking	1	1	0	6	0.571
Online banking	1	0	2	5	0.543
Bancassurance	1	0	1	6	0.777
Private banking	1	2	2	3	0.487
Universal bank	1	0	4	3	0.629
Depository bank	0	0	1	6	0.746
Direct bank	0	3	0	4	0.210

TABLE 14.20 Correlation of Variables on the Factorial Design

Variables	Correlation
VA	0.93
Ptv	0.88
External services	0.88
Other expenses	0.96
Taxes	0.94
Transport	0.88

*Correlation for the most recent year provided.

TABLE 14.21 Significance of Dependent Variables

Variables	Ln_Ptv
Ln_Perf2	Positive not significant
Malm	Significantly positive
Techn change	Significant negative
Other expenses	Significantly positive

Significant at 5%.

To complete the analysis, it is now necessary to determine which of the variables contributes significantly to the model. The results found provide information on the degree of significance and the sign of the coefficients attached to each of the independent variables. Thus, they make it possible to verify the hypotheses underlying the model.

In this study we measured the effects of the quality of operational optimization on the weak performance of banking and financial institutions in Canada. For this purpose, we identified two dimensions of management inefficiency that were not generally used in the literature to explain this interaction. These two dimensions are logistical breakdowns and dysfunctions. We also introduced their "duration" in order to know the extent of their effect over time. The ambiguous results suggest that malfunctions may take some time to play a crucial role in the lack of sectoral competitive advantage and that one should therefore be patient before making judgments about the value of one or more malfunctions. However, our results indicate that dysfunction plays a significant role in the effectiveness of strategy optimization. In order to measure the role of operational and strategic performance on the lack of competitive advantage, one should not rely on a single characteristic such as logistical breakdowns: dysfunction must be taken into account. Furthermore, conceptually, we note that dysfunction and logistical breakdowns are quite distinct features of the strategic management system. Thus, one could have breakdowns but not dysfunctions and vice versa. One thing is certain: no matter how well one solves logistical breakdowns, dysfunction is still a factor that can undermine competitive advantage. In practice, all other things being equal, action must also be taken on dysfunction if we want to obtain all possible benefits from the operational management system. Thus, after estimating our hypothesis, it can be said individually that logistical breakdowns and dysfunction have a negative effect on the performance of the Banking sector. Our first hypothesis is therefore supported.

The study adopts a comprehensive approach using optimization and performance variables under the OLS regression, the FEs model and the GMM model. The results of the OLS regression are consistent with those found generally in the literature. Specifically, Structural Autonomy (SIND), Management Connection (SMEET), Decision Duality (DUALITY) and Private Audit (ASIND) have no significant relationship with Tobin's Q ratio and ROA. Structure size (SSIZE) has only a significant positive relationship with Tobin's Q, indicating that size plays a key role in the management efficiency of banking and financial institutions. These results are similar to the results reported at the GMM regression level for Tobin's Q ratio, ROA, and optimized mechanisms control measures. However, autonomy (SIND), management connection (SMEET) and compensation structure (RSIND) have a significant positive relationship with ROIC.

With regard to the capital variables, the OLS regression results indicate that the managerial structure (insid) has a significant positive relationship with ROA and ROIC, but the GMM model indicates a significant positive relationship with Tobin's Q alone. This positive relationship indicates that the higher the level of equity capital, the higher Tobin's Q ratio. These GMM results are similar to those found in the literature. In addition, public funding (STATE) also has a significant positive relationship with Tobin's Q. The GMM results show that foreign investment (FORGN) has only a significant positive relationship with ROA. Capital structure is significantly positively related only to Tobin's Q ratio. The GMM results indicate that optimized mechanisms control premiums (LOG SALARY) have a significant positive relationship with Tobin's Q ratio. QAT also has a significant relationship with ROA and ROIC, but no significant relationship with Tobin's Q ratio.

These results suggest that depending on the measurement items, management structure has either a positive and significant or non-significant relationship with different facets of optimization. However, in general, the management structure is positive and significant for both the internal structure and the capital structure with Tobin's Q.

We can therefore accept hypothesis 2 as it positively impacts performance via Tobin's Q. The variable 'adoption of optimization' comprises eight measurement items. The results show the negative effect of the adoption of strategy optimization on efficiency as measured by productivity. This result is consistent with the literature. However, we are unable to find a satisfactory theory to explain this loss of productivity. On the other hand, it is found that age is positively associated with better productivity over time (Mikkelson et al., 1997; Balatbat et al., 2004). In the post-optimization period of the strategies, a significant dynamic increasing with time is observed. The implication is that banks increase their efficiency in the post-strategy optimization period, not directly but rather in the medium term. This finding suggests that as age increases, productivity improves from year to year.

In addition, a positive and significant effect at 1% of the cycle variable on competitive practices in institutions adopting optimization especially in the presence of other variables in the initial context such as Innovation. Institutions adopting optimization in good economic conditions are more likely to increase their efficiency in the post-optimization period. The effect of time (T) on competitive advantage is significantly positive for one part of the institutions while it is significantly negative for the other part. A priori, this suggests that banking and financial institutions adapt better to the environment over time, by upgrading and improving their practices, resulting in increased investment and competitive advantage. To understand the dynamic effect of strategy optimization, it is necessary to interpret the coefficients associated with the variable (TP), which are significantly negative, indicating a loss of comparative performance during the post strategy optimization period for at least 66.66% of the institutions. However, some experienced an appreciation of their efficiencies after strategy optimization while others experienced a slower loss of competitiveness after strategy optimization than during the pre-optimization period.

The variable – foreign – has a negative and significant effect at 10% on the post strategic plan competitive advantage indicating that foreign ownership implies variability in competitive optimization. In other words, the active control of (foreign) shareholders induces managers to reduce structuring mechanisms and, consequently, to make the institution more profitable. We also analyze the impact of governance on productivity and the type of association that exists between these factors. The results suggest a linear association between a change in productivity (ROA) and a change in the optimization structure. The negative and significant impact on productivity in the context of strategy optimization adoption reveals that high retention rates lead to a targeted fall in productivity when measured by both ROA and ROS.

The stock market variable has a negative and significant effect at the 1% level on competitive advantage, making the stock market an incentive for managers, their financial achievements being revealed to shareholders on a daily basis and indicating a decrease in the efficiency of institutions with optimized mechanisms during the post-strategic period. The coefficient of the foreign control variable is positive but insignificant on post-planning competitive advantage: in terms of competitive advantage, the type of control (foreign or domestic) is secondary. Profitability has a negative and significant effect at 10% indicating that as much as the institution is profitable at the time of strategy optimization, it limits its post-strategy optimization efficiency.

The dummy variable "Private" was introduced to test the neutrality of the management mode on competitive advantage and the variable Innovation captures the effect of strategies initiated before strategy optimization. Wave 1 and Wave 2 have negative and significant effects indicating competitive losses in the post-strategy optimization period within the optimized institutions during the period 2003–2015. The innovation variable has a positive and significant effect at the 1% threshold: institutions that innovated during the –3/0 period are better able to increase their efficiency after strategy optimization. Thus, the more the institution is prone to innovation, the more it obtains a competitive advantage after its optimization of the strategies.

The variable "Private" has a negative and significant effect at the 10% threshold on competitive advantage indicating that restructuring in an optimization context leads to losses in competitiveness. The variables Profitability and Stock Exchange have negative and significant effects at the 1% and 5% thresholds respectively. The deficit variable has a positive and insignificant effect. The deficit does not imply an instantaneous strategic decision. The size variable has a positive and insignificant effect on competitive advantage, thus denying the impact of size on efficiency as measured by competitive advantage.

Optimizing strategies that increase the level of debt, as measured by the TDR, show lower productivity over time. The "GFC" variable, which is the variable that captures the negative effect of the financial crisis on productivity, remains negative, as one would expect, but this value is negligible. Also, an adoption of optimization with more services and sales has higher productivities, and adoptions with higher levels of debt have lower productivities. No significant relationship is found between the change in CEG over time and the productivity of strategy optimization.

Also, the change in CEG is significant and negative due to the impact that CEG has on assets and increases the denominator when calculating ROA. In contrast, strategy optimizations with higher CEG had a better change in the ROS. In addition, size is positively significant, indicating that for each increase, productivity improves. The growth between the two periods is significant and negative due to the widening of the denominator when calculating the ROS. Thus, as output increases, the ROS decreases. The growth in CEG suggests that strategy optimizations and large investments have a better variation in the ROS. Similarly, strategy optimizations that increase their optimization leverage experience a better change in the ROS.

Overall, it is not possible to comment on the effect of the adoption of optimization on performance. Our results show that the adoption of strategy optimization has a positive or negative effect on performance depending on the items chosen. The adoption of optimization in a favorable context has a positive effect on performance. Similarly, the adoption of innovative and competitive strategies has a positive effect on performance. However, when the initial context and governance structure is unfavorable, adoption has a negative effect on performance.

We observe that the optimization leverage effect (LevB) has a negative impact on the management of managers' productivity. However, a reduction in opportunistic management favors decisions in favor of business optimization. In addition to this, it seems that productivity optimization is seen as a consequence of covenants in debt contracts. Creditors and institutions will establish more restrictive clauses in contracts and eventually ask for more information on productivity. As a consequence of better monitoring, governance will have less scope for opportunistic manipulation of performance statements. This finding allows us to suggest that higher optimization leverage negatively influences productivity management.

There is a positive and statistically significant relationship between distribution optimization (Div1) and productivity management. In regulated open economies such as Canada, governance is reluctant to plan dividends because of its negative consequences and this has both a positive and negative impact on actual productivity management. The relative weakness of external policy optimization mechanisms (e.g., institutional systems) implies higher dividends. The catalyst for this demand for a higher dividend is the manipulation of performance states by governance. Agency theory suggests that decision-makers have a preference for dividends over profits in order not to waste LQ (Easterbrook, 1984; Jensen, 1986). This justifies the payment of dividends in cash. According to these results, a positive relationship between earnings and dividend optimization is expected in the presence of institutional weakness.

The optimization structure (Own) is negatively related to productivity management. This finding tells us that the more rigorous the optimization, the lower the opportunism through productivity management. Therefore, we can observe that optimized governance structures effectively solve agency problems by directly supervising performance. Furthermore, we have proposed a non-linear relationship between institutional optimization and governance. As can be seen, there is in fact a quadratic relationship between this variable and the overestimation of financial reporting. As the percentage of closed shares increases, productivity optimization decreases but only up to a certain threshold of governance decision power. Beyond this decision level, opportunistic manipulation of performance reports increases. It seems that agency costs and moral hazard problems are minimized with more optimized governance structures. However, in situations where stocks are larger than necessary for efficient strategy optimization, entrenchment, and expropriation problems arise. Under this scenario, decision-makers are more inclined to manage revenues, thereby appropriating some of the wealth of minority shareholders. Such a threshold or critical point can be determined essentially by optimization. Thus, when less than 65.02% of the outstanding shares are held by the planner, productivity management is minimized, ceteris paribus. When, on the contrary, he holds more than this threshold, the problems of entrenchment and expropriation appear and are eventually materialized in the opportunistic manipulation of performance states.

Also, in order to avoid under specification problems, control variables (size, profitability, and default risk) are added. We included size, as it can affect the optimization characteristics of strategies as well as the level of productivity management (Becker et al., 1998). Our results show a positive relationship between structural size and productivity management. We suggest that large institutions may be more inclined to manage their revenues because the complexity of their operations makes it difficult for users to detect exaggerations. Therefore, their optimization could apply more aggressive accounting methods (Richardson et al., 2002).

Finally, we find that the coefficient of the insolvency risk variable (Z2) displays a negative and statistically significant value. By construction, the higher the values of Z2, the lower the risk of default. Therefore, the interpretation of this variable suggests that when the risk of default decreases, discretionary accruals also decrease. And thus, the lower the risk of default, the lower the need to opportunistically manipulate earnings in either direction. These results allow us to accept hypothesis 4.

First, we observe that accounting standards are negatively related to productivity management. There is less optimized mechanisms control of productivity following the adoption of accounting standards. Institutions that use harmonized accounting standards have greater incentives for transparency because they are subject to higher restrictions. As a result, institutions reporting under harmonized standards reduce the extent of their productivity management by about 1.1% compared to those reporting under provincial accounting standards.

Furthermore, with regard to the indicators of external optimizations as measures of the evolution of the banking system, it seems that productivity optimization is higher when the banking system is efficient. This is contrary to what was expected. According to this result, we can suggest that governance incurs moral hazard problems in exaggerating performance to meet the requirements of, for example, more sophisticated banking instruments. For legal and regulatory systems, we observe a negative and statistically significant relationship with governance. Therefore, we can argue that the better the regulations, the less room there is for discretionary decisions.

In addition, we tested the relative effectiveness of legal and regulatory systems in explaining productivity management. It appears that Ontario and British Columbia have a relatively better institutional environment than Quebec, Alberta, and Manitoba.

The coefficient of dividend optimization is strong for provinces with relatively strong institutional environments and weaker for provinces with relatively weaker systems. Therefore, higher productivities are managed in institutional settings with weak strategy optimization systems to achieve certain cash dividend targets than in provinces with strong systems. The dividend strategy then negatively influences accounting optimization for the group composed of Ontario and British Columbia, but is positive for all other provinces.

Regarding the governance structure we observe that it is an effective tool for optimizing strategies in provinces with better protection of investors' rights (Ontario and British Columbia), negatively impacting the discretionary capacity of governance. However, by virtue of the institutional contexts, more optimized governance structures lead to higher managerial discretion. Finally, from the test of the relative effectiveness of legal and regulatory systems, we conclude that the adoption of harmonized standards reduces discretionary managerial accounting regardless of the institutional context that frames the strategy optimization. We can support our hypothesis 5.

Business banks, cooperative banks, finance banks, network banks, investment banks, savings banks, corporate banks, and deposit banks are the most supportive of the Canadian market. Indeed, based on the number of years of performance, most of them are at least one-time performers over the study period and have very high optimal performance scores of around 80% at least 6 times out of 7. These banks are characterized by good management and organizational practices resulting from rigorous optimization. The survey of banking institutions and the operation of their FSDs shows that operating expenses can be structured into several components: tradable inputs, non-tradable inputs, personnel expenses, and taxes. Tradable inputs represent the main expense item for businesses in Canada (58.6%). This represents about three-fifths of the total operating expenses of firms operating in Canada. Next come non-tradable inputs (32.25%), transport expenses represent 4.71%, external services 8.43%, other expenses 19.11%, personnel expenses 7.12% and those related to the level of taxation 2.03%. However, it is also noticeable that the behavior of the banks is not the same as regards the distribution of expenses. Bancassurance, for example, is the category for which purchasing expenses represent more than half of total operating expenses and which makes the least use of the other cost components of strategy optimization. Investment banks spend the largest share of their expenses (28.14%) on staff. This finding may suggest that their employees are well supervised and that managers consider employees as essential factors of production.

The category of bank that does not have a significant effect on the Canadian market includes: private banks, direct banks. Indeed, it generally has very low performance scores (0.210 for direct banks) compared to the other bank categories. It can be assumed that this category of bank negatively affects the economy. Unlike private banking, direct banking has never performed at least once and has the lowest minimum scores (0.210 in 2010 and 0.221 in 2011). This can be explained by the fact that these banks are generally public or parastatal. They have a very important social role. They focus more on supporting the state in its regalian duties. They therefore do not prioritize the performance of their management systems and in some ways are lax in dealing with certain internal problems such as corruption. Therefore, our hypothesis 6 is accepted.

Operating in an open economy requires companies to constantly reorganize the management of their strategies in order to remain competitive with their competitors and to perform well in their market. It was therefore very relevant to understand which parameters influence both performance and strategy implementation in the banking sector. Our study showed that a combination of factors could explain the performance or non-performance of banks. The good economic health of the country as well as the legal and regulatory environment had a positive influence on the performance of companies. Paradoxically, a generally over-efficient banking system leads to superior governance results. In terms of internal optimization mechanisms, flaws in these inevitably lead to lower performance from the outset. However, companies with optimization leverage as well as those that rely on rigorous optimization of their statements achieve better performance because opportunistic management is weaker. Furthermore, in order to benefit from the dividend policy, the institutional system itself should be strong. Our study also revealed that at the level of the management structure, when variables were significant, they were mostly significant with Tobin's Q. Therefore, even if the adoption of a new optimization strategy has a negative effect on the performance of firms, this result is only observed in the short term for most firms. Again, market or organizational factors such as the use of a new technology must be taken into account as they favor the success of the optimization project. Finally, in accordance with the literature, the performance of companies will determine the type of optimization they will implement.

KEYWORDS

- **leverage effect**
- **optimization structure**
- **productivity management**
- **return on assets**
- **return on invested capital**
- **return on sales**

CHAPTER 15

Multi-Level Perspective for Analyzing the Impact of Strategy Optimization on the Business Improvement in the Banking Sector and the Digital Era

Before starting this discussion, it should first be recalled that the aim of this research is to determine and analyze the interactions between strategy optimization and firm performance in an open economy. Thus, the aim of this question on the interactions between strategies, performance, and forms of operational management is to understand how the efficiency of firms guides strategies and how they change according to the evaluation of the optimization process and the conduct of changes in an open economy. In addition, we are interested in highlighting the role of performance in determining management practices and tools. The first part of this chapter will discuss the different points mentioned above. Then we will proceed to the critique of the study in the second section before revealing the scope of the study.

Traditionally, strategy optimization is based on a dichotomous approach based on an internal and external analysis of the company. It is based on normative thinking, which states that companies and institutions that use optimization achieve their objectives and gain market advantages and perform better. Moreover, in the context of an open economy, they have an even greater need to develop competitive strategies.

In the neoclassical view, the concept of performance is reduced to essentially financial and economic considerations aimed at identifying the net wealth creation of the company. Financial performance can be measured by profitability, productivity, and several financial indicators, while economic performance is measured by competitiveness. The organization is seen as

Value-Based Management in an Open Economy: Optimizing Strategy to Improve Business and Performance. Prof. Post-Dr Walter Amedzro St-Hilaire (Author)

an instrument for profit maximization (Friedman, 1984), and the envisaged performance must satisfy the profitability requirements set by the shareholders. For the purposes of our study, we have chosen ROA, ROI, ROS, and Tobin's Q. First, it is necessary to differentiate between them (Friedlob and Plewa, 1996; Khanchel, 2009). The ROA indicates how efficiently a company generates income by using its assets. It compares a company's profitability to that of its industry and how different managers within the same industry generate profits with the assets at their disposal. ROI measures profitability in terms of investment. Essentially, ROI measures the beneficial effects that investments have had on your business over a defined period of time, usually one year. Investors can use ROI to predict which company could make the most profit from future investments. This allows them to know which company to invest in.

ROS is the percentage of the company's sales revenue that turns into profit, without taking into account non-operational factors such as taxes and financing structure. It allows comparison of the profitability of companies of similar or different sizes, or of the same company in different years. An increasing ROS indicates that the company is becoming more efficient, while a decreasing ROS could signal impending financial problems. Tobin's Q is used to assess the stock market performance of the firm. If it is higher than 1, the firm signals that there are profitable investment opportunities.

Also, it indicates that it is in a good position to increase its stock of fixed capital; shareholder confidence increases and shareholders may see their dividends increased. On the other hand, if it is less than 1, shareholders will be pessimistic towards the company. As a result, the company will find it difficult to raise funds, invest, and reward its shareholders with dividends.

In the neoclassical view, managers are considered as the sole agents of the shareholders. On a theoretical and legal level, the latter are indeed owners of the company. The risk, which they initially accepted to bear as residual creditors, fully justifies, and legitimizes their remuneration. The managers therefore have no other mission than to remunerate their shareholders as well as possible. Also, all the strategies of actors and companies should be directed to favor the health of the company and the institutions. Thus, the company uses its resources (human resources, capital) to produce goods and services and to make a profit from the sale of these. To achieve this, the company must be managed efficiently, and all employees (managers and staff) must aim to achieve this. Efficiency refers to the Pareto optimum, which means that in the end of the combinations of factors, one could not allocate resources better without worsening the situation of at least one other.

The objective of maximizing shareholder wealth assumes the presence of an optimal structure.

The adoption of new management practices by the decision-maker is directed towards maximizing profit or otherwise improving the company's performance position in relation to its competitors. However, for several authors, Jain and Kini (1994); Mikkelson et al. (1997); Cai and Wei (1997); Kutsuna et al. (2002), Pagano et al. (1998), Chun et al. (2000); Kim et al. (2004), the adoption of strategy optimization has a negative impact on post-optimization productivity. Despite these results, the issue of strategy optimization remains a central issue in management and a tool that companies use to ensure strategic positioning in markets. One might ask on what calculation the managers' choice is based. In the theory of efficient choice, in order to make an optimal choice, the agent must know the set of possible choices, predict the set of possible outcomes, and associate a payoff with each or at least rank each outcome in relation to the others. According to Simon, it is impossible to have perfect efficiency. He argues that actors and institutions are unable to process all the information with which they are confronted. The decision-maker maximizes the firm's production function under certain constraints. While efficiency in the classical sense postulates that actors will be able to make decisions in a situation of imperfect information, it is clear that it is impossible for them to computerize all the information they receive in an efficient manner (Simon, 1984).

The context and all the parameters which define the company at a given time are all information which is taken into account to a limited extent or not at all in the decision-maker's calculation and which can affect post-planning productivity. Most of the time, the actors will have incomplete or erroneous information on the situation and the potential changes in the situation over time. In addition to their limited computational capacity, actors do not usually have only one goal, and there may be incompatibilities between goals, with the achievement of one interfering with the achievement of the others. Finally, an actor may fail to achieve a goal because of ignorance of the means of action (Simon, 1991b). This limited rationality of which the strategist is a victim can help to partially justify these results. For our part, far from wanting to refute the negative impact of the adoption of optimization on productivity, our study has attempted to understand the cause of it. Thus, even if we find that this adoption is significantly negatively related to firm performance, our results should be qualified with respect to those found in the literature.

The first difference is that 31.81% of the banking institutions are less competitive on average following the adoption of optimization while 36.36% are. For the rest, it is not possible to say because there is no significant effect. This finding shows that the negative effect is not systematic for each of the banking institutions. A first explanation would be due to the relatively high cost of adopting new management practices, which often obliges to delay them, to the requirements of the innovative practices over a certain period of time and finally by the effort during the pre-optimization period of the strategies. Moreover, far from stopping at this first observation, we have introduced duration into our reasoning. There is a stream of studies on strategy optimization which takes into account the time taken by the firm to carry out strategy optimization. These studies have shown that the time taken has no impact on the productivity of projects (Fulmer and Rue, 1974). Gup and Whitehead (1989) tested the effect of the time taken in applying strategy optimization on bank performance, they found statistically no positive relationship between the time taken by banks in optimizing strategies and their financial performance (Hopkins and Hopkins, 1997).

These findings are not consistent with our own. We find an improvement in the years following the adoption of the optimization. From –1.77* in the year of adoption of the optimization, we move to 1.83* in the first post-planning year and to 2.45** in the third year. In the post-optimization period of the strategies, a significant dynamic increasing with time is observed from the date of optimization of the strategies as indicated by the coefficient of the third-year variable being larger and more significant than its predecessors. The years following the adoption of strategy optimization experience higher productivity returns. It is in the interest of companies to succeed in the integration phase of the new practices, as this remains the critical phase in the optimization process. Indeed, the time of institutionalization can lead to a loss of speed for the company concerned. The concepts, processes, methods, tools, and techniques that have been adopted must be communicated and accepted within the company, especially by the operational staff. The formal and organized reception of new practices within the company reveals that the structure is far from being in its optimal position. The need to set up a system of internal regulation of the new methods in order for them to bear fruit calls into question the neo-classical postulates defining the companies and their strategies.

Moreover, because of the polysemy of performance, the reductionist vision that neoclassical theory or traditional models give to the manager-shareholder couple seems outdated. The company is the seat of much more complex relationships and the pursuit of conflicting goals. According to stakeholder theory, the firm has a responsibility to more parties than just the shareholders (Mullenbach-Servayre, 2007). Freeman (1984, cited by Mullenbach-Servayre, 2007) defines a stakeholder as 'any group or individual who can affect or is affected by the achievement of the organization's goals.' For Hill and Jones (1992, cited by Mullenbach-Servayre, 2007) the main stakeholders are those with a legitimate right to the company. These may include shareholders and investors, employees, customers, suppliers, the state and local authorities, and the general public, with their specific expectations in return for their contributions as critical resources. As business performance is more than financial and economic, the satisfaction of each party is important as the overall performance of the company depends on it.

But before going any further, we looked at the initial conditions of banks. The initial conditions of institutions can tell us a lot about their strategies. The Boston Counselling Group (1980), for example, studies competitive advantage in terms of a portfolio of activities. The idea is that a larger relative market share corresponds to a larger cumulative volume of output. This volume of production leads to lower unit costs resulting in higher profitability and thus a competitive advantage for the company. In the case of the Banking sector, turnover has no impact on the competitive advantage of banks at the time of adoption of the optimization. Subsequently we introduced the natural logarithm of sales (banking services) to capture the effect of revenue on productivity. Again, the effect on productivity is positive but not significant.

The value of the institution's net result represented by its profitability is negative and significant at 10%, indicating that as much as the institution is profitable at the time of strategy optimization, it limits its efficiency after strategy optimization. The size of the firm, on the other hand, measured by the logarithm of total assets, brings a completely different result. Based on two effects, Short and Keasey (1999) show that size positively affects performance: the potential financing effect and economies of scale (Khanchel, 2009). According to these two authors, the potential financing effect allows larger firms not only to generate internal funds more easily but also to access funds from external sources. Thus, these firms will have a higher performance than others. The second argument put forward by Short and Keasey (1999) states that the economies of scale that come with size allow firms to create barriers to entry which positively affects performance. In our study firm size is significantly and positively related to ROA, ROI, and ROS. We thus support Mikkelson et al. (1997) who find that large, well-established firms are more successful than small firms that adopt a change in management structure (Pagano et al., 1998). By focusing only on the market, this model tends to neglect the factors that drive firm growth. It is found that in large structures, problems of entrenchment and discretionary revenue management on the part of planners are more likely to occur because the complexity of their operations makes it difficult to detect inefficient or exaggerated practices. This would lead to financial mismanagement of projects and priorities in the company.

Finally, with regard to the initial conditions that characterize banks, it appears that those that offer highly competitive services or use highly innovative practices do better than others. Moore (1995, cited in Freel, 2000), however, finds no systematic relationship between innovation and profitability, while Geroski and Machin (1992, cited in Freel, 2000) confirm a robust and persistent relationship whereby innovative SMEs have higher profit margins than non-innovative ones. Bismush and Oliveira (1986) state that for a firm or an economy to be successful, it must be able to capture market share in both the local and external markets. The open economy context implies an intensity of competition and the five forces of Porter's model help us to understand the external context of the company. According to him, differentiation is one of the bases that the company can use to face the five competitive forces. Thus, the fact that banks offer competitive services allows them to differentiate their products from those of their competitors and to best meet the needs of their customers. The technological factor taken into account allows them to be more competitive. In line with Schumpeter's theses, they should invest in R&D in order to design products and services with a high degree of novelty.

However, a company today cannot just satisfy its customers and seek to improve its business performance alone. The resources at its disposal can also be used to strategies for a higher level of performance.

Competitive advantage can also be understood as a portfolio of resources and in this case, the advantage is to be sought internally. Resources are tangible (material, financial) and intangible (patents, brands, customer databases, supplier relationships, management systems) assets held by a firm that enable it to design and implement strategies to improve its performance. Hofer and Schendel (1978) classified resources into five categories: financial, human, physical, and technological. To these categories, Grant (1991) added a sixth, reputation.

In the neoclassical view, the ruling elite is responsible for reconciling the interests of each stakeholder and integrating them in a more or less contractual way into decision-making. It manages the relations and interactions between stakeholders in order to serve a unitary and transcendent socio-economic goal: the maximization of profit and the well-being of the company and its institutions. However, for postmodernism, the failure to take into account the unequal power of the different stakeholders when confronted and the consideration that each stakeholder acts efficiently and is ready to make some sacrifices for the good of the companies and institutions is one of the main flaws of this type of management. For the classical theory, stakeholders put the interests of the company before their own interests and do not show opportunism. Indeed, a rational economic agent will tend to maximize his or her utility before that of others. It is in this light that postmodernism advocates the establishment of a system within the firm to limit opportunistic strategies on the part of stakeholders. If the aim is to maximize profit and enable the company to perform well in its context, the structures put in place and the actions of the implementers should reflect the optimal strategies for the company and not for their own benefit.

Moreover, the strategies formulated will have to reconcile diffuse goals. Political science sees the context as a field of power struggles. It sees the exercise of power as a game in which actors with very different interests pursue their interests through specialized structures and power relations. It also focuses on the actors of this strategy, the pressure groups, by examining how they exercise their power. Political models integrate into the analysis the actors' strategies to influence the mobilization of contextual resources to their advantage. Thus, there are power relations underlying the adoption of different strategies. Thus, the impact of optimization since its adoption on performance depends on several organizational, institutional, and market factors, the judicious combination of which makes it possible to achieve higher returns. According to postmodernism, a more participative management should be advocated. This vision is shared by the optimization model, which calls on a 'strategic planning' team to replace the manager, who has only marginal intervention. It is a question of encouraging dialog and conversation between stakeholders by decentralizing the managerial vision which imposed a privileged unitary industrial goal. It is a question of building together a relationship that suits everyone, and of determining together the objectives worth pursuing. This vision thus emphasizes the interdependence of the stakeholders and the company.

According to institutionalism, institutions highlight the importance of routine in cognitive processes. As actors evolve in a constantly changing context, their conduct overtime is influenced by institutional relationships and institutional change. Companies need to develop a corporate or organizational culture based on these systems and beliefs with the aim of mobilizing and involving stakeholders. The ultimate goal is to reduce the divergence of stakeholders and make them converge towards the overall efficiency of the company. In reality, shareholders, managers, employees, and customers have both divergent and convergent objectives. An employee can be a customer and a shareholder at the same time. They are ubiquitous beings (Martinet, 1984, cited by Mullenbach-Servayre, 2007) and the reconciliation of their preferences can only be temporary. When capital is not concentrated, shareholders can use several tools to exercise their power within the company.

The most traditional means is the board of directors and its various committees which are responsible for representing the interests of the shareholders. It has the power to hire, fire, and compensate the CEO and to ratify and control important decisions (Fama and Jensen, 1983). The board of directors is a key element in the understanding of strategy optimization and performance. The law states that a board of directors must consist of at least three and not more than 18 members (Godard, 2001). According to Jensen (1993), beyond 7 to 8 directors, agency problems may be encountered. In the Banking sector, the data reveal that the average size of the management structure is between 7 and 8 decision-makers, 59.7% of which are independent directors. Cheng, Evans, and Nagarajan (2008) state that there is a significant relationship between small optimized structures and better management effectiveness. For their part, Eisenberg, Sundgren, and Wells (1998); Jensen (1993); and Yermack (1996) have shown a significant negative relationship between market value and structural size of optimized mechanisms value-based mechanisms. They note that smaller size can improve the effectiveness of the firm's structural optimization.

The fact that the coefficient of the management structure is positive and significant with Tobin's Q implies that a board size of 7 to 8 decision-makers or less improves stock market performance. According to Hermalin and Weisbach (2001) the composition of the board structure has no influence on financial efficiency. This probably explains why the presence of outside directors does not influence the value of the firm when one might expect just the opposite. Furthermore, Pass (2003) points out that external structures do not always have the necessary industry knowledge to adequately support or guide the optimization of the firm's strategies. Brown and Caylor (2004) found that firms with a higher proportion of structural independence had higher returns on equity, higher profit margins, higher dividends, higher returns, and greater share buybacks. In addition, we note the importance of decision-making autonomy in the Banking sector. Having already noted that more than half of the board of directors is made up of outside directors, it turns out that 82.2% of the members of the compensation structure are independent and that an average of 9 to 10 meetings are held per fiscal year. The significant positive effect of autonomous structures suggests that non-executive decision-makers in open economy banking and financial institutions contribute positively to optimization management. Decision-making connections have a significant positive relationship with ROI, indicating that banking and financial institutions with decision-making forecasts often tend to generate higher returns. This finding is consistent with agency theory. Structural autonomy of payment is positively related to ROI, suggesting that it is composed of only independent decision-makers, and contributes to making recommendations for the optimized mechanisms control of the company. Thus, it mitigates conflicts of interest and improves management efficiency.

However, the overriding oversight role of outside directors may diminish the effect of their presence on the development of viable strategies for improving corporate effectiveness (Roberts et al., 1998; Kapopoulos and Lazaretou, 2007). The presence of audit mechanisms has the advantage of partially relieving them of this role so that, together with the other board members, they can focus on finding the best strategies. External auditing is an important factor that can positively affect corporate performance (Francis and Krishnan, 1999; DeFond, Francis, and Wong, 2000; Hay, Knechel, and Ling, 2008; Wahab, Haron, Lok, and Yahya, 2011; Fooladi and Abdul Shukor, 2012). More importantly, audit quality has a significant impact on management effectiveness (Adeyemi and Fagbemi, 2010; Willenborg, 1999). It appears that the independence of the audit structure has no significant effect on the performance of banks in Canada. However, in line with the literature, external auditing and its quality improves accounting performance indicators. This type of audit ensures that there will be good financial, service, and inventory management by the management. In Canada, 80.8% of companies in the banking and financial sector use the services of an external audit firm. Given the difference between the two types of audits, it is desirable that companies in the banking sector in general use the services of external auditors and integrate this practice into their strategies. Indeed, the second look that they bring allows to evaluate the quality and the efficiency of the internal control implemented by the internal auditor and to identify possible anomalies that could remain.

Finally, in line with the design model, we find no significant relationship between decision duality and firm performance.

Furthermore, the board and its committees will be at the heart of the development of systems for motivating and involving employees. These tasks are recommended by the optimization model. Motivation is the result of a stimulus to engage, while engagement is the result of the employee's own willingness to engage in a given activity. When the company does not provide energy to develop a sense of ownership among its employees, the implementation of the optimization process will be ineffective in achieving its objectives. We note that the "involvement" dimension is the least correlated of all in terms of logistical breakdowns and malfunctions. Indeed, given its character, it is much more difficult to get a person to become sincerely involved in the company's vision. However, there are many ways to help mobilize employees and improve their performance.

The resource and competence approach states that it is the resources and competences that the company possesses that determine the strategy. In our study, the various dimensions of logistical breakdowns are related to the aspect of societal optimization that concerns employees. We also note that dysfunction and logistical breakdowns are distinct features of the strategic management system. Therefore, both must be addressed in order to maximize competitive advantage. For Edvinsson and Malone (1997) human resources are at the heart of the optimization of the company and are the primary determinant of this performance. Our analysis shows that logistical breakdowns and malfunctions contribute to lower productivity and higher production costs. Although at the outset nothing can be said about profitability, all things being equal, taken individually they have a positive effect on the lack of competitive advantage of the Banking sector. A closer look at the determinants of logistical disruptions may help to understand this result. The engagement dimension is composed of performance appraisal, briefings, succession plan, strategic activities committee and incentive plan. All its variables suffer from dysfunction but the variable 'performance appraisal' suffers the most as it is correlated to it at 0.82.

Performance appraisal through performance management is a human resource management activity that allows you to step back and look at the development of both the employee and the manager. Evaluating performance means analyzing the performance of staff against a performance standard, focusing on maintaining job satisfaction while promoting professional and personal development. It is normal then that if this evaluation is not done regularly or is done poorly, the company's productivity suffers. Furthermore, the judicious use of incentives can increase production and reduce control costs. One way of retaining employees and preventing them from being tempted to accept another company's offer is to implement an incentive plan. This plan consists of giving a key employee or an executive the option to buy shares in the employer for a specific period of time and at a pre-determined price. This procedure allows the employee to share in the success of the company. Succession planning, which is part of personnel and career management, is at the heart of human resource optimization. It aims to ensure that the necessary talent and skills will be available when needed and that essential knowledge and skills will be retained after the departure of employees in key positions. Succession planning is an area that is often overlooked by companies that are preoccupied with other immediate issues. Ensuring succession is like making a good long-term investment.

The dynamic capabilities model on this point deepens the resources and skills model by eliciting the improvement of internal capabilities in the strategic plan. However, it suffers from the same limitations as its counterpart in focusing only on internal resources to explain the link between strategy optimization and performance. The strategic activities committee is responsible for studying all questions relating to the company's strategies (investment, alliance, partnership, etc.). It generally assists the board of directors in its mission and helps to improve the operational management of the company. In order to strengthen the sense of belonging and increase the capacity for mobilization, it is necessary to give the employee a sufficiently clear vision of the company so that he or she understands how his or her efforts contribute to the company's success (Capraro and Baglin, 2002). In this sense, information meetings contribute to the dissemination of information on the life of the company, its markets, its difficulties, and its projects.

Moreover, information given quickly and without distortion creates a climate of trust favorable to dialog. This point supports the importance of information management in achieving higher levels of performance.

The same reason can be used to explain why logistical breakdowns have a positive effect on costs with regard to the determinants of its 'mechanism' variable. This is composed of variables such as work mechanism or skills which are correlated with dysfunction at 0.83 and 0.75, respectively. Within this dimension, the review committee is responsible for collecting and monitoring the various complaints. In addition, it can assist the Board of Directors in some of its tasks. The training committee's aim is to ensure the ongoing training of staff. It identifies the company's needs in terms of skills and fills them internally. It allows the company to gain in efficiency and helps to retain its employees. The measurement items of the 'mechanism' variable in general contribute to the improvement of operational management through effective personnel, career, and performance management.

Finally, focus groups are a key tool in the optimization process. They involve a small group of people, which can be stakeholders, answering questions about the business in short sessions. They allow the company to collect valuable information that will be incorporated into the strategic plan. Depending on the nature of the group, the information obtained will be of a different nature. However, to substantially impact the lack of competitive advantage, it is necessary to play on several dimensions and to identify logistical breaks that reinforce and complement each other over time. Canadian banks and financial institutions should therefore improve the human resources component so that the resulting management practices improve their competitiveness in the context of open markets. In the context of an open economy, banks must not only be competitive in the provision of goods and services but they must also renew their skills. For Cao (2002) the dynamics of capabilities are essential for the firm to establish a long-term competitive advantage. It is therefore essential to focus on the ability to change, evolve, and learn, i.e., on achieving dynamic capabilities (Johnson et al., 2005).

Several authors have studied the capital structure and its effects on the performance of companies. Capital structure briefly refers to the way in which the firm finances its activities using different sources of finance. One of the powerful and widely used strategic tools to achieve this objective is debt. Debt is part of the various elements that resource theory uses to explain the performance of firms. Both theoretically and empirically, the debate about the effect of debt on firm performance does not allow for clear-cut results. The debt ratio indicates the percentage of a firm's assets that have been financed by creditors. Depending on the financial health of the company and its objectives, it will affect its debt level. In the long term, this ratio is essential to assess the ability of banks to meet obligations. The level of debt and equity is fundamental for shareholders and debt holders. For shareholders, the level of debt is an indication of the risk of non-payment of dividends, while for lenders, the level of debt can be used to assess the facilities for obtaining loans. A lower ratio suggests a better financial position based on the company's assets. In the context of adopting optimization, the debt ratio has a significant negative effect on ROA, indicating a high level of debt. However, higher leverage implies higher optimization levers.

In terms of the leverage position, we find that institutions have a debt level of about 48.65% of their total assets. High levels of debt reduce the power of opportunistic managers by reducing the cash flow available to spend on non-discretionary assets, such as benefits or negative net present value projects (Frank and Goyal, 2009; Harris and Raviv, 1991; Jensen, 1986).

Our results show that institutions with more structured and optimized leveraged capital reduce opportunistic management of their productivity. Indeed, higher levels of debt imply large disbursements of cash flows that could otherwise be used opportunistically by governance. Debt thus contributes to some extent to better financial management as revenues will be injected into priority and profitable projects. Moreover, creditors and institutions that establish more restrictive clauses in contracts and demand more information on productivity prevent governance from opportunistically manipulating performance statements. This is also confirmed by the improvement of the ROS in the second period. Strategy optimizations that increase their optimization leverage, experience a better change in the ROS. However, the negative relationship between optimized mechanisms control efficiency and the financial LR suggests that the increase in debt may lead to more borrowing costs from banking and financial institutions, which absorbs the firm's profits. On the other hand, CEG has a significant and negative relationship due to the impact that CEG has on assets and increases the denominator when calculating ROA. Their growth on the other hand suggests that strategy optimization and large investments have a better variation in the ROS. Half of the banks surveyed have a LQ ratio of 0.873 or less, indicating that in the short term they may not be able to meet their commitments. This finding implies that suppliers will be reluctant to deal with these banks if they do not solve their solvency problem. However, on average, it is 1.209. Having a LQ ratio above 1 is equivalent to having working capital. Our results show that the LQ ratio is positively correlated with the different forms of performance studied. The risk of firm failure can have various effects on managerial behavior. Managerial opportunism can be explained either by the search for utility satisfaction or by the desire to show satisfactory and stable accounting results. However, when the risk of bankruptcy in banks is low, the need to opportunistically manipulate earnings in either direction is lessened.

The authors of the agency theory and those of value-based management (Charreaux, 1997; Shleifer and Vishny, 1997) assume that the ownership structure can be an effective means of controlling the management of directors, because it makes it possible to bring together, when certain conditions are present (concentration of capital and nature of the shareholders), the bases of an efficient control system. For a long time, the shareholder was seen only as the bearer of financing and a claimant of dividends. However, the political school sees him as a major influence in the elaboration of the strategy (Laroche and Nioche, 1998). In a firm with a widely dispersed shareholder base, a single shareholder has no incentive to commit resources to exercise control over management. Under these conditions, each shareholder has an incentive to become a free rider and the managers can give free rein to their opportunism in the management of the firm. On the other hand, a shareholder with a significant share of the firm's capital has a strong incentive to invest in controlling the management of the firm, as he will appropriate a significant share of the additional profits thus made (quote). Moreover, he can control more effectively than minority shareholders, as he holds enough voting rights to influence the management or possibly convince other shareholders to support him in case of opposition with the management of the firm.

It appears that on average 27.13% of shares are in the hands of the majority shareholder. The results show that the concentration of ownership in the hands of a majority shareholder is negatively related to performance management. Thus, the higher the ownership concentration, the less room there is for opportunistic behavior through optimization management. Therefore, we can observe that concentrated ownership structures effectively solve agency problems by directly supervising managers' performance. This result provides insight into the effect of the board of directors and its composition on firm efficiency. The interests of the majority shareholder drive him to monitor the actions of the manager himself. Banks and institutions that will base their long-term strategic calculation on an increasingly concentrated structure will experience superior performance by reducing the opportunistic management of their results. Also, we find that the importance of foreign investors reduces the ability of companies to compete, as there is variability in competitive optimization. In other words, the active control of (foreign) shareholders encourages managers to reduce structuring mechanisms and, consequently, to make the institution more profitable. When the control of the institution is a function of the shareholding, a similar effect on competitive advantage is achieved, indicating that restructuring in the context of optimization leads to losses in competitiveness. The aim is to influence management decisions to maximize the return on their investment and thereby improve project productivity La Pointe (2000).

On the other hand, in the Banking sector, we find that on average 15.2% of the shares are held by managers, 25.2% by institutional shareholders,

0.5% by the government and 28.1% by foreign ownership, which represents the total shares owned by foreign individuals and institutional investors. Hyeon Cho (1998) showed that there was no relationship between capital structure and management efficiency. Rogers, Dami, Ribeiro, and Sousa (2007) report the lack of influence of capital composition on the effectiveness of optimized mechanisms control. Gurbuz and Aybars' (2010) study reveals that foreign capital improves firm performance, while (Bai, Liu, Lu, Song, and Zhang, 2004) indicate that issuing shares to foreign investors has a positive and significant impact on market practices. Jensen and Meckling (1976) argue that capital diversity has different effects on firm efficiency. Using Pearson's correlation for all variables of the Canadian bank and financial institutions, the result is that the number of shares held by management capital or managerial ownership is positively correlated with ROA and ROI only, indicating that the more integrative the structure, the more effective the optimized mechanisms control. This result is consistent with agency theory. Own capitalization cannot reduce agency costs but improves management efficiency. The share of institutional funding has no significant relationship with either ROA or ROI, but a significant positive relationship with Tobin's Q ratio. There is a significant positive correlation between foreign investment or foreign ownership (FORGN) and outcomes measured by ROA and ROI.

There is a significant positive relationship between managerial ownership and Tobin's Q indicating that equity capital increases managerial performance. In addition, taking into account the decision capital structure as an endogenous variable, indicates that it has a significant positive relationship with Tobin's Q ratio. Although this result does not corroborate previous studies that failed to find a significant relationship between capital structure and management performance using the GMM model (Pham, Suchard, Zein, and 2011; Schultz et al., 2010), this result is consistent with that of Jensen and Meckling (1976) who pointed out that the agency conflict between decision and management structures could be mitigated through the convergence of interests between managers and shareholders.

These authors note that managerial ownership can reduce the tendency of managers to take advantage of their position, expropriate shareholder wealth and engage in decisions that do not maximize the value of the firm as they themselves will suffer the consequences. The convergence of interests then suggests that the value of the firm increases if the percentage of capital held by managers increases. This confirms the work of Randall et al. (1988); and McConnell and Servaes (1990), who found a significant positive relationship between these two parameters. The participation in the company's capital is a source of motivation for the employees, as we pointed out above, so that their interests and those of the shareholders converge in the same direction. Again, one might expect the disciplinary role of the board to be less important in firms with high managerial ownership.

The institutional share has a significant positive relationship with Tobin's Q ratio, suggesting that institutional shareholders provide the monitoring capacity necessary to increase Tobin's Q ratio. These agents also have privileged access to information due to their activity and the numerous investments they make, which implies a better knowledge of the performance of companies in the sector, abundant information on the environment and consequently a better appreciation of the manager's optimization (Alexandre and Paquerot, 2000). This result is similar to studies by Shleifer and Vishny (1997); and Henry (2010) while previous research is inconsistent with Tobin's Q ratio (Wei et al., 2005; Xu and Wang, 1999). Government funding has a significant positive effect on Tobin's Q ratio. The percentage of government action in banks suggests two findings. The first is that the Banking sector is largely privatized. The low level of equity participation shows that the ultimate goal of banks and financial institutions is not the maximization of collective utility but that of stakeholders. The second finding concerns the regulatory control exercised by the presence of the political authority within them.

In the same vein, we studied the link between ownership structure and the optimization of financial statements. It appears that on average 58.71% is in the hands of the majority shareholder and the managers. When the shares are in the hands of the majority shareholder and the managers, a non-linear relationship between ownership concentration and overestimation of financial reports is found. As the percentage of closed shares increases, performance management decreases but only up to a certain threshold of ownership concentration. Beyond this level of ownership concentration in the hands of the majority shareholder and managers, opportunistic manipulation of financial statements increases. It seems that agency costs and moral hazard problems are minimized with more concentrated ownership structures.

However, whenever the majority shareholder and the managers hold more stock than is necessary for efficient strategy optimization, entrenchment, and expropriation problems arise. Under this scenario, managers are more inclined to manage revenues, thereby appropriating some of the wealth of minority shareholders. Such a threshold or critical point of ownership concentration can be determined essentially by optimization. Our results indicate that when the majority shareholder and the managers hold no more than 65.02% of the outstanding shares, performance management is minimized, ceteris paribus. When, on the contrary, they hold more than this threshold, problems of entrenchment and expropriation appear and are eventually materialized in the opportunistic manipulation of financial statements to the detriment of minority shareholders. Strategic plans should therefore determine the critical threshold before allocating shares to management and the majority shareholder. Poor consideration of valuation could lead to lower returns. This and other aspects of the ownership structure are essential to explain the interaction between strategy optimization and performance.

The establishment of restrictive clauses in contracts or the presence of a majority shareholder are effective ways to combat discretionary management and protect investors' rights. Otherwise, the institutional system could control interests, reduce opportunistic approaches to productivity optimization, and improve the quality of accounting information (Ball et al., 2003). According to Welfare Economics, when the context impacts badly, the state must simply act to prevent spillovers because there is no other recourse. Institutionalist analyzes claim to offer a more refined explanation of the presence of the socio-political within the economic.

In Porter's model, the state is an element that will determine competitive advantage. Although his model only focuses on the external environment of the company, we have used it and found an interaction with the internal environment of the company. We find that at this point, we need models that focus on everything external to the company and resource-based models to formulate a more relevant strategy.

Although some theoretical models have noted the importance of the external environment in its impact on project productivity, most have focused only on economic variables. Our study broadens the field of reflection by introducing new variables.

The institutional system can influence dividend levels. Anglin et al. (2013) state that planners manipulate earnings upwards through actual activities to mitigate the shortfall (of pre-managed earnings over previous years' dividends) when these are lower than the dividends paid for the previous year, suggesting that dividend levels are an important benchmark for earnings. They provide evidence of a strong relationship between changes in pre-generated earnings and actual productivity management by suggesting that dividend optimization impacts actual productivity management.

We confirm this result as we find a positive and statistically significant relationship between dividend optimization and productivity management. However, depending on the institutions in place, this relationship will be more or less attenuated. Indeed, the legal origin of the country could also determine how governance uses its discretionary capacity to manage outcomes (Ball et al., 2003). Although Canada is a regulated economy, it appears that Ontario and British Columbia have a relatively better institutional environment than Quebec, Alberta, and Manitoba. As a result, banks there have less productivity management in their dividend policy and a more effective strategy optimization structure.

For Canada, we can argue that the better the legal and regulatory systems, the less room there is for discretionary decisions. The importance of contextual regulation to protect investors exposes the imperfections of the market and the economic process, but operates in the absence of other incentives on management behavior. In addition to the legal and institutional system, the economic environment must be taken into account when considering the external factors that explain bank performance. It appears that banks that adopt management practices in a performing economic environment (GDP growth) are likely to increase their performance post-strategy optimization. On the contrary, when the economic environment is bad or subject to a crisis such as the one the world experienced in 2008 with the financial crisis, any change in practice may not bear its positive fruits on the company's performance. Particularly concerning the efficiency of the banking system, we observe however that overly sophisticated banking or regulatory systems push planners to manage results to meet their requirements.

Regarding the type of optimization used, we find that formal optimizations show better results than non-formal optimizations. Although most schools and schools of thought recognize this concept as a means of reorienting the evolution of a company, not all agree on the importance to be attached to it, on the means of conceiving it and then implementing it, nor on the usefulness of optimization in a decision-making process. The explicit formulation of strategies is a necessary element for the success of companies. Explicit formulation must take place in a context of uncertainty, because this uncertainty puts the future of the company at the mercy of the hazards of the future. This explicit formulation makes it possible to take account of these hazards and would prevent the firm from deviating from its objectives in an unfortunate manner. It argues in favor of a rigorous optimization of the strategy, but recognizes the place that must be given to experience on the one hand and to emergent strategies on the other, because they are part of the explicit optimization, just as unforeseen expenses are included in a pre-established budget.

However, a strategy that constantly adapts puts the company at a disadvantage. The adaptation time, the waiting time during the realization of the experiments is a hindrance for the company in a market where competitors have already optimized their advance. Optimization can only objectively be carried out in the short term, due to budgetary constraints, the inability to predict the future, and the lack of objectivity of optimizations, which are generally biased by the vision and desire of their designers and the hierarchy. With environmental constraints, the needs of the company are constantly changing. They cannot be defined after an assessment of the company's strengths and weaknesses, but rather gradually. Optimization, based on the needs of the company, must therefore respond to these needs, and must take account of their gradual evolution.

The consideration of the expectations of internal and external stakeholders in the management of the company by the managers has put an end to a performance-focused solely on profitability and has broadened the change of considerations to the level of the social responsibility of the company. The development of the best corporate strategies is therefore based on taking into account the different realities that the company faces in its environment on the one hand and the power relations that exist within it on the other. The evolution of optimization from a purely economic aspect to a multidimensional one is reflected in the advent of the concept of corporate social responsibility (CSR), which is "a concept that designates the voluntary integration by the company of social, societal, environmental, and governance concerns into its strategy, management, and relations with its stakeholders. CSR is a way for the company to strengthen its profits in the short, medium, and long term" (MEDEF, 2012). Performance tends to be approached from a more global perspective than simply assessing profitability for the company or for the shareholder.

Performance now takes into account the company's social responsibility towards all its stakeholders. Corporate responsibility is no longer limited to shareholders, but includes other stakeholders (associations, NGOs, trade unions, customers, suppliers, etc.). These new stakeholders demand to be heard and this listening becomes a vital target for the performance and sustainability of companies. Performance in its entirety mobilizes several sub-types of performance (commercial, economic, financial, organizational, etc.), which gives it a heterogeneous status but which must nevertheless be taken into account when configuring the best management practices. This observation limits all market- and product-oriented models such as Porter's model, the positioning model, the BCG model, McKinsey's model, and the ADL model, because the elaboration of optimization will only be done with reference to considerations that do not integrate the multidimensional character of optimization.

From an economic point of view, all economic agents are rational and concerned with maximizing their utility or satisfaction in their relationship with the company. The multiplicity of utility functions and their interactions make it difficult for interests to converge in the same direction. Moreover, several theories have demonstrated the limited rationality of agents on the one hand and the power they have to take maximum advantage of the benefits of the company on the other. In this sense, the theory of regulation admits that the actors will ensure the regulation of the context by creating rules to which all will agree. These rules can be made at the external level (e.g., imposed quality standards) and at the internal level. The development of strategy is the development of a "hardcore" of rules, which may be economic, financial, or organizational. In this perspective, they are guides for action and decisions in order to ensure that the latter conform to the general orientations of the company.

Developing a strategy and putting it into action is a long process. The management team needs to have a good understanding of the company's potential and its environment before committing to it. The LCAG model allows goals to be set by the owners after an internal and external review of the company has been done. Also, the understanding of the changing environment becomes a guarantee for the survival of the company. Several analytical tools can be used by managers to help them decipher their environment and use the information gathered in the formulation of strategies: the BCG industry matrix, the McKinsey model, the ADL model, and Porter's five competitive forces dynamics.

In a constantly changing competitive environment, companies will have to adapt their strategies to keep up with the times and remain successful. In order to do this, they need to monitor their performance. According to the optimization model, giving a clear direction to the strategy means being able to evaluate it ex-post through value-based management. In practice, steering consists of providing the company's management with a limited number of varied financial and non-financial, short-term, and long-term indicators, often grouped together in the form of a dashboard, in order to help managers in their strategic decision-making. The task of steering performance is partly entrusted to value-based management. In order for value-based management to be effective, it is first necessary to discern and prioritize the areas of performance sought. This is why a value-based management system must be built specifically for each company. The basic dynamics of value-based management include two main phases: optimization and analysis of results (Giraud et al., 2004). The optimization phase serves to prepare for action. The role of optimization is to determine targeted objectives (the expected level of performance) and to plan the means that the company will use to achieve these objectives. The results analysis phase makes it possible to assess the degree of success of the actions undertaken and to orient future actions accordingly. Bouquin (1997) defines value-based management as the set of systems on which managers rely to control the decision-action-results process.

In the context of an open economy, the role of value-based management must be to identify new market needs, to conquer new strategic segments, but also to check whether the objectives of the strategy optimization implemented have been achieved and to encourage the implementation of new strategy optimization processes. In addition, value-based management must ensure the steering of global performance, which can allow the maintenance and control of the company's resources and key skills in the sense of the resource-based view (RBV), and oriented towards the detection and control of the company's dynamic capacities. The strategy optimization allows companies to control their objectives and how to achieve them, and is only important if the company makes the effort to implement it. But the only way they can check whether they have achieved their goals is through monitoring. There is indeed a correlation between the control systems implemented, the type of strategy developed and performance (Simons, 1987). Strategy and control are therefore linked in that the nature of control systems and processes cannot be separated from strategic considerations.

There are two types of dashboards. The balanced scorecard or BSC and the intellectual capital scorecards. Initiated by Kaplan and Norton (1992, 1998, 2001), the BSC allows for the evaluation of the overall performance of the company, through which elements related to financial and non-financial performance are taken into account. Kaplan and Norton believe that in today's competitive environment, financial indicators are no longer appropriate to guide the company's strategy. They advise managers to focus on improving customer satisfaction, quality, employee skills and motivation, and production cycle times. The BSC is a tool for top management. It translates the mission and strategy into objectives and measures along four axes: financial, customer performance, internal processes, and organizational learning. A company will only choose one top goal to measure its long-term success. The second type of dashboard is the Skandia navigator and the Sveiby monitor.

Their distinctive feature is that they analyze the intangible capital of the company and are designed on the basis of a resource approach to strategy (Roos and Roos, 1997).

The navigator takes into account the aspect of societal optimization that relates to employees and customers (human resources). It breaks down intangible capital into two categories of banks: human capital (knowledge, know-how, attitude, behavior, and agility) and structural capital (organization and partner relations). It is divided into five orientations: finance, customers, processes, renewal, and development, and human resources. Edvinsson and Malone (1997) see human capabilities as the basis for optimizing each objective for each axis. Human resources are positioned at the center of the value creation system and feed into the other four axes (Cappelletti, 2006). Human resources are therefore at the heart of the optimization of the company and constitute the primary determinant of this performance. The human resources axis "corresponds to the skills of employees and the commitment made by the company to maintain their level" (Germain and Trebucq, 2003).

The Monitor (Sveiby, 1997), like the Navigator, proposes that people are the basis of the company's profit. According to the author, it is the people, considered as intangible assets, that enable the generation of income. The Monitor develops three indicators (competences, internal, and external) of the asset that allow the competence of the personnel to be evaluated according to three criteria (growth, efficiency, and stability). This tool shows the importance of competencies within a performance management system, but does not link competency management to the strategic vision of the company. According to the competence literature, Grant (1991) determines that the most important thing in a strategic approach is the ability to integrate individual competences for organizational performance. In this sense, the navigator is similar to the idea proposed by Grant. Indeed, this performance management tool attempts to enhance the value of human capital through indicators of involvement and development of individual competencies within a company.

Against the backdrop of the ideological debate around the issue of strategy optimization, using value-based management to rethink the company's strategy in relation to the desired performance is in line with the argument that optimization is an adaptive process, which evolves progressively and emerges from the different orientations that the company follows in relation to its context. This challenges the models that claim that optimization is a formal process. It is this emergent process that will lead the company to success. The company has no need for explicit optimization which would be a brake on its expansion. It is impossible to measure the unpredictability of an environment in order to adapt an explicit strategy to it. Moreover, a company cannot be sure of its strengths and weaknesses until it has tested them experimentally. A rational approach cannot lead to business success. He therefore advocates those strategies should not be explicit but rather emerge as they are developed, based on the conditions, experiences, and mistakes experienced by the company. They should be the result of a learning process and non-predetermined actions. Thus, the manager should focus more on the evolution and optimization of the company's capabilities and efficiency, rather than on optimization, which is likely to be limiting. The real role of performance, he says, is to develop and articulate the consequences of a pre-existing strategy: optimization does not create strategy. This view is best understood in a context that is subject to constant change and the interaction of both internal and external forces that can challenge its stability overnight. Optimization plans should therefore not be rigid but should be responsive to their environment so as to implement strategies that will meet the desired level of performance.

Furthermore, in line with resource theory, the company must think about developing organizational or collective competences in order to meet the expectations of its context more effectively. This configuration of optimization is opposed to that of market-product models which find that strategy is developed in response to industry performance, in relation to competitors and the place the company wants to obtain in the sector. This new approach to strategy for achieving overall performance will not happen without the establishment of an integrative structure. Structural integration reflects the extent to which a company ensures the compatibility of its decisions in one field with those taken in others and the extent to which participatory decisions characterize decisions taken at the highest level of the hierarchy (Miller, 1988; Miller and Droge, 1986). Moreover, the heterogeneity of internal and external optimization at the individual level, which is not accounted for by any of our models that address the relationship between strategy optimization and performance of Canadian banks and institutions, implies that banks should not design their strategies in the same way unless they belong to the same category. In order to answer our question, we studied the performance of banks and financial institutions based on certain indicators. The positioning school is the development of strategy as an analytical process. By analyzing the industry, the sector and thus the competition, the strategy to be adopted will be determined.

First, it appears that some types of banks contribute more to the Canadian economy than others. According to the scorecard and the navigators, to improve their performance, companies need to assess their financial and non-financial indicators, and their human capital, respectively. In a first step, we studied turnover and value-added, taking as inputs purchases of inputs, transport, wages, and external services. The study of turnover and value-added places bancassurance and general banks at the top of the league in terms of performance. We were also interested in their efficiency, whether technical, pure or of scale. Technical efficiency reflects the level of quality of resource management and organizational capacity of the production unit (pure technical efficiency) where it reflects the optimal scale or not. Focusing instead on the efficiency score, corporate banks, cooperative banks, finance banks, network banks, investment banks, savings banks, corporate banks, and deposit banks are the most supportive of the open market with very high optimal performance scores of around 80%. They then operate at optimal scales, i.e., by proportionally increasing the level of all their factors, they achieve the best possible and expected situation. They therefore have increasing marginal effects. On the other hand, we find the banks with the lowest efficiency scores. The direct bank, for example, has never been at least once efficient and has the lowest minimum scores (0.210 in 2010 and 0.221 in 2011).

The type of banks is based on their legal status, the type of ownership or the type of activities they carry out. The successful banks mentioned above are private banks or banks with a strong shareholder presence. They are subject to a year-end performance requirement. Their main goal is to make a profit and satisfy shareholders. In order to minimize the risks of opportunism on the part of governance and to increase their competitiveness in the competitive market, they will tend to opt for rigorous optimization upstream. However, depending on the changing environment and their objectives, they will adopt emergent strategies. Concerning the least performing banks such as direct banking, their performance can be explained by the fact that these banks are generally public or parastatal. They have a very important social role. The concept of social responsibility is much stronger in this type of enterprise. While Friedmann (1970) defines CSR as making a profit, Carroll (1979) defines the notion of societal responsibility as what society expects of organizations in open economic, legal, ethical, and discretionary matters. He was the first to formulate the idea that social and economic objectives were not incompatible, since they would participate together in a common objective: 'global social responsibility.' For Germain and Trebucq (2004), the overall performance of companies is formed 'by bringing together financial optimization, social optimization and societal optimization.' As far as direct banks are concerned, they focus more on supporting the state in its regalian duties. They therefore do not prioritize the performance of their management systems and, in a way, are lax in dealing with certain internal problems such as corruption.

By analyzing the behavior of banks in relation to optimization and optimization, multiple factor analysis allows us to split our sample of banks into two sets. Atypical categories of banks that are factorially isolated, with distinct characteristics and behavior; and homogeneous categories with homogeneous behavior and close to each other. This second category includes the maximum number of banks indicating that the overall situation in the Canadian economy has not changed over the years of the study interval.

According to institutionalists, there is a set of values, norms, laws, or organizational models outside organizations that will influence their structures and management methods (Meyer and Rowan, 1977). The institutional context is therefore composed of cognitive, political, and cultural dimensions. It constructs structures, rules, and practices that give meaning and guide the behavior of agents in ways that are both enabling and constraining (Scott, 2001). Neo-institutional theory (NIT) explains that the firm's environment is characterized by social and cultural rules and requirements to which it must conform in order to gain legitimacy from its environment (DiMaggio and Powell, 1991; Meyer and Rowan, 1977). Stakeholders exert pressure on companies to follow institutional rules (Boussoura and Ben Mlouka, 2008). These coercive, normative, and mimetic constraints are likely to lead organizations towards increasingly similar practices (DiMaggio and Powell, 1983). With regard to the grouping of atypical banks, these are the ones that have contributed most to the formation of the factorial plan. They include universal banks, bancassurance, general banks, online banks, investment banks and merchant banks. They are among the banks that perform well according to turnover, value-added or efficiency scores. This result confirms the findings of Louizi (2011) in the banking sector. High-performing banks are optimized differently from low-performing banks.

In addition, the survey of banking institutions and the operation of their FSDs shows that operating expenses can be structured into several components: tradable inputs, non-tradable inputs, personnel expenses, and taxes. Tradable inputs represent the main expense item for firms in Canada (58.6%), i.e., about three-fifths of the total operating expenses of firms operating in Canada. This is followed by non-tradable inputs (32.25%), transport expenses (4.71%), external services (8.43%), other expenses (19.11%), personnel expenses (7.12%) and expenses related to the level of taxation (2.03%). These elements that appear in Porter's value chain have a positive and significant impact on the productivity of banks. Thinking like Porter, it is expected that the level of performance and the desired positioning will determine how banks manage their services and stocks. The first finding of our study is that banks do not behave in the same way when allocating expenses. Bancassurance, for example, is the category for which purchasing expenses account for more than half of total operating expenses and which makes the least use of the other cost components of strategy optimization. Investment banks spend the largest share of their expenses (28.14%) on staff. This finding may suggest that the employees of this bank are well supervised and that the managers consider the employees as essential factors of production. It would seem that this type of bank favors the resource approach to strategy (Roos and Roos, 1997) in order to remain competitive.

Around 62.06% of the variability in output at Canadian banks is explained by a combination of performance, productivity, technological change, and other operating expenses. Performance has a positive impact on output. Both variables fluctuate in the same direction. Indeed, increasing optimization by 1% leads to an increase in production of 0.054%. In other words, if planners can influence the organizational and managerial capabilities of firms to increase their performance by 1%, they would increase, for the same quantities of inputs, the quantities produced by 0.054%. Similarly, this positive relationship is also observed with regard to efficiency and production. The Breusch Pagan test confirms that the null hypothesis of non-significance of random effects (RE) is rejected (p_value less than 5%). Thus, each category of bank has a significant random effect over the period which, all other things being equal, distinguishes it from the others.

It also appears that the overall performance of the banking system can influence the optimization mechanisms. Discretionary management is indeed higher when the banking system is efficient. According to this result, governance faces moral hazard problems in exaggerating performance in order to meet the requirements of, for example, more sophisticated banking instruments. Therefore, it seems that, at least with respect to the efficiency of the banking system, discretionary management is a reaction to more sophisticated financial instruments and needs. The variables used to measure this impact are i) the size of the assets of the central bank and ii) other financial institutions relative to national GDP, and iii) the size of private credit by deposit banks relative to GDP. To appear competitive, some planners will act on their financial statements.

Finally, the study of the internal stakeholders of some banking companies reveals that the lack of performance is due to logistical breakdowns and dysfunction among employees. The negative impact on performance can be revealed as a result of value-based management using the Skandia browser for example. According to Johnson et al. (2005), the development of strategic capabilities enables long-term competitive advantage. Capabilities are the managerial skills needed to coordinate, harmonize, and strategically deploy available value-creating resources. According to the dynamic capability model, companies have an interest in renewing their competencies in order to be in line with a context that is evolving more and more rapidly, where technologies lead to increasingly frequent innovations, accelerating the obsolescence of offers and facilitating substitutions. The dynamics of capabilities are then essential for the firm to establish a long-term competitive advantage (Cao, 2012). Building and developing competitive advantage requires the firm to exploit its current specific internal and external capabilities but also develop new ones (Wernerfelt, 1984). Teece et al. (1997) have established those dynamic capabilities allow the firm to integrate, build, and reconfigure their resources and competencies and therefore maintain performance in a changing context.

The overall performance of the industry or of the most competitive firms, as well as failures in the management system, will cause the less successful firms to rethink their strategies and adopt better ones. NIT is relevant to explain the organizational changes that take place when a new system is introduced. And this restructuring can take place through benchmarking. Benchmarking is a process of systematically seeking out best practices and innovations in a given activity with the aim of adopting, adapting, and applying them for greater business performance and to ensure superiority over the competition (Hermel and Achard, 2007). The results achieved are then evaluated by comparing them to desired results or benchmarks (Bouquin, 2004). Benchmarking is a lever for optimizing innovation in the company, because it is not only a method of analysis that enables benchmarking by taking inspiration from the best references, but also a state of mind and a management style. In order to improve their results, banks can choose between several types of benchmarking. The first is internal benchmarking, where the comparison of processes, products or services is done within the same organization. The expected improvement in performance is about 10%. When it is functional, the comparison is between non-competitive organizations in the same industry to identify innovative techniques. The expected improvement in performance can exceed 35%. Generic Benchmarking is the comparison between organizations in different sectors on processes or working methods. It is the most powerful and beneficial but is more difficult to implement as it compares with practices that are not specifically adapted to the sector to which one belongs. This benchmarking does not concern the competition and is therefore a source of innovative ideas (Costa, 2008). The improvement in performance can also be as much as 35% or more. Competitive benchmarking allows a specific comparison with competitors on the product, method, process. It is therefore necessary to compare with the best of the competitors on the market and for this type of comparison between two directly competing organizations, it is often more difficult to obtain information. The expected improvement in performance is about 20%.

A number of variables could not be included in our analysis because of their unclassifiable nature. These variables are mostly control variables or additional variables. We find the variable MVBE, BFRISK, SSIZE (t–1), SIND (t–1), INSIND (t–1), ROA (t–1), ROI (t–1), Tobin's LogQ (t–1). Concerning the lagged profitability variables, they have a positive and significant effect on present profitability. So how can we explain that the present profitability in the context of the adoption of optimization negatively influences the competitive advantage of the firm. In any case, it is important to find a way to fully integrate them into a model that addresses the relationship between strategy optimization and performance so that it is comprehensive.

KEYWORDS

- **corporate social responsibility**
- **neo-institutional theory**
- **profitability variables**
- **return on assets**
- **return on investment**
- **return on sales**

CHAPTER 16

Key Determinants for Operational Interactions Between Strategy Optimization and Business Performance

In the previous chapters, we discussed the impact of strategy optimization on project productivity and the performance factors that play a role in the development of corporate strategy. This part will now focus on the interaction between our different performance variables and strategy optimization. The strategy optimization is a means that the company uses to meet its long-term objectives. However, in the process, using value-based management, it makes adjustments to improve its strategy in the light of its capacities and the evolution of its environment.

The literature very often presents the manager as an efficient agent who works for the interests of the owner. However, we have seen that there can be agency problems between shareholders and managers within the firm. Since the effectiveness of the manager in managing the company has been questioned, Simon tells us that his choices and the decision-making process he adopts are strongly dependent on the context in which this individual evolves. To this fact, he adds the emotional context of the agent. He shows that emotion can also guide an agent's choice. Indeed, emotion can draw an agent's attention to one contextual aspect rather than another, and thus alter the agent's perception of it. In order to channel the executive, the shareholder can use several internal value-based mechanisms, which are: the board of directors; internal and external audit; ownership structure and capital concentration; and debt. These important elements in strategy development can also be driven by a given level of performance. Also, the legal, regulatory, and institutional systems will need to be taken into account.

Value-Based Management in an Open Economy: Optimizing Strategy to Improve Business and Performance. Prof. Post-Dr Walter Amedzro St-Hilaire (Author)

The traditional control system par excellence is the board of directors. We found that with optimal size and an autonomous decision-making and payment structure, profitability was increased by limiting the opportunistic management of the manager. Similarly, verification mechanisms and the quality of external auditing had a positive effect on performance indicators. We also found that other tools could limit opportunistic management of results. Also at the internal level, an integrative structure that allows the manager to benefit from the company's profit encourages him or her to manage the company's accounts well. This has the effect of improving the company's performance. This integrative structure consists of managerial ownership and profit-sharing. Also, the concentration of capital in the hands of a major shareholder could allow him to mobilize the necessary control to ensure the profitability of his assets.

Furthermore, the debt policy, by increasing the levers of optimization, reduced the management of a portfolio of unprofitable projects. At the external level, the legal, regulatory, and institutional systems have proven to be effective in reducing opportunistic account management. The mechanisms used by the shareholder as well as the performance of the political, legal, and regulatory system lead to an improvement in the optimization of the overall performance of the company. The aggregate improvement of all the performances of the banking sector will have the effect of improving the efficiency of the banking sector. Our results show that an over-efficient banking sector had the effect of increasing the opportunistic management of results. An increase in discretionary management has an impact on the financial and accounting indicators of the firm. A company that is not known to be performing well in the industry will find it difficult to attract investors. Similarly, an increase in opportunistic management implies that the manager takes more account of his or her preferences than those of other stakeholders.

At the employee level, for example, this can lead to logistical disruptions and malfunctions that will have repercussions on the company's overall performance, as the employees who represent the company's human capital are a key factor in improving its institutional and financial performance. A decline in the overall performance of the company again sends a bad signal to the industry. These declines due to poor workforce management can lead to a decline in customer satisfaction and therefore business performance. A decline in business performance through creditworthiness can cause three effects. The first is that suppliers will fear not being paid and may limit the supply of products and services to the banks. The second effect could be a decrease in customer satisfaction due to the decrease in competitive services and products. We have seen in the course of the study the importance of providing such technology-rich products. The direct consequence is the decline in commercial profitability. The second effect is lower profits and therefore lower shareholder satisfaction. The third effect is that in a business context of voluntary or involuntary indebtedness, opportunistic management is reduced because it is important for the manager at that moment to make the company profitable. However, it can be seen that the lower the risk of bankruptcy, the lower the level of opportunistic management. When the accounting results are not good, managers tend to manipulate the results to attract investors.

The presence of well-performing firms may lead to shaping the institutional landscape so that harmonized accounting standards are adopted within the sector. Successful firms will tend to maintain their position and less successful firms to improve their position. As noted above, an integrative structure will result in an increase in the overall performance of the firm, which in turn will result in the strengthening of the integrative structure following a diagnosis using value-based management tools. Strengthening the integrative structure involves improving mobilization factors such as performance evaluation, profit-sharing, succession planning, information meetings, etc., mechanism factors such as work mechanisms, competences, focus groups, training committees, and involvement factors such as involvement in committees. The strong sense of ownership and corporate culture that is developed within such a structure increases the overall performance of the company and improves the efficiency of the banking system.

We also found that dividend optimization could increase opportunistic earnings management. Institutional efficiency is negatively correlated with dividend management. This has the effect of protecting the rights of investors more and thus encouraging them to invest in regions where the institutional system is efficient. It is expected that the ROA will increase. This increase may lead to an increase in opportunistic management if the capital structure is not concentrated. This would lead to a decrease in the optimization of the firm. However, if the ownership structure is efficient, or the capital is concentrated, the strengthening of value-based mechanisms to effectively combat discretionary management of accounts and ensure the profitability of the firm is secondary.

The size of the business increases the efficiency of the business through economies of scale by reducing costs. However, larger firms may be subject to opportunistic account management because of the complexity of their operations. The consequence will be a decline in business performance. To counter this, the strengthening of verification mechanisms such as external auditing could be of great help. The quality of the country's economic environment will have an impact on improving the optimization of the company. The effect on the company will be in the same directions as explained above. The size of the company is also decisive for the composition of the number of members in the board and its committees. The larger and more profitable a company is, the more shareholders will want to bind themselves with explicit contracts and covenants in order to reduce transaction costs and protect their dividends.

Finally, the possibility of opportunistic management of accounts will lead to preventive measures such as the introduction of restrictive clauses and contracts in the relationship. At the level of the external environment of the company, again as a preventive measure, this threat can lead to the shaping of a legal and regulatory framework and, within the company, to the setting up of mechanisms that take into account the interests of all as far as possible.

Some management models have focused on the link between strategy optimization and performance. In this section we will present the models that were used to discuss our results in the previous section and then for each of them we will show their limitations in dealing with our study problem.

The LCAG model offers a logical reasoning in several phases. Through an external assessment, it identifies threats and opportunities in the context of companies and institutions, but also key success factors. The results of this first step will be confronted with those of the internal assessment, which consists of identifying the strengths and weaknesses of companies in relation to the competition (and in relation to time) and identifying distinctive competences in relation to the competition. Here the company's resources, activities, and performance are examined. Then all the possibilities for action are created and evaluated, and the contextual and managerial values are clarified. The contribution of this very first strategy formulation model is that it allows an understanding of the environment in which the company evolves. It also helps with an internal diagnosis of the company. It is based on a formal analysis. Our study has shown the importance of formal optimization in the success of companies. The model by stating that "all possible actions will be created" does not take into account the limited rationality of the manager which is even more important in an open economy. Moreover, the model behaves like a photograph taken at a given moment on which the manager bases his choice of strategy. It can be useful for the very first strategy but can show flaws in the medium term. Indeed, the context in which the company is situated is dynamic and unstable. Therefore, what is a strength or an opportunity today may become a weakness or a threat tomorrow (represented by the red crosses).

Furthermore, the model does not say anything about the type of performance that is put forward by the manager. The type of performance will determine what are called strengths and weaknesses, threats, and opportunities. It is assumed, given the age of this model, that the manager will do everything possible to maximize shareholder profit and this ultimate goal will determine what he calls strengths and weaknesses, threats, and opportunities. However, as our study has shown, taking into account the needs of all stakeholders in a model is important in an era that advocates CSR. Furthermore, there may be discretionary management of the company's accounts by the manager. The LCAG model, however, does not address this reality. Going further, the classical efficiency that guides this model has led to the belief that personnel management will be efficient by default. However, we note that human resources, which are an important pillar at the internal level, may or may not be an asset in the search for optimization. These findings show the bias of the model in defining what is a strength or a weakness and what is a threat or an opportunity. Our study has clearly shown that the boundary is not as watertight as that (Tables 16.1 and 16.2).

TABLE 16.1 Internal Variables

Internal Analysis	Forces	Weaknesses
Logistical breakdowns	–	x
Malfunctions	–	x
SSIZE	x	–
SIND	x	–
SMEET	x	–
DUALITY	–	–
ASIND	–	–
ASMEET	–	–
RSIND	x	–
LOG SALARY	x	–
QAT	x	–
Capital structure	x	x
ISIZE	x	x
LQ	x	–

TABLE 16.1 *(Continued)*

Internal Analysis	Forces	Weaknesses
LR	x	x
MVBE	x	–
BFRISK	–	x
IAGE	x	–
Innovation	x	–
Financial deficit	–	x
Governance	x	x
profitability	–	x
Ownership structure	x	x
Dividend policy	x	x
solvency	x	x
Sales	–	x
Sales growth	–	x
Capital expenditure	x	x
TDRC	x	x
Efficiency score	x	x

TABLE 16.2 External Variables

External Analysis	Threats	Opportunities
Cycle	x	x
Financial crisis	x	–
Exchange	x	–
Bank efficiency	x	–
Legal and regulatory system	x	x
Categorization of banks in the sector	–	x

Table 16.3 shows the variables from our study that are not taken into account by the LCAG model. The model, although making an internal and external diagnosis of the company, does not status on the impact of the year of optimization on the productivity of the projects. Since it is not dynamic, it does not address the impact of past optimization on the development of future strategies and their implementation. Our problematic, however, in its second part, seeks to understand the link from optimization to optimization.

Finally, it does not address, as mentioned, the possibility of opportunistic management of accounts or the fact that flaws in management mechanisms may become more pronounced over time.

TABLE 16.3 Other Variables

Other variables
Year of optimization
SSIZE (t–1)
SIND (t–1)
INSIND (t–1)
ROA (t–1)
ROI (t–1)
Tobin's Log Q (t–1)
Duration of logistical breaks
Discretionary management

It is also about identifying the strengths and weaknesses of companies and institutions in order to analyze their context to assess the opportunities and threats that may arise. Strategic decision making or strategy optimization is about ensuring that the firm's products and markets are well chosen. This emphasizes the relationship between the firm and the context. Administrative decision's structure the company's resources for optimal success and ensure the development of these resources: financing, equipment, personnel, raw materials. Finally, operational decisions aim to implement the operation under optimal conditions of capital profitability.

The model, like the previous model, is a model which offers a diagnosis of the company and its environment which is too rigid and static, whereas we are in an open economy and in full competition. However, this model is a little more precise than the previous one because of the market-product orientation it gives to the strategy optimization. Thus, performance should be achieved if the banks in our case offer products that meet the expectations of its target customers. This view is reductive because it ignores the other aspects of business optimization and the means to achieve them. Like the previous model, there is a bias in defining what is a strength or weakness and what is a threat or opportunity. The model further argues that: 'Management decisions structure the resources of the firm for optimal success and ensure the development of these resources' and 'Operational decisions aim to implement the operation under optimal conditions of capital profitability.' In this model, management decisions are efficient in the classical sense. Our results showed that operational and capital management were not always in the direction of optimal business success. On the contrary, personnel management could be such that it would lead to logistical breakdowns and malfunctions. These failures would result in higher costs and lower productivity. On the capital side, managers might be tempted to invest in projects that are not profitable for the company or to act opportunistically. Under these conditions, the allocation and management of resources cannot be optimal. As our study has shown, value-based mechanisms and incentives must be put in place to ensure that resources will be structured for the optimal success of the firm. Finally, as in the analysis of the previous model, the same conclusions apply here.

The design school is the development of strategy as a design process. It is based on the effectiveness of managers. This school bases strategy development on the notion of strategic diagnosis as in the LCAG model. It separates the strategy development phase from the strategy implementation phase. As a result, it denies some fundamental aspects of strategy development, such as incremental development, emergent strategy, the influence of the existing structure on strategy, and full stakeholder participation. Since this school is based on the same assumptions as the LCAG model, the criticisms of that model apply here. In this school, the manager is efficient and makes all strategies. While his bounded rationality is demonstrated by Simon, our results clearly show that the manager can behave opportunistically in the allocation of resources, which runs counter to the profitability of the firm. Regarding the separation of the strategy development and implementation phases, although the literature remains mixed on this point, our results reveal a non-significant effect of decision duality on the performance indicators of Canadian banks. As a result, the model could not be used to advise Canadian banks in this sense. Furthermore, the design model, by denying the emergent strategy and the influence of the existing structure on the strategy, shows its limitations with respect to our results because the link between strategy optimization and bank performance is bidirectional. We have seen throughout the study that strategies must emerge as the process unfolds to respond to the instability of a porous environment. Using this approach, we would only be answering one aspect of our problem. Finally, the study of stakeholders teaches us to involve them all as much as possible in the development of strategies so that their implementation is facilitated. Non-involvement of stakeholders can lead to what we have defined as logistical breakdowns and dysfunctions.

The optimization model is the development of strategy formation as a formal process. It essentially takes up the hypotheses of the design school by breaking down its development into distinct stages and giving it a more rigorous formalization in the form of checklists. The 'strategic planning' team thus replaces the manager, who is only marginally involved. Here, strategy is considered as an objective process, organized, and optimized in its smallest details (Marchesnay, 2004). This school gives a clear meaning to strategy and allows not only a good allocation of the company's resources, but also an ex-post control of the realization on the basis of what was optimized. Since it takes up the assumptions of the design school, the criticisms of the latter apply to the present case. However, its contribution is that it advocates a more participative management within the company. Formal optimization is much more understandable to everyone. This can lead to mobilization and involvement of employees who understand the projects and their goals. This model also relies on controlling to improve its strategy as it goes along. The trade-off is that it shifts the problem of possible managerial opportunism to the strategic planning team, which is also considered effective. Moreover, the dynamic that this model seems to take on because of the ex-post value-based management that it carries out masks its staticity. Indeed, it should be recalled that the optimization model has its origins in the SWOT matrix strategy development. Although it recognizes the place of emergent strategy and existing structure, it can only be properly applied in a stable framework. As a result, it does not lend itself well to our analysis because it is conducted in an unstable and dynamic framework.

The positioning model is the development of strategy as an analytical process. By analyzing the industry, the sector and thus the competition, the strategy will determine a positioning, choosing a place where the potential for development is highest. This school emphasizes the interactions with the environment on the study prior to action. The context is understood essentially in economic, market, and competition terms. Defining a strategy is above all positioning oneself in a competitive context. Its contribution to our study is that it enabled us to take an interest in the banking sector. We studied the evolution of turnover and overall added value. We also separated the banks into categories of banks and by studying their efficiency defined the best and worst-performing categories of banks. In this model it is the achievement of a high position that will drive the strategy. The balance of power that drives the behavior of firms could weaken the sector in the long run as the most efficient ones could be tempted to apply predatory pricing to drive their competitors out of the market. In the long run, the competitive structure would tend towards an oligopolistic structure.

In the study of our problem, we noted on the one hand the elements specific to the company which could help us to solve it and on the other hand the items of measurement of the external environment of this one. As a result, we can only suggest that one of the major limitations of this model in understanding our problem is that it only focuses on the external environment of the company to define its strategy. However, basing the strategy on the internal environment of the company is just as important because the knowledge of the ins and outs of the competition is limited. Moreover, since the context is understood essentially in economic terms of market and competition, it denies the heterogeneity of the external context of the company. The influence of the country's politics and regulations are therefore not factors taken into account in the search for this positioning. Presented in this way, this model seeks to understand how the performance of the sector in which the company evolves influences its strategic choices. Furthermore, this model implicitly suggests that the manager will always invest in projects that ensure that the company acquires the highest development potential. Again, this denies the possibility of opportunistic behavior on the part of managers. By using this model, companies will not be able to put in place adequate measures to remedy this internally. Finally, such an analysis cannot take into account the expectations of all stakeholders, especially the staff. Inadequate personnel management, as we have shown in the results, leads to a decrease in the company's performance.

The Boston Consulting Group Matrix 1 (Boston Consulting Group, 1980) uses two strategic variables: market growth rate and relative market share. The basic idea behind this matrix is that it would be better for the company if a product had a larger market share and the market for goods grew faster for the company. Market growth measures market attractiveness and is based on the concept of the business life cycle. Relative market share is measured against the nearest competitor or leader and indicates the competitive strength of the company as a result of a significant experience effect. The BCG model 1 method consists of dividing the company into a number of strategic business areas (SBAs). In response to criticism of this model, BCG has developed a second model that takes into account the factors that can influence the competitive position of the company. It is based on two variables, price advantage (differentiation) and cost advantage (height of the strategic barrier), and takes profitability as a performance indicator.

In its first version, this BCG model, by dividing the company according to its activities, seeks to develop strategies according to these activities and the optimization of the sector or its closest competitor. As a result, it only deals with one aspect of the problem. Furthermore, in this model a successful company is one that knows how to increase its market share. Accounting and commercial performance are only the two types of performance that will direct the strategies. However, performance is multidimensional and it is essential in order to talk about global performance that the analysis takes into account the performance of employees, suppliers, customers, shareholders, and all internal and external stakeholders. When looking at the variable 'adoption of optimization' we found that sales had a positive but insignificant relationship with ROA. Managers cannot then use only product-market analysis to improve their performance. On the other hand, in its second version, the BCG model attempts to achieve a level of performance based on a strategy relying solely on price advantage (differentiation) and cost advantage. By using both versions of the BCG model in the analysis of our problem, we seem to be able to explain the reciprocity of the relationship between performance and optimization. However, the weakness of the strategy indicators (only two) to explain a performance still based solely on the profitability of the company leaves us wanting more, as the results have allowed us to identify no less than 30 strategy indicators that influence profitability as much as other performance indicators.

The second matrix is that of McKinsey. This model is based on the approach proposed by the Boston Consulting Group but has the advantage of offering a wider range of strategic choices for the company. It is based on the representation of the company's SBAs in relation to two dimensions: medium-term market attractiveness and competitive strength or position. The establishment of its matrix requires the identification and analysis of the external factors that control the attractiveness of the sector to which the activity under consideration belongs; it also requires the identification and elucidation of the internal factors whose degree of control is the basis of the company's competitive strength. These two dimensions are assessed on a three-position scale (strong, medium, and weak).

Although presenting a wider range of strategic choices than in the BCG model, this model only addresses the impact of sector and company optimization on strategy development. The strategy is based only on the SBAs. Market attractiveness is defined by growth, number of competitors, accessibility, profitability, volume… This medium-term assessment of the external context does not make this model as dynamic as one would wish in the context of an open economy. Moreover, the diagnosis of the external context does not take into account its heterogeneity. While it is true that we have studied the efficiency of the banking sector, we have integrated other variables of the external environment. We also note the tight boundary between the internal and external diagnosis of the company. Yet we have seen how, for example, the institutional and regulatory context can have internal repercussions on the opportunistic behavior of managers or the impact of the company's profitability on the efficiency of the banking system, to name but a few. Moreover, by focusing only on the SBAs, the analysis of optimization itself in its influence on strategy development remains incomplete. Focusing only on strategic areas does not allow for adequate operational management of the company. Finally, the McKinsey model suggests: 'the identification and elucidation of the internal factors whose degree of control underpins the competitive strength of the company.' This suggestion is guided by the same spirit that drives the model, which is that the company should base its strategy on the most profitable activities in order to acquire a position in the sector. In terms of our study and our results, the McKinsey model is not sufficient because it only deals with certain aspects of overall performance.

The strategic analysis is structured on the basis of two variables: the competitive position (company strengths or competitive position) and the maturity of the business (attractiveness of the sector). The model adopts the same principle of DAS but uses the notion of key success factors of the company. These are three factors: supply factors, production factors and marketing factors. These three operations constitute three systems with different mechanisms, in which the company may or may not have advantages over its competitors. The sum of the advantages or disadvantages allows it to have a level of profitability that is lower or higher than that of its sector and to have greater or lesser possibilities of choice. In this model, the strategy is developed according to commercial optimization, productivity, supplier optimization and sector optimization. Recalling that our study seeks to capture the effect of optimization in its global aspect on strategies, we note that this model, like the two previous ones, does not allow us to understand the influence of strategy optimization on performance. Furthermore, the ADL matrix acts as a snapshot of the competitive position and market attractiveness. It does not lend itself well to our case, which is dynamic. At the internal level, the ADL model does not take into account certain aspects that we have identified, such as human resources management, capital management, debt, and its corollary, the optimization lever, opportunistic management, and performance at period t–1. The mastery or otherwise of these indicators may or may not lead to higher levels of performance. At the external level, this model, by focusing only on the attractiveness of the sector in economic terms, denies the influence of the institutional and regulatory context as much on the attractiveness of the sector as on the company itself. This aspect of the external context is however apparent in our results.

Finally, project productivity is defined by its level of profitability in relation to the sector. This view of optimization is reductive in the sense of our study, which has sought to highlight the overall performance in these results.

According to Porter, the productivity of projects is determined by the structure of the industry in which it operates and by the competitive advantages obtained by the company in a given sector (Grant, 1991; Porter, 1991). Porter's competitive model is based on four elements. The first (Porter, 1980) concerns the intensity of competition. The competitive game results from five forces (competitors in the sector, potential entrants, suppliers, customers, substitutes) whose intensity determines an average level of profitability influencing the attractiveness of the company. For him, there are three basic strategies for dealing with the five competitive forces: overall cost dominance, differentiation, and concentration. The company must make only one choice among the basic strategies if it wants to gain a competitive advantage. Furthermore, he introduces the notion of international competitiveness which results from four elements determining the competitive advantage of a nation (Porter, 1980). These are the factors of production, demand, upstream, and related industries, and the strategy-structure-rivalry of the firms in the sector. To these elements Porter adds chance and STATE. All these factors determine the chances that firms in a given country and in a given industry have to build a competitive advantage over competing firms in other countries. Finally, Porter (1985) analyzes the sources of a firm's competitive advantage using the value chain. The value chain highlights the activities that have a real impact in terms of costs or differentiation from competitors. The value chain thus makes it possible to analyze the strategic potential of the company at the origin of the greatest value, i.e., the key factors of success that explain a competitive advantage.

Porter's model tells us that the performance of a company is determined by the structure of the industry in which it operates. Our analysis showed that the Banking sector is open and very competitive. Moreover, the Banking industry, which includes several categories of banks, is very successful overall. According to Porter, this should be a catalyst for the less successful banks to become more successful. However, we have seen that bank efficiency, other things being equal, could encourage more discretionary management.

Its competitive model is based on four elements, the first being the intensity of competition. The analysis of the banking industry by categorization in our study again showed us that Canadian banks are highly competitive and have maintained this trend for the most part over the years. The table shows how a company can decrease (x in red) or increase (x in green) Porter's five forces.

Porter's model identifies five forces that the company must master using only three strategies. All these forces are outward looking. However, our study has shown that in addition to these forces, other strategies must be put in place to control the forces concerning human resources (employees and managers), otherwise we would end up with dysfunctions and opportunistic behaviors that would call into question the profitability of the company. He also introduced the notion of international competitiveness. For our part, this could be manifested by the presence of the company on the stock markets, the share of institutional investors in the ownership structure and in the capital. Indeed, without wanting to refute Porter's model, we found that the stock market and foreign investors could be a source of competitive advantage at the international level. In a more recent version of his model, Porter analyzes the sources of a firm's competitive advantage using the Porter value chain (1985). This new analysis of competitive advantage includes more of the company's own factors that ensure its positioning. However, he only lists them without studying in depth the factors that can influence them positively or negatively. In support activities:

1. **Corporate Infrastructure:** We found that for this function to play its supporting role properly, sufficient integrative, incentive, and monitoring mechanisms had to be put in place to reduce the bounded rationality and opportunism of actors;
2. Human resources were, in our study, subject to logistical breakdowns and dysfunctions that made them more of a handicap for the company.
3. **Technology Development:** We found that technology can have a significant negative impact on the variability of output in Banking firms. Innovation most certainly has beneficial effects on performance. But a technological change in the situation of adoption of optimization may for some time have the opposite effect which corresponds to the adaptation time.
4. **Purchasing:** The negotiation of contracts with suppliers, for example, could be called into question if the company embarked on a policy of indebtedness, even though this was necessary to control the behavior of its managers.

At the core level, we found, as he predicted, that taxes, transportation, external services, and operating expenses were positively correlated with bank productivity. Porter's model, like the product-market models, seeks to understand how industry or firm performance leads to the best mix of strategies. As a result, it only allows us to understand one direction of our problem given the factors it uses.

In the resource and competence model, the company is no longer conceived as a portfolio of products and markets but as a portfolio of resources. Competitive advantage is to be sought internally. The resource theory is based on the observation that performance is heterogeneous within an industry (Depeyre, 2005). It is based on the fact that firms compete on the basis of their resources and capabilities (Peteraf and Bergen, 2003) and focuses on these as a determinant of competitive advantage and optimization. In analyzing the sources of competitive advantage, the RBV approach is based on two fundamental assumptions (Barney, 1991; Peteraf and Barney, 2003): heterogeneity and immobility. However, the assumptions of heterogeneity and immobility are not sufficient conditions for sustainable competitive advantage. According to Barney (1991), firm resources must have the following properties to be a source of sustainable competitive advantage: value-creating, rare, inimitable, and non-substitutable (VRIN). Competence refers to the organizational ability to deploy resources in combination to achieve a goal.

The contribution of this model to our problem is that it allows us to identify, as far as possible, all the internal variables of the company that are likely to have an influence on the productivity of projects. The resources concern the human, financial, and material aspects. We can assume that, given the hypothesis of resource heterogeneity, performance by category of bank is only a consequence. However, the theory does not tell us anything about the origin of the difference in resources and capabilities, although Barney (1991) notes that the firm's resources must have the following properties to be a source of sustainable competitive advantage: VRIN. The question is whether heterogeneity derives its strength from the natural characteristics of the resources or from a development of them. In surveying the resources and capacities of firms, we have seen that their very management ensures their heterogeneity from one firm to another. We have noted logistical breakdowns and dysfunctions in the management of human resources that make some companies perform better than others. In addition, the distribution of capital and monetary incentives affects the profitability of firms. The study of human resource management factors showed that certain internal practices such as training, know-how, and culture could give resources a VRIN character. With regard to skills, our study has specified the type of organization needed to manage resources adequately. The organization must be inclusive, incentive-based, and have good monitoring mechanisms. Although this model addresses both directions of our problem, it is insufficient in that it does not address the external environment of the company or the influence of its initial context. However, we have identified a number of variables in the methodology section of our analysis that are of great importance.

Capacities are the managerial skills needed to coordinate, harmonize, and strategically deploy the available value-creating resources. These capabilities can be sources of competitive advantage (Teece et al., 1997). According to Johnson et al. (2005), the development of strategic capabilities leads to long-term competitive advantage. Grant (1991) states that resources are the source of a firm's capabilities while capabilities are the source of its competitive advantage. The element of 'dynamics' concerns the company's ability to renew its skills in order to keep pace with an increasingly fast-changing environment, where technologies lead to increasingly frequent innovations, accelerating the obsolescence of offerings and facilitating substitution. The dynamics of capabilities are then essential for the firm to establish a long-term competitive advantage (Cao, 2012).

This model, like the previous one, focuses more on the internal resources of the company to explain the strategy of companies. The disadvantage of this model, however, is that it only focuses on the manager's contribution to explaining the link between performance and strategy. Throughout this study, we have shown that managers can suffer from limited rationality and opportunism in the management of the company. The hypotheses of our study therefore set out to identify the internal and external factors that reduce this handicap for the company. These factors concern all the variables of supervision, incentives, and integration of other employees in the company. We have seen that the absence of such variables in the dynamics of the company has an impact on its competitive advantage and profitability. Capacities will underpin strategy by managing resources to create value. There is no feedback from optimization to strategy so necessary to explain our problem. Moreover, while it is true that we have seen, for example, that the absence of a succession plan in a company or the weakness of work mechanisms or skills could play on the comparative advantage of companies, using this model limits the analysis of our problem. The advantage of this model is that it is part of a dynamic that fits with the context of an open economy. It therefore adapts to its environment as it goes along. However, our results have shown us the importance of taking into account the global environment of the company as well as all its stakeholders in the strategic calculation.

KEYWORDS

- **BCG model**
- **human resources**
- **LCAG model**
- **stakeholders**
- **strategic business areas**
- **strategy optimization**

CHAPTER 17

Modeling the Organizational and Structural Determinants of the Value-Based Management

Strategy optimization is a concept that has acquired a strong position in corporate performance management, even if disagreements on how this variable affects the competitive advantage of institutions persist. Two sets of factors contribute to the understanding of the interactions between strategy optimization and the performance of the banking sector in the face of internal conditions and exogenous shocks: the first, which emphasizes the concept of transaction and market costs, allows for the micro-mechanisms affecting sectoral growth in the context of an open economy to be taken into account in a model of its own, while the second integrates the organizational and macro-structural aspects affecting the application of the optimization of the strategic decision. This research has succeeded in establishing the framework for the implementation of strategy optimization with parsimonious factors, quantified the impact of strategy optimization on bank performance, and established the key factors involved in the role that performance plays in the dynamics of multifaceted management practices.

The place of situational and decisional factors or how does strategy optimization and its variables enter into the functions related to the efficiency of banking institutions in the context of an open and digital economy?

The first step towards a specific approach to explain the nature of the relationship between optimization and banking and financial sector performance is to identify how strategy optimization enters into the production function of banking institutions in an open economy context. In this dynamic, it should be noted that banking and financial institutions incur transaction costs. Consequently, they use (optimization) mechanisms in their management to

Value-Based Management in an Open Economy: Optimizing Strategy to Improve Business and Performance. Prof. Post-Dr Walter Amedzro St-Hilaire (Author)

reduce transaction costs. This particular type of management tool facilitates operations: easier and less costly transactions (less transaction costs, in the presence of a mediator).

Considering the function Q(w, y), where w is the profit vector for input x and y is the output ($Q(tw, y) = tQ(w, y)$ for all $t \geq 0$; $Q(w, y) \geq 0$ for $y \geq 0$ and $w \geq 0$; $Q(\acute{w}, w) \geq Q(w, y)$ for $\acute{w} \geq w$). By integrating the costs of the technology $V(y) = \{x \geq 0 : wx \geq Q(w, y), \forall\ w \geq 0\}$ to the cost function and if the vector w_i represents the task control vectors or decrease costs, the principle of duality mapping will be:

$$x_i\,(w,y) = (\partial Q(w,y))/(\partial w_i) i = 1, \ldots, n$$

In which, *x* represents the conditional demand function for input *i* (optimization factor). Thus, each pair (*w* and *y*) will refer to a particular *x* that minimizes the cost of production. This optimal status is called the conditional factor demand function. In general, the production function is the sum of the conditional demand functions for the inputs. For the production function *y* that uses *n* inputs for production, we have:

$$y = f(x_n) n = 1, \ldots, i, \ldots, n$$

Banks and financial institutions, in addition to the factors of optimization, enjoy physical and human factors so that the cost function is as follows:

$$C(w,y) = yw^{\alpha}\, r^{\beta}\, \varnothing^{\varphi} \tag{1}$$

In this equation, *w* is the human factors profit, *r* is the cost for physical quality, and $\varnothing$ is the cost for optimization determinants and *y* represents the output of institutions. Based on the duality of the mapping principles we will have:

$$x_i\,(w,y) = (\partial c(w,y)) \,/\, (\partial w_i)\ i = 1, \ldots, n \tag{2}$$

In which x_i is the conditional demand function for input *i*. Therefore, in this function, for each pair of *w* and *y*, we have a choice of *x* that minimizes the cost of production. Thus:

$$k_p = (\partial C)/(\partial r) = \beta y w^{\alpha}\, r^{\beta-1}\, \varnothing^{\varphi} \tag{3}$$

$$k_h = (\partial C)/(\partial w) = \alpha y w^{\alpha-1}\, r^{\beta}\, \varnothing^{\varphi} \tag{4}$$

$$k_s = (\partial C)/(\partial \varnothing) = \varphi y w^{\alpha} r^{\beta} \varnothing^{\varphi-1} \quad (5)$$

Relationships (3), (4), and (5) will represent the conditional demand for the optimized mechanisms, human factors, and control factor, respectively. From this, we obtain:

$$(k_p\, r)/(\beta w^{\alpha} r^{\beta} \varnothing^{\varphi}) = (k_h\, w)/(\alpha w^{\alpha} r^{\beta} \varnothing^{\varphi}) = (k_s \varnothing)/(\varphi w^{\alpha} r^{\beta} \varnothing^{\varphi}) \quad (6)$$

Taking into account the cost of production and the human determinants in the mechanisms of strategy optimization, we will have:

$$w = (k_s \varnothing \alpha)/(\varphi k_h) \text{ and } r = (k_s \varnothing \beta)/(\varphi k_p) \quad (7)$$

By integrating Eq. (7) into the conditional demand of the optimization, we will have:

$$k_s = y k_s^{\alpha+\beta}\, k_h^{-\alpha}\, k_p^{-\beta}\, \varphi^{-(\alpha+\beta)}\, \alpha^{\alpha}\, \beta^{\beta} \quad (8)$$

Solving Eq. (8) for production, we will have:

$$y = [(\varphi^{\alpha+\beta})/(\alpha^{\alpha}\, \beta^{\beta})]\, k_p^{\beta}\, k_h^{\alpha}\, k_s^{\varphi} \quad (9)$$

Equation (9) is the production function of the banking and financial sector in an open economy in which optimization mechanisms are one of the factors of production. There is then an input related to the optimization of transactions that leads to a decrease in the costs of banking and financial institutions. Thus, the control in services leads to the establishment and strengthening of relations between factors. These services are nothing more than determinants of the optimization mechanisms of banking and financial institutions. In fact, by lowering the fraction rate and thus controlling the transaction and lowering the costs, institutions in the open and digital economy are basically willing to pay certain costs in their optimization to control the transactions. This optimization can be the result of the activity of control factors or the formation and increase in the level of optimization in banking institutions. Thus, the control factors include a set of factors that help to increase the effectiveness of strategic management and therefore the optimization of transactions. We can say that the performance in the banking and financial sector is a function of the service and the determinants of the optimization of the transaction. Hence the following equation on the quality of optimization and control of services:

$$SC_{micro} = H(x_i) \rightarrow \partial H / (\partial x_i) \geq 0 \tag{10}$$

This equation indicates that quality at the micro-level is a function of service control, so that with an increase in the factor – control – the level of optimization in the banking sector increases.

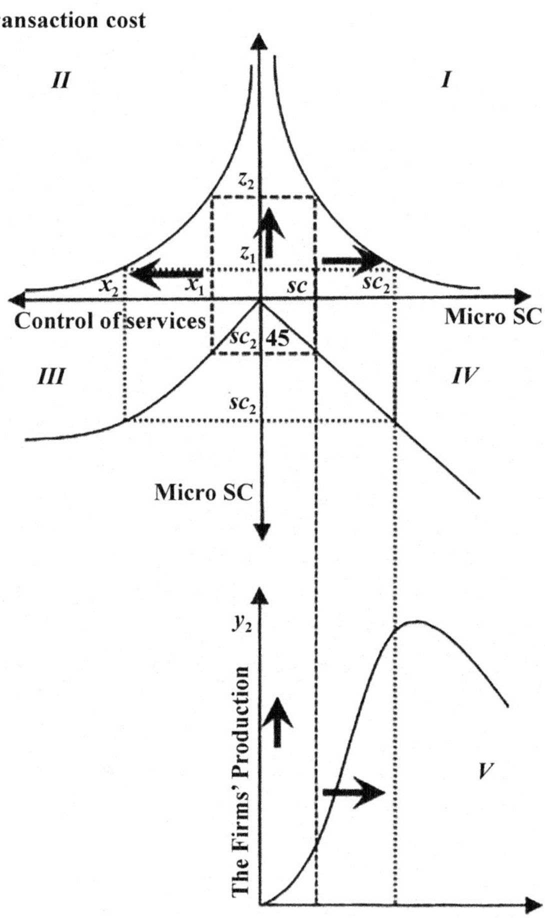

Micro SC

FIGURE 17.1 Effect of strategy optimization on production.

Let's assume that banks employ ξ_1 many control tools (input) (*see* Figure 17.1). This input of facilitators creates $\sigma\chi_1$ unit of optimization and also ζ_1 unit of transaction costs. This value of input causes ψ_1 unit of output of the institutions. Now suppose that the banks will use more control measure input, ξ_2. In this case, the level of optimization increases to $\sigma\chi_2$. By increasing the quality of the optimization, the transaction costs are reduced to ζ_2. Drawing the general equilibrium lines shows that in this case, the output of the banks increases. Thus, increasing the level of optimization leads to an increase in output for each bank.

To generalize to all banks and financial institutions, we must assume that the net supply functions of banks are well-defined functions if we have m banking institutions then the total supply function will be equal to:

$$y(p) = \sum_{j=1}^{m} y_j(p) \tag{11}$$

Now, if the institutional net supply function is a well-defined and continuous function, the total net supply function will also be well-defined and continuous. We can then write:

$$x_i \in y_i \rightarrow x_i \in Y \tag{12}$$

Equation (12) indicates that since the input to the optimization is one of the inputs to the banks and financial institutions, the open economy production function (which is the result of the horizontal aggregation of the production function) will also have a level input to the optimization. Therefore, given that there is a quality vector in the cost-minimizing and profit-maximizing production program, in the total production program (which is the result of all production programs), there is also a vector as the optimization-level input. This vector leads to the total cost to be minimized and the total profit of the economy to be maximized. Considering the set of production possibilities y, this set is representative of the set of practical net output vectors. The set of total production controls is the sum of the set of production control tools, which we can write:

$$\mathrm{Y} = \sum_{j=1}^{m} y_j \tag{13}$$

Here, each production program belongs to *Y*; *y* and *i* belongs to *Y*. Thus, *Y* represents the set of production programs that can be obtained from the distribution of production among institutions $j = 1, 2, \ldots, m$. Since the program of each production maximizes profit, therefore the whole production program also maximizes total profit. The production function of the economy by the horizontal aggregation of efficiency institutions, the variables existing in it are also aggregated horizontally. If we assume that, in addition to the role of optimization, only the inputs bank performance and physical factors play a role in the production function and in the overall economy, there are m efficient banking and financial institutions. Then for the aggregate function of the digital economy, we will have:

$$Y = f(\sum_{j=1}^{m} k_{jp}, \sum_{j=1}^{m} k_{jh}, \sum_{j=1}^{m} sc_{j\,Micro}) \tag{14}$$

This function indicates that aggregate output is a function of the quality of total physical factors of all banking institutions efficiency $\sum_{j=1}^{m} k_{jp}$. And there are also the total human factors function of all institutions $\sum_{j=1}^{m} k_{jh}$ which is equivalent to human factors. This function will also depend on the total performance of all banks. But the total micro efficiency does not correspond to the total optimization level. But, most of the control measures of the institutions are in opposite directions from each other and thus neutralize each other. Thus, the vector aggregation of micro efficiency will not equal the level of macro-optimization $\sum_{j=1}^{m} sc_{j\,Micro} \neq sc_{Macro}$.

The micro and macro levels of optimization are not two distinct packages that could easily be divided into two types depending on the banking network. This distinction exists although the application of it in its true concept is not easy. The macrostructure of the banking economy is made up of systems, organizations, and different banking networks that are distinguished by the actions of its actors. In the relationship between the micro and macro aspects of optimization, it seems that the formulation of macro mechanisms is a function of the amount of accumulation of micro factors. It is possible to consider in the framework of micro-optimization a border point so that an increase up to it has positive effects on the increase of the level of optimization and the expansion of the level of optimization from within the networks and the transformation of specific trust into general trust causes the formation of macro efficiency. But with the increase of the level of micro-optimization beyond a border point and the expansion of cross-sectoral prejudices, the intranet work connection goes out of the optimal amount and causes oppositions and hostility, which means a decrease of the level of macro-optimization. Thus, the relationship between macro and micro efficiency can be shown in Figure 17.3. We can conclude the magnitude of a positive relationship between micro and macro performance up to the point of frontier. Before this point, we can consider the micro performance in proportion to the macro efficiency. Therefore, we can write in the case of financial and banking institutions in the context of an open and digitized economy:

$$sc_{Macro}\begin{cases}\cong \sum_{j=1}^{m} sc_{j\,Micro} & \text{if} \quad \sum_{j=1}^{m} sc_{j\,Micro} \le sc^{*} \\ < \sum_{j=1}^{m} sc_{j\,Micro} & \text{if} \quad \sum_{j=1}^{m} sc_{j\,Micro} \ge sc^{*}\end{cases}$$

Thus, if we assume that the level of global micro-optimization is lower than sc*, the result we can be sure of is that there is a level of global micro-optimization in the global production function. It is under this condition that the factors affecting the application of optimization decisions will be able to find a place in a model specific to the case of the banking and financial sector.

Organizational and structural determinants of the interaction between strategy optimization and performance of banking institutions in open economies.

The fundamentals to establish a parsimonious explanatory model of the relationship between strategy optimization and performance of banking and financial institutions also relies, in addition to its market component to integrate the [1]organizational components and factors that influence the effectiveness of the optimization process. The first aspect is the use of market mechanisms to build and determine innovative practices of strategic management of the performance of banks and banking institutions. The determination of management practices, to be effective, is based on the combinatorial factors such as familiarity, resource allocation, flexibility, responsive culture, priority, support, and process efficiency. The many issues explored in the "content" and "process" flows must be integrated at the level of our case.

The second determinant is the strengthening of management tools for effective strategy optimization because the strategic resources of banking and financial institutions consist of a set of idiosyncratic capabilities such as equipment, facilities, procedures, and routines. This bundle of resources is

[1] These components contribute to the institutional capacity to support the coordinated deployment of resources in a way that helps achieve goals. Coordinated deployment of these resources includes processes that leverage and strengthen open economy planning. Leveraging is applying existing mechanisms to existing or new market opportunities in a way that does not require qualitative changes in assets or capabilities of the form. Component planning is the process by which banking and financial institutions make qualitative changes to their existing stocks of assets and capabilities, including new capabilities to help them achieve objectives.

of several kinds (assets and human development, capabilities, and structure, processes, and routines, and behavior and value). Thus, the effectiveness of optimizing strategies in the banking and finance sector depends on the constant adaptation of new management practices, and the qualitative change of organizational performance. The two aspects of integration are illustrated in Figure 17.2. The factors that impact the determination of innovative management practices, and the performance results are based on the effectiveness of the optimization process.

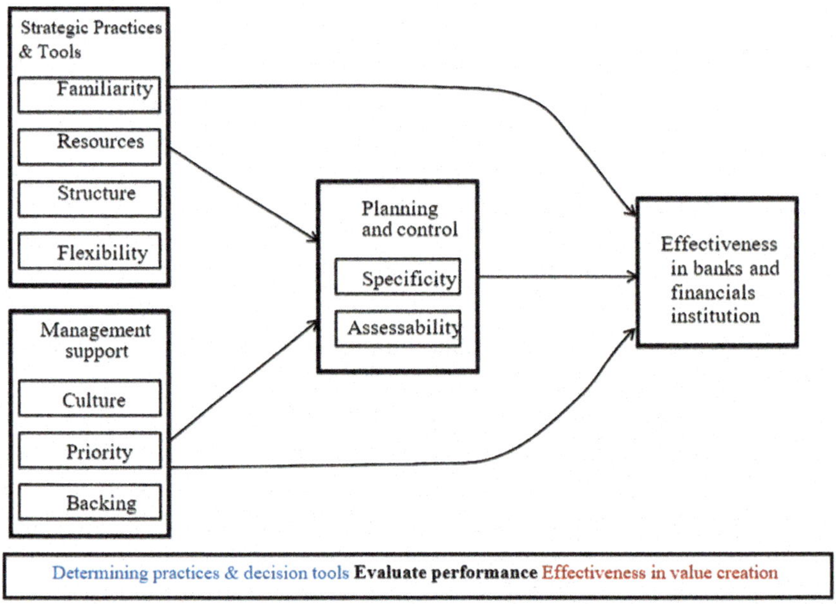

FIGURE 17.2 Integrative framework for implementing strategy optimization.

Strategic management practices (familiarity, resource allocation, structural optimization, and flexibility) are treated as independent variables because these variables influence the effectiveness of strategy optimization in the banking and finance sector. Experienced strategic management (familiarity) ensures successful optimization by providing sufficient resources (experienced personnel, time, equipment, know-how, and funds) to ensure effective optimization. The management thus wants to put in place a system to clearly define productivity standards, and to provide feedback on optimization progress. As a result, there is flexibility in management that allows optimization to maneuver around unforeseen events. Management support (creating the responsive culture, priority, and support) is also considered an independent variable so that governance can tailor the content and tools to influence the effectiveness of strategic performance. Indeed, by instilling a culture receptive to strategy optimization, management mechanisms influence the values of institutions. This leads in an open economy context to a significant power addition to management effectiveness. The (priority) setting thus improves the visibility of optimization and allows management support to remove obstacles to the successful execution of strategic plans.

Optimization and control are treated in the model as control variables for the following reasons. First, the explanatory model uses optimization and control as a generating mechanism through which profitability and output influence the effectiveness of the management process. Governance anticipates visions and actions so that performance mechanisms are based on a method, plan, or logic. Bank optimization thus puts in place control systems to evaluate results, provide feedback, identify productivity gaps, and redirect strategic mechanisms, so that optimization can be properly carried out without having to waste time and ensure that the management practices carried out are in fact enabling the institutions to achieve performance. Second, open-economy, and digital banks require optimization to ensure that the plan is based on sound strategic logic and put in place management control systems to redirect mechanisms and monitor performance. Therefore, strategic management practices and management support drive the need for optimization, and control in turn affects the effectiveness of profitability mechanisms. The dependent variable in our model is process efficiency and is assessed in terms of completion, achievement, and performance. Completion is the degree of finalization. Achievement is used to assess inventory reduction, operational efficiency improvement, and stakeholder satisfaction (including the bank's customers and shareholders). Acceptability is used to assess the adoption of the strategy. Thus, it can be considered that through the implementation of innovative management practices, the components of performance and the factors that influence the effectiveness of the process provide a parsimonious explanation of the role of performance in the implementation of management tools. From this, the determinants of strategic management performance in the banking and financial sector each have a different role in influencing the effectiveness of innovative management tools. Familiarity has a direct influence on effectiveness ($b = 0.19$, $p < 0.05$) and relevant marketing/sales experiences contribute to successful productivity by elaborating the "construct" of the business unit's strategic mission. It should be noted that these banking practices are used to generate and modify resources and management performance to integrate them together and recombine them. At the level of banks in the context of an open and digital economy, performance conveys the exchange and combination necessary in the development of new management practices. It allows for a greater capacity to process information in a more meaningful way, and for relevant, deep, and principled analysis. Therefore, it provides a better foundation for assimilating and interpreting opportunities and more effective in developing and applying new and innovative practices.

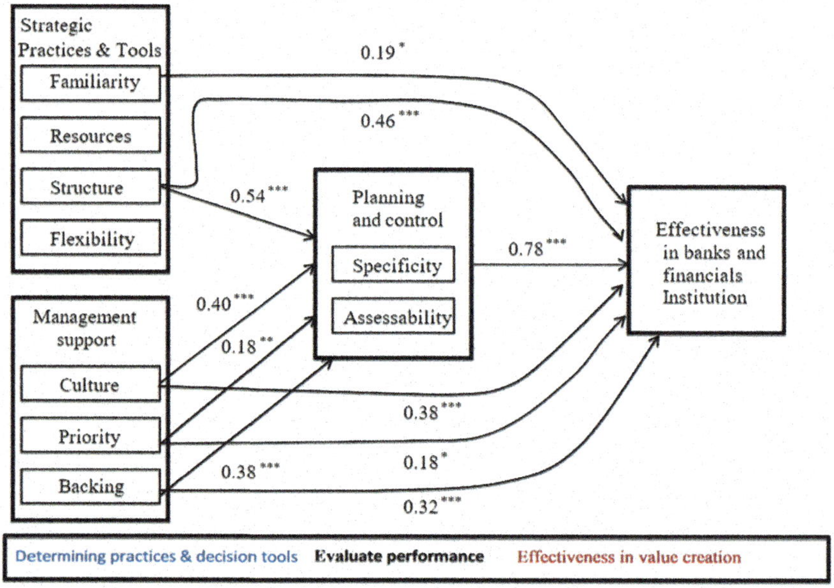

FIGURE 17.3 Summary of the relationship between strategic management practices, management support and performance of banking and financial institutions.

Notes: N = 171, factory managers = 95, Bank, and financial managers = 76. * $p < 0.05$; ** $p < 0.01$; *** $p < 0.001$.

With the profitability of the management and the different mechanisms available, the banking institutions have developed a more acute awareness in the framework resolutions of the problems and their anticipations: a reduction and an alternative to the risk. As a result, the practices experienced by the banking and financial sector incorporate original and useful ideas in the performance resources and capabilities to improve the interaction between the factors and improve the quality of resources and capabilities. Thus, performance has a direct influence on the efficiency of the process in an open economy context. In complex institutions, management practices have an important role regarding optimization and control, which in turn lead to the successful implementation of strategy optimization. Conversely, in branch offices, the low level of optimization coupled with the close relationship between performance practices is positively associated with management effectiveness.

As for the place of decision architecture, which is the degree to which the organizational structure is conducive to optimized performance, it aims to facilitate the effectiveness of optimizing critical business strategies. In open and digitized economies, banking, and financial institutions have set up separate agencies for mass customization to foster innovative management tools. Thus, in the implementation of structural optimization practices, a weekly meeting, information on product quality, past productivity status, and budget variances to facilitate weekly discussions are granted by the central bank. The decision-making architecture of institutions is thus positively related to management efficiency, because by providing a structure for optimizing critical business strategies, banks are able to exercise their ability to do their best to achieve efficiency. In addition, structural architecture is positively related also to optimization and control, which in turn influence management effectiveness. Resources such as experienced practices and support staff, time, equipment, know-how, and funds are necessary for institutional productivity. Banking institutions allocate resources in line with strategic business requirements and strong resource provision leads to successful strategy implementation. Thus, it is clear that resources are necessary for effective strategy optimization but do not provide the capacity to create innovative management practices. Architectural performance is required to reconfigure, mix, add, or modify existing resources to create new innovative management practices. Therefore, the provision of sufficient resources supports in the banking and financial sector in the open and digital economy the effectiveness of strategy optimization without being able to achieve performance (profitability and productivity).

Flexibility is the extent to which the banking and financial optimization process can be adapted to accommodate changing circumstances. The effectiveness of institutional optimization then involves uncertainty, and intentional and unintentional changes are inextricably linked like two sides of the same coin. Thus, in open economy banks, a certain degree of flexibility is necessary to allow a course of action to be modified in response to an emerging event that has been capriciously deviated from previous expectations. Adapting the process of optimization mechanisms to unexpected conditions requires innovative management tools, great commitment, effort, and sacrifice to prioritize optimization, reorganize the human factor, reallocate resources and despite time pressure, produce the desired results in the midst of uncertainty. As can be seen, extra effort and time are needed to adapt the mechanisms of optimized management to the evolution of open economies. As a result, flexibility improves the reliability of decision-making tools in banks and financial institutions, but does not contribute to process efficiency.

In this model of the interactions between strategy optimization and performance of banking and financial institutions in a digitized context, three supporting attributes of management also contribute directly to management effectiveness and indirectly through optimization and control to ensure institutional effectiveness. The set of beliefs and practices about how banking and financial activities are conducted directly influence banking efficiency. The banking and financial sector in open economies, through the creation of an appropriate performance culture, provides, on the one hand, a system of informal rules and group pressures on the role of performance factors and, on the other hand, the structures, norms, and a value system within which to operate. Thus, the culture has a positive influence on the optimization and control which in turn influences the effectiveness of the process and the institutional performance. The prioritization factor is the extent to which optimization takes precedence in banks and has a direct influence on process effectiveness: this priority is communicated up and down the institutions, so the strategy becomes observable and prominent. Prioritization allows institutions to put together shared priorities to allow for quick agreements and has a positive influence on optimization and control which in turn influence institutional effectiveness. Institutional support, on the other hand, has a direct influence on process efficiency and takes the form of uniform sanctions (strict discipline for under-productivity) and adjustment policies. In addition, support has a positive influence on optimization and control, which in turn influence process efficiency (Figure 17.4).

Thus, this model coherently explains and successfully establishes the framework for the effectiveness of strategy optimization with parsimonious factors, quantified the relationships between the factors and established the key determinants of success in the effectiveness of strategy optimization. The model perfectly integrates with relevance, the problems of the operational design of banking and financial institutions. In addition to identifying the relationship between strategy optimization and bank performance, it traces the conditions under which this relationship is negative, neutral, or positive, as well as the crucial role of design in extending the performance frontier. By applying operational principles, the model shows that problems related to the competitive advantage of banking institutions can be analyzed successfully like all other business problems, using traditional tools of economics, finance, and strategy (provided that the dynamics of industry-specific interactions are considered). He thus determines, finally, that the strategic means necessary for the banking sector to improve their performance while offering superior value depends on the structure in which it competes and on its own industrial capabilities. The use of economic models and the accompanying optimization of analytical reasoning processes are aimed at solving institutional problems in order to remedy organizational and market failures. Based on cross-sections and panel data estimations, this model indicates that the reorganization of optimized management practices has a significant interaction effect on the performance of banking institutions in the context of a digital and open economy.

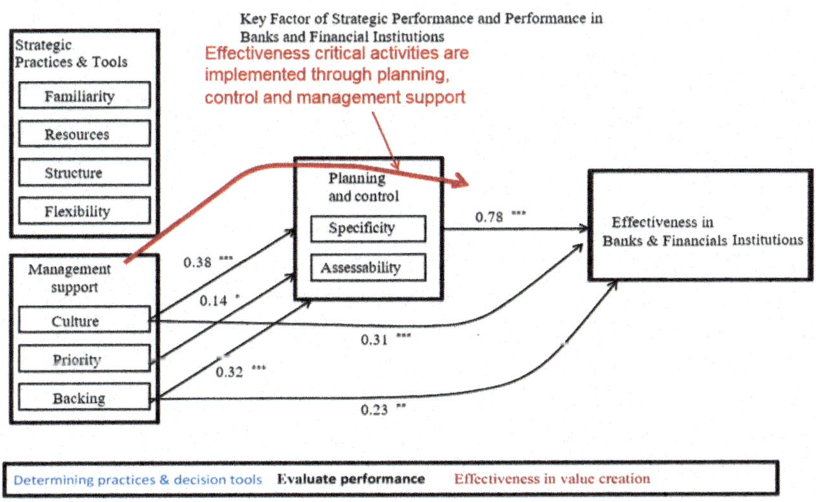

FIGURE 17.4 Determinant of the relationship between bank efficiency and strategy optimization.

Notes: N = 171, factory managers = 95, bankl managers= 76. * $p < 0.05$; ** $p < 0.01$; *** $p < 0.001$. Betas are standardized coefficients.

At the theoretical level, this research has made it possible to identify a set of organizational and market factors involved in understanding the interaction between strategy optimization and the performance of banking and financial institutions in an open economy context. Through the combined use of different models and methods, it has been possible to integrate into a global research process dimensions that, most often, have been considered separately. The approach used is in line with the recommendations of various authors to the effect that it is important to consider more than one dimension when studying the performance of companies and its impact on decision-making tools.

Moreover, our model used features rarely considered together in theoretical models, namely dysfunction and logistical breakdown. Although an important innovation, the significance of the results of this research is difficult to assess due to the limitations of current knowledge. Another contribution is the use of three interconnected dimensions of competitive advantage (production, cost, and profitability), which further corroborates the findings here. While it is true that many studies have focused on the internal and external factors of optimization in explaining the performance of firms, our study stands out by the multitude of parameters it uses and by the secondary consideration of the effect of performance on optimized management tools. It thus considers optimization as a continuous process that is part of a competitive context and competitive advantage not as a given but as a status that must be preserved and maintained. It would then be interesting to propose and test a framework integrating all of the variables measuring, on the one hand, the optimization of strategies and its inefficiency in performance and, on the other hand, the efficiency of firms in developing strategies, while noting how they change according to the evaluation of the optimization process and the conduct of changes in an open economy. Our research study proposes avenues for the development of an integrative conceptual framework that allows for the simultaneous consideration of the effects of optimization on firm performance and the impact of performance on the strategy formation process. No theoretical model has been developed to date that takes this interaction into account.

This research is the first, to our knowledge, to analyze the interactions between strategy optimization and the performance of banking and financial institutions using an integrated approach and total triangulation. Other studies that have focused on this aspect have used other frameworks that had several limitations mentioned in the theoretical section. Our approach provides a relevant model for analyzing the nature of the relationship between optimization and performance in the banking and finance sector. However, additional studies are required to validate the possibility of extending the model and its methodology to other economic contexts, other sectors of activity and other themes. In addition, weaknesses in the measurement of some of the constructs in this model indicate that special attention should be paid to the development of measurement tools in future work.

Moreover, the exploration of a double question (the first concerns the impact of strategy optimization on firm performance while the second concerns the role of firm performance in the productivity of strategy optimization tools) to understand how, at the industrial level, firm efficiency guides operational strategies and how they change according to the evaluation of the optimization process in an open economy represents an original theoretical contribution of this research. Indeed, previous work has not employed a framework that considers both the internal characteristics of financial institutions and market factors. However, the significance of the results of this research is difficult to assess due to the conceptual limitation. The theoretical and methodological conceptualization of the study of decision characteristics should therefore be encouraged. Regarding the theoretical aspects, it would be important to propose and test a framework that integrates all the variables associated with the problem identified to date. To do this, more qualitative research is needed to further explore the dynamics of the interaction between the different variables. In addition, further quantitative analysis would allow for an assessment of the importance of the associated factors in different contexts.

In order to evaluate the interaction, the use of theoretical bases is essential. In fact, the use of explanatory models has made it possible to propose a set of hypotheses that have guided the collection of data and the analysis of the relationships between the influencing factors observed and the performance of banking management practices. Furthermore, this research highlights the importance of developing measurement tools adapted to specific contexts. To do so, the combination of qualitative and quantitative approaches seems to be a strategy to be favored. In this research, qualitative approaches allowed for the adjustment of theoretical constructs, while quantitative approaches facilitated the validation of research hypotheses as well as the possibility of comparing and generalizing results.

Finally, this research proposes guidelines for the development of an integrative conceptual framework that would allow us to simultaneously consider new dimensions of our problem as well as the interrelationships between these determinants. No theoretical model has been developed to date to explain these two issues at these two levels. This research has also highlighted the relevance of considering the influence of different management factors in future research. For example, it would be relevant to conduct longitudinal studies comparing the effects of different factors on profitability levels over time. The integration of new theoretical perspectives could contribute to proposing an explanatory model of this interaction taking into account the particularity of the context, the influence of the characteristics and the environment in which they evolve.

On a practical level, this study provides information that can be directly transferred to all stakeholders interested in operational banking management in open economies. Indeed, the banking and financial sector constitutes a vast network that provides a basic model for its operational management and decision-making. The results of this research provide stakeholders with information to assess the factors related to our research question. This research also provides indications on the elements to consider for the extension of the mechanisms understudy to other management practices and other sectors.

This study provides information that can be directly transferred to all stakeholders interested in improving the performance of the banking sector. One important player could be the Bankers Association. The association is a vast network of banks and foreign affiliates operating in an open, digital economy. The findings from this research provide banking stakeholders with information on the position of bank categories in terms of efficiency. They also indicate the factors that positively or negatively impact the performance of banking institutions and how well and poorly performing banks design their strategies. This last point is important in that it may allow struggling banks to revise their strategies in order to catch up on their performance. This research also provides insights into what should be considered when studying this sector in other countries or regions.

At the national level, the responsible authorities will need to ensure that the country's economic environment is conducive to the development of the banking sector. In addition, they will need to minimize threats such as financial crises that undermine post-optimization performance. In addition, the legal and institutional system must be improved to ensure that investors' rights are better respected.

At the level of banking institutions, we have identified various elements in the management structure that could constitute strengths or weaknesses for them. At the level of internal structure, ownership structure or capital structure, we find different impacts whether we are interested in accounting returns or in the value of the bank on the markets. In general terms, the ownership structure has a positive effect on the value of banks, but beware of the concentration of capital in the hands of the managerial ownership, which could lead to problems of entrenchment, or in the hands of the majority shareholder, which could lead to the marginalization of the rights of minority shareholders. Decision-making independence increases the performance of banks and is therefore to be encouraged. Leveraged banks are better able to solve opportunistic management problems, as are banks with low default risk. We also find that innovation supports post-planning performance and improves returns. Banking institutions would benefit from investing in research and development in order to offer highly technological products and services. We also find that the size of the institution has a positive effect on returns and the size of the board on market value. In addition, human resource deficiencies are detrimental to the objectives that banks have set for themselves. It is therefore in their interest to invest in the evaluation and involvement of their employees in order to increase their societal performance. This study is also relevant in that it will allow banking companies to know which aspects to focus on to improve a particular aspect of their performance or to improve their overall performance in general.

KEYWORDS

- **macro-structural aspects**
- **micro-mechanisms**
- **micro-optimization**
- **opportunistic management problems**
- **strategic management**
- **strategy optimization**

CHAPTER 18

Transforming the Value-Based Strategy: A Reinterpretation of Structural Constraint in Action Coordination

At the heart of the discussion on the structuring of the value-based of firms in the face of digital transformation lies the weak disjunction between an approach to value creation organization as a structure, as a concrete form, and that of value creation organization as a process, as a system of intentional activities. This double facet expresses the duality of the concept. It is a duality, not a dualism. In contrast to dualism, which puts forward the existence of two essentially irreducible principles, duality refers to the existence of two essentially complementary principles. Dualism often organizes the optimization strategies of the value creation: thus, the oppositions between micro/macro, local/global, internal/external, static/dynamic, etc., are often used as a basis for the development of the value creation.

In this analytical framework article, we propose to go beyond this disjunction and to move resolutely towards a renewal by going beyond traditional dichotomies. The objective is to situate this perspective in the current debate in order to show its heuristic aspect and its operational scope for the governance of the value creation of firms at the level of digitalized and economy. The duality of the value creation structure means that the rules and resources mobilized in the action of value creation actors are at the same time the result of the action and the condition of it: the study of the structuring of the operational systems of the value creation is that of the modes by which these systems, which are anchored in the activities of competent actors, situated in time and space and making use of rules and resources in a diversity of action contexts, are produced and reproduced in the interaction of these actors (Giddens, 1987) of the value creation.

Value-Based Management in an Open Economy: Optimizing Strategy to Improve Business and Performance. Prof. Post-Dr Walter Amedzro St-Hilaire (Author)

This conception leads to a reinterpretation of the notion of structure and structural constraint of the value creation: the structures of the value creation, because they are produced and reproduced, are simultaneously constituted and constitutive. On the one hand, the structure of the value creation is not external to the actors of the value creation; it is constitutive of their actions. On the other hand, it is the framework that allows this action. The structure of the value creation is not only a constraint, it is both constraining and enabling (Giddens, 1987). This approach rejects both the over-socialized and under-socialized vision of the actors in the value creation. The value creation of firms in digitalized and economy are built by digital transformation actors, whose actions are both facilitated and constrained by the value creation structure and available resources of the operational networks in which they are embedded.

1. **The structured and structuring nature of the action relevant to the analysis of digital transformation:** Due to the original methodological posture adopted and the heuristic nature of the construct, the structuring model offers a relevant and robust framework for analyzing the operational interactions of firms in digitalized and economy. It allows us to put forward a concept of value creation organizations that designates two essential properties of any system of activities: the structured and structuring nature of the action. The model is based on three fundamental statements. The first is the assertion that, in any operational system, the interactions between the actors in the value creation are structured and structuring. This differentiation between the structuring and the structured in the operational interaction makes it possible to propose a relevant distinction between structure and system of the value creation. The vision thus leads to a concept of digital transformation structure as a set of rules and resources that enable interaction between actors in the value creation.

   The second statement indicates that the structural aspect of the value creation of firms is made up of three dimensions: a semantic dimension that allows operational actors to give meaning to their actions and their professional relationships. It helps reduce the uncertainty linked to any technique. The rules allow mutual understanding between the actors of the value creation. They are the support for operational communication in digital transformation; they induce common patterns of interpretation. Interpretation schemes are the modes of representation and classification that are inherent to the reservoirs of technical knowledge of the actors in the value creation and that they use reflexively in their communications (Giddens, 1987).

It is also this semantic requirement of interaction in digital transformation that seems to be put forward when the concept of interpretation is introduced to designate the objective of all interaction in digitalized and economy: to achieve a common understanding by the actors in the value creation of the situation and its meaning (Salais, 1994). A dimension of power and domination at the value creation level. Because it is inherent to any organizational action, power is the ability to continuously deploy a battery of causal capabilities, including the ability to influence the causal capabilities deployed by other stakeholders (Giddens, 1987) in digitalized and economy. Concretely, this capability is expressed in the development of artificial intelligence rules and in the control of operational and strategic action resources. The rules, which are more or less codified structures for digital transformation, have a strong normative content. They influence to a certain extent the strategies of the value creation' actors. A dimension of legitimization of operational action. Compliance with the rule becomes a means of legitimizing the digital transformation action.

The third statement takes into account the interaction between the technical actors of the value creation; this interaction is constituted according to a double perspective. A structural component that refers to the development in space and time of regularized patterns of relationships that reproduce technical activities. While the system designates the space of interaction of the actors in the value creation, the structural designates the set of rules and resources organized in a recursive manner (Giddens, 1987) that the operational actors concretely implement to produce and reproduce their interactions at the level of the value creation. A systemic component that designates the operational system as the organized space of recurrent relations between actors in the value creation. These relationships, reproduced, and organized as regular strategic practices (Giddens, 1987), are built on the activities of the value creation actors. In other words, an operational system (organized digital space of interactions) only appears through the activities of the value creation actors, activities that produce and reproduce the system.

Thus, the operational analysis grid inherent in structuring processes makes it possible to apply the theoretical elements previously put forward to any place and place of interaction in the value creation; whether, therefore, at the level of operational action constructs in general or, more precisely, on the occasion of a more localized transformation of action structures. In these terms, we will successively consider the ins and outs of the structuring features characterizing at the level of digitalized and economy, on the one hand, firms as a structure and, on the other, digital transformation as a process.

2. **The organization for the digital transformation of the value creation: forms and dynamics of activities:** The operational system of the value creation for the digital and open economy does not depend on the activity of a particular technical actor or digital device, and yet it would cease to exist if the operational and technical actors in the value creation disappeared. This reality can be approached in two equally legitimate ways: either we look at the activities of the operational actors who carry each block, their technical relations, etc.; or we bracket these activities to apprehend only the overall figure in its form and persistence. The fact remains that these are two sides of the same reality that are thus apprehended (Eraly, 1988).

The fundamental idea we infer here is that value creation structures, a set of rules and resources, organize operational and strategic activities just as much as technical activities organize them and give them meaning and purpose at the level of digitalized and economy. Just as they are the condition of technical activity, the structures of the value creation do not exist independently of the activities of the technical actors who invest in them. The theory thus leads to a concept of value creation structure as a set of digital rules and resources that allow interaction between operational actors. Digital rules provide a good illustration of the duality of the structural in the value creation. In this operational communication, the respect of technical and digital rules is a constraint. At the same time, they are what make this operational communication possible. The action contains the structure of action at the same time as it receives its obligatory form and possibility (Eraly, 1988).

The structure-activity-technical and digital feedback is at the heart of the duality of the structural and the systemic paradigm of the value creation. It expresses the fact that in the value creation of Digitalized and economy firms, a system of actions, technical or digital actors mobilize rules and resources to produce and reproduce their operational activities in particular time and space contexts and for shared technical purposes. The introduction of time and space to specify the role of technical and digitized actors allows us to conceive in digitalized and economy as a non-homogeneous whole; as a plurality of spatial and temporal situations in which different specific actors located at different levels can implement different rules and resources. In fact, the use of differentiation is much stronger. In the value creation, the concept of positioning is introduced to signify the existence of several technical and digitized processes of structuring according to the different technical positions and in the connection of this with the different environments. This refers to the notion of rule in its operative content. We will come back later on to the conception of the rule that is best able to account for the constraint-enablement duality in the value creation. Let us note, already, that this approach contrasts with the status of the rule-constraint in traditional organizational models.

Moreover, it seems to us that the junction we want to show between the concept of structure and that of action at the level of the value creation of firms helps to resolve the analytical difficulty that lies behind the generic use of the term. The concept refers to two essential properties of a system of activities: the structured and structuring nature of action in digitalized and economy. This duality of the organized and the organizing in the digital transformation of the value creation of firms simultaneously allows us to account for the dynamics of digital transformation, through the study of the activity processes of the value creation; and the relative invariance of its forms, the structures of the value creation. The duality of the value creation designates both a state and a dynamic. The state refers to an object, i.e., technology in all its forms in digitalized and economy. Dynamics refers to the digitized processes by which technical actors adjust their strategies and coordinate their methods in the pursuit of operational and strategic action (Friedberg, 1993) in the value creation optimization. The structuring of the value creation in digitalized and economy calls for the development of a concept of strategic action, especially in relation to the necessary coordination.

3. **Coordination of action in the value creation: another emergence of organizations?:** The theory of action and coordination is characterized by the extreme heterogeneity of the concepts used. This

heterogeneity overlaps with divergent traditions of thought and epistemological options. The divide between holism and methodological specificities is also significant in terms of the distribution of approaches. While most authors only refer to the notion of action, in order to speak of the actions of several actors, if the conjunction of their acts makes it possible to observe a certain order, a certain coordination (Livet and Thevenot, 1994), the explanations for the emergence of the resulting order are largely different. Two main explanations organize the emergence of action figures: figures emerge as the technical result of the aggregation of decisions. This is the normative approach of the strategies of the standard model. It is the explanation by the invisible hand. The normative model is a model of reconstruction of reality to which the adequacy or otherwise of the logic described as rational is related. Rationality constitutes, in this case, an ex-post conformity to what emerges from the strategies of the actors in the value creation. These figures are intentional modes of coordination that serve as a reference for strategies and are the results of these strategies.

These two explanations are obviously antinomic. However, a bridge must be found between them if we want to constitute a theory of action, that is to say, a theory that accounts for both strategic and operational action. For Thevenot, for example, the overcoming of dualism encourages us to consider actions as part of a framework of common references. This framework is built around the imperatives of coordination of action. Indeed, coordination seems to be a relevant entry point in an attempt to apprehend action simultaneously. While this statement is obvious, it requires some explanation in the case of strategic action. In what way does strategic action induce a coordination load in the value creation in digitalized and economy?

Two elements may provide the answer to this question. One of the founding hypotheses of conventionalist analysis, which marks a break with the orthodox problematic (Quere, 1993), is that strategies are not established between actors who are free of all external references. In the most trivial of daily actions, actors constantly refer to norms, rules, and routines. This reference consists of a constant adjustment of actions in relation to these references. The second element of the answer lies in the teleonomic dimension of action. An action contains the search for coherence between the ends pursued and the means used: we develop the action that suits the ends pursued according to the capacities and the context. This reflexive character, including of strategic action, induces, at least, a coordination between the cognitive domain and the conative domain of the actors in the value creation; between the domain of reason and their tendencies to act. In other words, strategic action in the value creation presupposes a dose of coordination in so far as it is always carried by a teleological rationality of adequacy between an end and the means.

From the point of view of artificial intelligence, the fact that this coordination involves only one actor does not change anything: coordination is an artificial process. The involvement of other actors in the value creation, the interaction in operational strategies, only introduces a higher degree of coordination between technical actors. As soon as the action involves a large number of actors, as soon as it requires the use of more or less formalized, more, or less stabilized digital techniques; but, in all cases, coordination does not go without saying and calls for systems of conventions, rules, standards, procedures, routines. In short, it calls on organizations. The value creation of firms thus becomes the natural place for implementing these technical and digital figures.

4. **Two mechanisms for coordinating action in digitalized and economy: the rule and the procedure:** The model of the duality of the structural would make the value creation a system of finalized and hierarchical activities carried by collective figures of coordination of the action. Conventions, rules, procedures, and routines are the supports of this coordination. In the paradigm of conventions, there is a consensus widely shared by the different authors that rules, routines, and procedures are an essential pathway for building a theory of action. This consensus stops at the scope and, very often, the meaning to be given to these effectively polysemous notions. Rather than exploring this diversity, we will set out a few points of reference that seem fundamental to us in order to reduce this polysemy by drawing on the achievements of the duality of the structural.

First of all, a rule is a guide for possible actions. As a result, it is inseparable from the context of the action, and is only effective in particular situations. Its statement can therefore only be general and abstract. This generality also indicates its incompleteness: it cannot establish, nor foresee in a precise way, all the real situations of the interaction. Rather, through a rule, actors possess a capacity for intervention in an indeterminate set of circumstances (Giddens, 1987). This capacity to intervene makes it possible to construct a set of strategic actions and possible decisions (Dupuy and Kechidi, 1996). It is important to note that the rule does not set all the terms of the action, it is an interpretative framework for the action. This implies, and this is probably also where its generality lies, that a rule can take on a plurality of operating modes, a plurality of forms of concrete action. As a result, a rule is transferable without this transferability altering performance through errors or interpretations.

At the same time, and because it is based on a logic of applicability, a rule must be practicable (Reynaud, 2001). A statement that is too general and therefore interpretable at will is not a rule. It is simultaneously a limitation of its generality and its interpretability that makes the rule a practicable process. In other words, to be practicable, a technical rule must stabilize the universe of action of the actors in the value creation. This stabilization concerns both the context of the action and the domains of the action. A practicable rule is a rule that participates in the reduction of the complexity of the action contexts. It is selective in the representation of the variables of the numerical environment to be considered, it simplifies the field of possible strategies. As a result, the existence of rules saves the mobilization of the cognitive capacities of the operational actors in the action. Not only is it no longer necessary to consult each other in order to act (Midler, 1994), but it is no longer necessary to have an exhaustive representation of the context of the action in order to act. This, in any case, is sensibly impossible. At the same time, by limiting the domains of possible actions (desired, prohibited, obligatory actions), the rule reduces the participants' fields of action, or, more precisely, directs their actions towards the achievement of a singular goal. These two limitations on the deployment of the cognitive and conative capacities of the actors are, it seems to us, the basis of the economy of resources that coordination by rules induces.

As an enabling intelligence structure, a rule is also binding. As a model of artificial intelligence, it exerts a certain power of obligation on the strategies of the actors in the value creation. It presents itself as a prescription to which it is possible to confirm. It also indicates, in relation to determined contexts, which technique is required, preferred, or prohibited (Daraut, 2004). It is this dimension of regulation of strategies that makes Favereau write that the rule is not just

> the statement of a simple regularity: it is a normative statement. This normative character of the creation of values distinguishes the rule from a routine, which has a stronger prescriptive power. For example, for a robot X, going into functional mode each time it detects a command is the statement of a regularity. Because this statement is not a prescription to which X would submit, but especially because it does not include the idea of sanction, it cannot be assimilated to a rule. It is a routine operating practice. In order to be a rule, a provision of action must include the idea of sanction. However, we must not attribute to the sanction a coercive content from the outset. The degree of sanction is linked to the formalization of the rule. In this aspect alone, going into functional mode every time it detects a command can turn into a strong prescription in particular situations and where its transgression would lead to sanctions. The more a rule is presented in the form of a formal prescription, the stronger the charge associated with the sanction.

Finally, a rule is always finalized; it only makes sense in a given situation and for a more or less stated purpose. Beyond the general function of any rule (to allow and ensure the coordination of actions), a rule always relates to a particular result or at least to a desired consequence. It is precisely the intentionality contained in the rule that makes it exist and determines the concrete modalities of its implementation. The finality is a property revealed by the application of the rule. From this point of view, to act intentionally is to follow a rule in order to achieve its goal. This statement should not lead one to think that the rule contains all the action, determines the artificial action for the value creation. There are two main reasons for this, which we borrow from Livet (1994): firstly, because action is, to a certain extent, revisable in the course of action: it is the nature of action to have to adapt to changing environments; and secondly, because at the start of any action, the goal is never fully formed.

The purpose of an action is never totally saturated: the work of specifying the conditions of the action can never go so far as to define strictly what the action corresponding to such and such an end must be. At this point, it must be recognized that a significant portion of the activity in a digital firm's value creation proceeds from numerical patterns of execution, artificial executory and non-deliberative activities. These activities are aimed at achieving broadly identified goals. Because they are precisely executable, these patterns leave no room for revision in the course of action. Does this mean that some rules are interpretable and others are not? In fact, if we introduce the dimension of power and domination, the third dimension of the structural in Giddens' work, we can show that when rules take on the aspect of strong prescriptions that totally organize the behavior of the actors in the value creation, they take on the form of procedures, of deontic modalities of artificial action.

Let us first give content to the technical and operational procedure. The procedure has as its direct object the arrangement of a strategic action or, more precisely, of an operational conduct. It prescribes the strategy, in the sense of a description of artificial acts, corresponding to a given situation. Thus, the procedure is stated in the imperative: x must do a. The nature of the connection determines the normative content of the statement: it is prescriptive (must do), prohibitive (must not do) or permissive (can do). If x cannot do a, in a given situation, if therefore the artificial procedure cannot be respected, we have recourse to the rule, i.e., to an interpretative framework. This recourse concerns either the reinterpretation of the context or the reconsideration of the corresponding type of action. The rule can thus be seen as a recourse to a process of solving problems not explicitly codified by the procedures. In the value creation, the coordination of technical actions by rules, by interpretable artificial devices, is only ever ensured by default of procedures, non-digitizable devices. This vision leads to attributing to the procedure a completeness that distinguishes it from the incompleteness of the rule.

The completeness of the operational procedure should not be assessed in absolute terms. A procedure is only complete in relation to the purpose towards which it is directed. If the purpose of the action is saturated, then the action is also saturated. This means that there are modes of action where the action no longer lends itself to revision during the course of action (repetitive work, for example). It is transformed into numerical patterns of execution of a saturated purpose. The saturation of the purpose of a strategic action in the value creation by a digitized procedure is generally the result of the crystallization of an artificial learning process. It expresses the fact that previous practices have led to a precise knowledge of the concrete modalities of action in the face of recurrent situations. These concrete modalities are presented as formalized prescriptions.

The deontic vision of the process induces a re-examination of the position of the different actors in the value creation. Reconsidering this position does not mean giving up the status of actors in the operational process. This re-examination concerns the perception of the environment of the action and the constraining character of the procedures for its deployment. In digitalized and economy and its digital transformation, the perception of the environment of an action likely to take the form of a procedure at the level of firms is the responsibility of a particular category of actors in the value creation. In a technically organized relationship, two types of actors are in a situation: actors who have the power to develop procedures and enforce them, and those whose action is imposed and who therefore do not have control over the contexts in which they are activated.

Between these two actors, the relationship is asymmetrical: the power of the former is a power to organize the strategic action of the latter. This relationship can, for example, take the form of a strict authority relationship. Here we find a classic definition of power: power is an asymmetrical relationship that expresses the capacity to intentionally exert influence on the strategy of the actors (Dupuy and Kechidi, 1996). This notion of power and domination strongly influences technical interactions (Giddens, 1987). Contrary to the rule of digitization in the value creation, the procedure dispossesses the actors of their powers of technical interpretation. It turns them into technical operators. This dispossession concerns both the interpretation of the situation and the corresponding form of action. More precisely, the procedure solicits less cognitive and conative capacities than the rule. From this point of view, because it transfers the power of technical interpretation to a particular type of actor, the procedure saves cognitive resources. Therefore, a procedure can be defined as a program established at a central or local level of the value creation in digitalized and economy and consisting of a set of digitized execution patterns that drive the resolution of recurring operational problems. It is these procedures that feed the structural memory in the form of codes of conduct, execution patterns and formalized digital practices. This formalization can allow the strategy (March and Simon, 1999) of the operational actors in the value creation to be predicted in detail.

The preceding developments have attempted to propose distinctions between the operative rule and the technical procedure in order to make what is generally understood by these two conventional figures of coordination of strategic action practicable. Ultimately, because they are, to a greater or lesser extent, loaded with prescriptions, the numerical figures of coordination of action base action on common determinations that act as a framework of coherence and homogenization of strategies. As a result, technical facts can only be accounted for by: on the one hand, grasping them as unequally stabilized products of the composition of a set of actions and, on the other hand, by considering that the characteristics of the action situation constituted by the interdependencies between actors and institutions provide the intelligence of the motives and instruments that the actors in the value creation use to act (Paradeise, 1991). It is only from this perspective that an approach to strategic action calls for the notion of coordination and the technical figures of this coordination. Only because in a normative perspective, with the invisible hand, the norm-convention appears as the technical result of numerical actions.

The fundamental epistemological concern in structuring the value creation of firms is to go beyond the operational structure/strategic action divide in digital transformation. We seek to link the ideas of structure and action in an inseparable way, replacing the concept of structure by that of structural sets, the latter being spatiotemporally situated manifestations based on the set of technical rules and operational resources involved in the institutional articulation of digitized systems. The structuring model therefore succeeds in overcoming the dualism between the specific and holistic methodological postures through a definition of the value creation by the inter-actor relationships.

Two imperialisms are thus called into question: that of the operational actors, on the one hand, and that of the structural constraints according to the structuralist and functionalist approaches, on the other. Thus, the theory of structuration conceptualizes again in the context of the value creation of digitalized and economy the opposition between subjectivism and objectivism by proposing a new duality: the duality of the structural expressing the structured and structuring character of technical and digitalized interactions.

To construct the duality of the structure of the value creation, the author starts from the refusal of the opposition between the organization structure and the organization-action. He develops an approach that captures both the technical structure and the strategic action in a theory of structuring of and in the organization. The organization is, in the same movement, a set of structural properties and a set of action properties. This unitary theory of structure and action is a theory of the constraining and the enabling. The duality of the structure also means that the rules and resources mobilized in the action of the actors of the value creation are at the same time the result of strategic action and the condition of the latter.

It is this positioning that allows some actors of the value creation to situate the approach to the constitution of the digital, in a constructivist perspective, while for others, it is close to constructivist structuralism. This second viewpoint is based on the recognition of theoretical affinities between the two theoretical postures as well as on the same confrontations with the same issues and the same research questions. While structuration theory basically claims to go beyond controversies, it does not focus on epistemological aspects. We have seen that the undeniable contribution lies in a dual approach that accredits the idea that conation and digitalized artificial intelligence are two inseparable dimensions. It is this approach that provides the argument of the structuring and structured aspect of technical interactions for digitalized and economy and the digital transformation of the value creation.

This perspective will allow future research to examine how firms' engagement in a specific dimension of value creation activities facilitates the improvement of their position based on access to key resources or a relationship with specific actors on the digitalized and economy platform. This reorganization of the value creation structure among firms breaks down the barriers of space and time, which often characterize the disaggregated and geographically dispersed activities of the digitalized and economy value creation. These platforms as a value base can have complex implications for firms, their value creation activities, relationships, and participation or orchestration (Kano et al., 2020).

KEYWORDS

- **analytical framework**
- **asymmetrical**
- **deontic vision**
- **digitalized and economy**
- **goods and services**
- **mobilization**

Conclusion

Moving beyond partial empirical approaches has led to the choice of the specific factor model as the best suited to capture the interactions between firm efficiency and policy optimization in the open economy banking sector. While it is hoped that some aspects of this model can be extended over time, both in the past and in the future, it is unrealistic to assume that it can be applied systematically to other economies. At the very least, such a claim would have to be carefully substantiated. A more difficult task would be to formulate a less parabolic model in order to run simulations with it and obtain forecasts on different aspects of the interaction between business efficiency and strategy optimization: what kind of investments are needed and in which categories of banks?

In order to take into account, the relationship between strategy optimization and performance, especially in the context of an open economy, several sources of data and techniques were mobilized to bring out a common denominator of the various situations and points of view that would make it possible to explain this interaction. At the theoretical level, this study has identified a set of internal and external factors that influence the interaction between optimization and performance. Through the combined use of different theoretical frameworks and methods, it has been possible to integrate dimensions that have usually been considered separately into a comprehensive research process. The approach used is therefore in line with the recommendations of various authors who advocate considering more than one dimension when studying a phenomenon. Furthermore, our model has used characteristics that are rarely considered together in theoretical models, namely dysfunction and logistical breakdown. Although an important innovation, the significance of the results of this research is difficult to assess due to the limitations of current knowledge. Another contribution is the use of three interconnected dimensions of competitive advantage (production, cost, and profitability), which further corroborates the findings here. While it is true that many studies have focused on the internal and external factors of optimization in explaining the optimization of firms, our study stands out by the multitude of parameters it uses and by the secondary consideration of the effect of optimization on the tools of optimized mechanisms control. It thus considers optimization as a continuous process in a competitive context and competitive advantage not as an achievement but as a status that must be preserved and maintained. The models found in the literature are considered too partial or biased because, for the most part, they are too static or just focused on one aspect of the relationship (activity portfolios vs. product portfolios). It is, therefore, necessary to go beyond them by creating a model that will not only integrate all of their assumptions and "other variables" but will also be more dynamic, to explain the interaction between strategy optimization and performance.

Therefore, it would be interesting to propose and test a framework integrating all the variables measuring, on the one hand, the impact of strategy optimization and its ineffectiveness on performance and, on the other hand, the effectiveness of firms and industry in the development of strategies, while noting how they change according to the evaluation of the optimization process and the conduct of changes in an open economy.

Our study thus proposes avenues for the development of an integrative conceptual framework that allows for the simultaneous consideration of the effects of optimization on firm performance and the impact of optimization on the strategy formation process. No theoretical model has been developed to date that takes this interaction into account.

At the end of this study, certain lessons can be drawn. First, the concept of factor specificity and the hypothesis of the plurality of mechanisms seem to complement each other and enrich the analysis of the interactions between company productivity and strategy optimization. The notion of factor specificity allows to account for the heterogeneity of the design factor. Thus, the existence of several qualities of the mechanism is explained in particular by different investments in decision tools. These management qualities give each company a certain gain in profitability. Our own calculations have shown that Canada's exports seem to be rather intensive in skilled applied processes. The notion of factorial specificity takes on its full meaning when interpreted with the hypothesis of the plurality of mechanisms. Indeed, each quality corresponds, in theory, to a strategic plan. Our calculations also show that these plans, far from being transversal, are, on the contrary, specifically linked to certain groups of firms: those whose comparative advantage is revealed by the exchange.

Everything seems to indicate that the various goods produced are produced using techniques that require a certain specific quality of work. Under these conditions, the specific factors model seemed to be the most appropriate one to interpret the reality of the Canadian case. The hypothesis of the plurality of strategy optimization also appears to be fruitful in the analysis of the interactions between firm productivity and strategy optimization in models of imperfect competition. Thus, it seems that a category of banks characterized by monopolistic competition conditions in the product market has a homogeneous behavior on strategy optimization. Moreover, and our calculations confirm this point, the opening to trade has the effect of restoring competitive structures, not only in the product market but also in the factor market. There seems to be a strong basis for developing models that incorporate the notion of specific factors in imperfect market structures. In particular, the link between firm efficiency and plan failure rate should be further explored, using developments in the theory of the optimization process. In this respect, our calculations did not allow us to come to any conclusions. At most, we can consider that the failure rate of the plan linked to the efficiency of the firms comes, if we accept the analytical framework of the specific factor model, from the non-mobility of labor and the non-competitive functioning of the market of this factor. In this perspective, developments in the theory of firm efficiency tend to reduce, under certain conditions, the rate of plan failure. An intuitive explanation of this phenomenon would be to return to the explanation of exchange as an outlet for surplus, which MYINT discovered and highlighted in the work of Adam SMITH. The opening up to exchange, by increasing the value of goods, offers them an outlet. They can even be produced in greater quantities. From then on, labor, a specific form of labor that is underemployed, once again finds its use.

A second conclusion concerns the need to address simultaneously, in theoretical approaches and empirical work, the dual aspect of analyzing the interactions between firm efficiency and strategy optimization. As the models studied in the first part show, each set of hypotheses leads to a particular set of conclusions concerning both the effects of firm efficiency on strategy optimization and the role of strategy performance in determining value-based management and risk mechanisms.

For effective preventive measures, the designing/development and implementation of a strong value-based management system is a prerequisite. All efforts for minimizing the risk will go in vain if the Control system is weak and ineffective. Any loophole in this system itself invites misappropriation, misuse, unauthorized disbursement, and non-recovery due to either connivance or ignorance. An effective system ensures risk-free operation for the achievement of this objective, emphasizing upon preventive measures instead of a detective. The value-based management system is responsible for plugging these loopholes by identifying the discrepancies, irregularities, and malpractice, guiding the organization to improve its performance through strict vigilance, close cooperation among the various tiers of the organization, and developing the Human Resources to shoulder their responsibilities with all honesty, efficiency, and diligence. Conventionally banks rely on the periodical audit of accounts and inspection. This exercise is beneficial only to the extent of identification and detection of irregularities and discrepancies. However, the basic issue of risk remains intact because its elimination or, for that matter, mitigation preventive measures are required to be suggested and implemented, which audit and inspection fail to ensure. The observations in audit reports are, no doubt, dealt with through compliance but the process of compliance remains restricted to the extent of removal of the discrepancies detected. An independent value-based management system will cater to the overall efficiency needs of the organization. The system will not only be confined to the monitoring and supervision but give equal importance to the skill development and behavioral improvement of the Human Resources, which in fact, plays a decisive role in the risk-minimizing process of the organization.

We wanted to show that partial empirical approaches had to be overcome and that a single model, however imperfect, could be identified that would allow both aspects of the question to be answered simultaneously. This model can only be a first, very vague draft. More precise empirical studies, with more elaborate technical means, however, could make it possible to improve, if not the principle, at least the precision of the empirical results and thus give them a predictive value.

Throughout this book, we have tried to identify the different empirical and theoretical aspects of the interactions between firm productivity and strategy optimization. In doing so, we have remained consistent with the traditional definition of the open economy as a trade balance, i.e., the difference between exports of goods and services (EX) and imports of goods and services (IM), i.e., EX – IM, can be positive or negative. Without ignoring the fundamental role that interest rate parity can play in this respect, it seemed preferable to reserve this aspect for later studies.

Bibliography

Aghion, P., & Bolton, P., (1987). Contracts as barriers to entry. *American Economic Review, 77*, 388–401.

Aglietta, M., & Breton, R., (2001). Financial systems, corporate control and capital accumulation. *Economy and Society, 30*(4), 433–466.

Akeel, M. L., & Dennis, W. T., (2012). Governance characteristics and role effectiveness of audit structures. *Managerial Auditing Journal, 27*(4), 336–354.

Aldrich, H., & Mindlin, S., (1978). Uncertainty and dependence: Two perspectives on environment. In: Karpik, L., (ed.), *Organization and Environment: Theory, Issues and Reality*. Beverly Hills, CA: Sage.

Amabile, T. M., (1983). Social psychology of creativity: A componential conceptualization. *Journal of Personality and Social Psychology, 45*, 357–376.

Amedzro St-Hilaire, W., (2011). *Management of Telecoms in Changing States: Challenges and Strategies* (p. 187). Ed. Presses of the University of Quebec.

Amedzro St-Hilaire, W., (2011). *Organizational Adaptation in Managerial and Social Theories* (p. 167). Ed. Presses of the University of Quebec.

Amedzro St-Hilaire, W., (2011). Strategies with respect to the organization's business strategies in order to gain competitive edge. *Business Management and Economics European Union Journal, 17*(2).

Amedzro St-Hilaire, W., (2011). The strategic impact of the business dynamics in emerging countries on contemporary perspectives. *Business Intelligence Journal – BIJ (London), 4*(2).

Amedzro St-Hilaire, W., (2014). *Foundations and Methods in Applied Management*. L'Harmattan, Col. Pour Comprendre.

Amedzro St-Hilaire, W., (2018). *Industrial Relations Research and Analysis* (p. 475). Coll. Business & Management Ed. World Scientific Publishing.

Amedzro St-Hilaire, W., (2019). *Entrepreneurship Strategies & Policies* (p. 314). Coll. Business Management Ed. Apple Academic Press, CRC Press and Taylor & Francis Group.

Amedzro St-Hilaire, W., (2019). The strategic impact of performance appraisal on corporate governance dynamics. *Journal of Management Development, 38*(5).

Amedzro St-Hilaire, W., (2020). *Applied Managing for Entrepreneurship* (p. 314). Coll. Business Management; Ed. Apple Academic Press, CRC Press and Taylor & Francis Group.

Amedzro St-Hilaire, W., (2020). Key-synergies between strategic resources and operational tool orchestration for organization development. *Journal of Management Development, 39*(6), 811–822.

Amedzro St-Hilaire, W., (2021). Leading with digital technologies governance in the state-owned enterprises. International *Journal of Public Administration*. https://doi.org/10.1080/01900692.2021.1993898.

Amedzro St-Hilaire, W., (2021). Reconnecting governance configuration variables to enterprises contingency factors for optimizing productivity at the start-up level. *Journal of Enterprise Information Management, 34*(6), 1630–1648. https://doi.org/10.1108/JEIM-12-2019-0410.

Amedzro St-Hilaire, W., (2022). Disruptive leadership for organizations: strategies & innovations in the digitalized economy. *Coll. Management & Business Administration,* (p. 412). Ed. Routledge.

Amedzro St-Hilaire, W., (2022). Multidimensional and global indicators for government start-ups' risk decisions-making. *International Journal of Public Administration* https://doi.org/10.1080/01900692.2021.1998109.

Amedzro St-Hilaire, W., & Boisselier, P., (2018). Evaluating profitability strategies and the determinants of the risk performance of sectoral and banking institutions. *Journal of Economic and Administrative Sciences, 34*(3).

Amedzro St-Hilaire, W., & Boisselier, P., (2019). The coordinated strategy for the optimization of the interaction level of business model. *Journal of Economic and Administrative Sciences, 35*(2).

Amedzro St-Hilaire, W., & De La Robertie, C., (2018). Correlates of affective comment in organizational performance: Multi-level perspectives. *Australian Journal of Career Development, 27*(1).

Amedzro St-Hilaire, W., & De La Robertie, C., (2018). The empirical impacts of risk-management on the enhancement of value-based mechanisms in organizational studies. *Research in Management Sciences*, 123.

Amedzro St-Hilaire, W., & Prasad, B., (2018). The governance correlation matrix of the financial institutions. *Management & Applied Economics Review, 26*(1).

Amit, P. L., & Schoemaker, J. H., (1993). Strategic assets and organizational rent. *Strategic Management Journal, 14*, 33–46.

Anderson, R. C., Mansi, S. A., & Reeb, D. M., (2004). Structure characteristics, accounting report integrity, and the cost of debt. *Journal of accounting and Economics, 37*(3), 315–342.

Ansoff, H. I., (1968). *Business Development Strategy*. Men and techniques edition, Paris.

Ansoff, H. I., (1991). Critique of henry Mintzberg, the design school–reconsidering the basic premises of strategic management. *Strategic Management Journal, 12*, 6.

Ansoff, H. I., Avner, J., Brandenburg, R. G., Portner, F. E., & Radosevich, R., (1970). Does planning pay? The effect of planning on success of acquisitions in American firms. *Long Range Planning, 3,* 1–7.

Ansoff, I., (1995). *Corporate Strategy: An Analytic Approach to Business Policy for Growth and Expansion*. New York: McGraw-Hill.

Anthony, M. S., (1997). *Commercial Bank Risk Management: An Analysis of the Process* (1st edn.). The University of Pennsylvania: The Wharton Financial Institution Center.

Aram, J. D., & Cowan, S. S., (1990). Strategic planning for increased profit in small business. *Long Range Planning, 23*(6), 63–70.

Archambeault, D., & DeZoort, F. T., (2001). Auditor opinion shopping and the audit structure: An analysis of suspicious auditor switches. *International Journal of Auditing, 5*(1), 33–52.

Armour, S., (2014). *Bank Risk Management* (1st edn.). Washington: The Wall Street Journal.

Armstrong, J. S., (1982). The value of formal planning for strategic decisions. *Strategic Management Journal, 3*(3), 197–211.

ASX, (2003). *Principles of Good Strategic planning Practice and Best Practice Recommendations.*

Atuahene-Gima, K., & Li, H., (2004). Strategic decision comprehensiveness and new product development outcomes in new technology ventures. *Academy of Management Journal, 47*(4), 583–597.

Baden-Fuller, C., & Volberda, H. W., (1997). Strategic renewal in large complex organizations: A competence-based view. In Heene, A., & Sanchez, R., (eds.), *Competence-Based Strategic Management*. John Wiley & Sons.

Banker, R. D., Field, J. M., Schroeder, R. G., & Sinha, K. K., (1996). Impact of work teams on manufacturing productivity: A longitudinal field study. *Academy of Management Journal, 30*, 867–890.

Baysinger, B. D., & Butler, H. N., (1985). Strategic planning and the structure of directors: Productivity effects of changes in structure composition. *Journal of Law, Economics, & Organization, 1*(1), 101–124.

Becker, B. E., & Gerhart, B., (1996). The impact of human resource management on organizational perform: Progress and prospects. *Academy of Management Journal, 39*(4), 779–801.

Bédard, G. M., & Simard, G., (1996). *At the Dawn of the 21st Century: Challenges for Management Sciences*. Academic Ed Guérin.

Belleflamme, P., & Peitz, M., (2010). *Industrial Organization: Markets and Strategies*. Cambridge University Press.

Ben-Shakher, G., Bar-Hillel, M., Bilu, Y., & Shefler, G., (1998). Seek and ye shall find: Test results are what you hypothesize they are. *Journal of Behavioral Decision Making, 11*, 235–249.

Ben, A. L., Dubrocard, A., & Tudor, H., (2010). *Productivity, Innovation and Perceived Effects in Luxembourg Companies, Economy, and Statistics*. STATEC working papers and Competitiveness Observatory.

Benartzi, S., & Thaler, R., (2001). Naive diversification strategies in defined contribution saving plans. *American Economic Review* (Vol. 91, No. 1, pp. 79–98). American Economic Association.

Berck, P., & Perloff, J., (1988). The dynamic annihilation of a rational competitive fringe by a low-cost dominant firm. *Journal of Economic Dynamics and Control, 12*, 659–678.

Bergh, D. D., & Lim, E. N., (2008). Learning how to restructure: Absorptive capacity and improvised views of restructuring actions and execution. *Strategic Management Journal, 29*, 593–616.

Bergh, D. D., & Lim, E. N., (2008). Learning how to restructure: Absorptive capacity and improvisational views of restructuring actions and productivity. *Strategic Management Journal, 29*, 593–616.

Bertrand, M., & Mullainathan, S., (2003). Enjoying the quiet life? Strategic planning and managerial preferences. *Journal of Political Economy, 111*(5), 1043–1075.

Black, J. A., & Boal, K. B., (1994). Strategic resources: Traits, configurations and paths to sustainable competitive advantage. *Strategic Management Journal, 15*, 131–148.

Black, S. E., & Lynch, L. M., (2001). How to compete: The impact of workplace pracoutils de prise de décisiones and information technology on productivity. *The Review of Economics and Statisoutils Stratégiques, 83*(3), 434–445.

Blau, P. M., (1982). Structural sociology and network analysis. In: Marsden, P. V., & Lin, N., (eds.), *Social Structure and Network Analysis* (pp. 273–279). Beverly Hills, CA: Sage.

Boddy, D., Magbeth, D., & Wagner, B., (2000). Implementing collaboration between organizations: An empirical study of supply chain partnering. *Journal of Management Studies, 37*, 1003–1017.

Boron, R., & Kenny, D., (1986). The moderator-mediator variable distinction in social psychology research: Concept, strategic, and statistical considerations. *Journal of Personality and Social Psychology, 51*, 1173–1182.

Boudon, R., (1977). *Perverse Effects and Social Order*. Paris: PUF.
Bourdieu, P., (1980). *Le Sens Pratique*. Paris: Éditions de Minuit.
Bourgeois, L. J., (1980). Execution and consensus. *Strategic Management Journal, 1*, 227–248.
Bourguinat, H., (1992). *Finance International*, PUF (eds.), Paris.
Boyd, B. K., & Reuning-Elliott, E., (1998). A measurement model of strategic planning. *Strategic Management Journal, 19,* 181–192.
Breidbach, C. F., & Maglio, P. P., (2016). Technology-enabled value co-creation: An empirical analysis of actors, resources, and practices. *Industrial Marketing Management, 56*, 73–85.
Bresnahan, T. F., Brynjolfsson, E., & Hitt, L. M., (2002). Information technology, workplace organization and the demand for skilled labor: Firm-level evidence. *Quarterly Journal of Economics, 117*(1), 339–376.
Bresser, R. K., & Bishop, R. C., (1983). Dysfunctional effects of formal planning: Two theoretical explanations. *Academy of Management Review, 8*, 588–599.
Brews, P. J., & Hunt, M. R., (1999). Learning to plan and planning to learn: Resolving the planning school/learning school debate. *Strategic Management Journal, 10*(20), 889–913.
Brock, D. M., & Barry, D., (2003). What if planning were really strategic? Exploring the strategy-planning relationship in multinationals. *International Business Review, 12*, 543–561.
Brown, L. D., (1983). *Managing Conflict at Organizational Interfaces*. Addison-Wesley Publishing Company.
Bulow, J. I., Geanakoplos, J. D., & Klemperer, P. D., (1985). Multimarket oligopoly: Strategic substitutes and complements. *Journal of Political Economy, 93*, 488–511.
Bunge, M., (1998). *Social Science Under Debate: A Philosophical Perspective*. University of Toronto Press; Toronto.
Cai, J., & John, K. C. W., (1997). The investment and operating productivity of Japanese initial public offerings, *Pacific-Basin Finance Journal, 5*, 389–418.
Cameron, A. C., & Trivedi, P. K., (2009). *Microeconometrics Using Stata* (Vol. 5): Stata Press College Station, TX.
Carroll, G. R., Delacroix, J., & Goodstein, J., (1988). The political environments of organizations: An ecological view. *Research in Organizational Behavior, 10*, 359–392.
Carroll, J. S., (1998). Organizational learning activities in high-hazard industrial: The logics underlying self-analysis. *Journal of Management Studies, 35*, 699–717.
Charles, V., Puthraya, N. R., & Kavitha, S. I., (2007). Simple Kanban technique for better inventory management. *The ICFAI Journal of Operations Management, 4*, 60–73.
Chen, Y., & Jermias, J., (2012). Business strategy, executive compensation and firm performance. *Accounting & Finance, 53*(3), 1–22.
Chen, Z., (2003). Dominant retailers and countervailing power hypothesis. *RAND Journal of Economics, 34*, 612–625.
Chen, Z., & Ross, T., (2000). Strategic alliances, shared facilities, and entry deterrence. *RAND Journal of Economics, 31*, 326–344.
Cheng, J. L. C., (1983). Interdependence and coordination in organizations: A role-system analysis'. *Academy of Management Journal, 26*(1), 156–162.
Cheng, S., Evans, I. J. H., & Nagarajan, N. J., (2008). Structure Size and firm performance: The moderating effects of the market for corporate control. *Review of Quantitative Finance and Accounting, 31*(2), 121–145.

Chiang, H., & Chia, F., (2005). An empirical study of strategic planning and corporate performance. *Journal of American Academy of Business, 6*(1), 95–101.

Choi, J. J., Park, S. W., & Yoo, S. S., (2007). The value of outside directors: Evidence from strategic planning reform in Korea. *Journal of Financial and Quantitative Analysis, 42*(4), 941–962.

Chun, H., (2000). *The Role of the Stock Market in Korea: Determinants and Effects of Initial Public Offerings*. PhD Thesis, The George Washington University, United States—District of Columbia.

Church, J., & Roger, W., (1999). *Industrial Organization: A Strategic Approach*. Boston: Irwin McGraw-Hill.

Church, J., & Ware, R., (1999). *Industrial Organization: A Strategic Approach*. Boston: Irwin, McGraw-Hill.

Collier, P., & Zaman, M., (2005). Convergence in European strategic planning: The audit structure concept. *Strategic Planning: An International Review, 13*(6), 753–768.

Collins, J. C., & Porras, J. I., (1994). *Built to Last-Successful Habits of Visionary Companies* (p. 123). Harper Business.

Conein, B., (2003). The social dimensions of distributed artificial intelligence. *Paper Presented at the CNRS Thematic School Information and Communication Technologies and Structuration of Collectives*. Carry-le-Rouet.

Conyon, M. J., (1997). Strategic planning and executive compensation. *International Journal of Industrial Organization, 15*(4), 493–509.

Conyon, M. J., & Gregg, P., (1994). Pay at the top: A study of the sensitivity of top director remuneration to company specific shocks. *National Institute Economic Review, 149*(1), 83–92.

Corbin, J., & Strauss, A., (1990). Grounded theory research: Procedures, canons, and evaluative criteria. *Qualitative Sociology, 13*(1), 3–20.

Corcuff, P., (1995). *Les Nouvelles Sociologies*. Paris: Nathan.

D'Aspremont, C., Gabszewicz, J. J., & Thisse, J. F., (1979). On Hotelling's stability in competition. *Econometrica, 47*, 1045–1050.

Daraut, S., (2004). *On Technical-Organizational Learning or the Role of Rules in the Structuring of Contexts of Interactions*. PhD thesis, University of Social Sciences of Toulouse.

Darnall, N., & Edward, D. Jr., (2007). Predicting the cost of environmental management system adoption: The role of capabilities, resources and ownership structure. *Strategic Management Journal, 27,* 301–320.

David, F. R., (2009). *Strategic Management*. Prentice Hall International, Inc., Upper Saddle River, New Jersey.

De Bondt, W. F. M., Muradoglu, G., Shefrin, H., & Staikouras, K. S., (2008). Behavioral finance: Quo Vadis? *Journal of Applied Finance*, 7–21.

Dean, J. W., & Sharfman, M. P., (1996). Does decision process matter? A study of strategic decision-making effectiveness. *Academy of Management Journal, 39*, 368–396.

Defalvard, H., (1990). The invisible hand: Myths and reality of the market as a spontaneous order. *Revue D'économie Politique, 6*, 870–883.

Delery, J. E., & Doty, D. H., (1996). Modes of theorizing in strategic human resource management: Tests of universalistic, contingency and configurational performance prediction. *Academy of Management Journal, 39*(4), 802–835.

Demsetz, H., (1983). Structure of ownership and the theory of the firm. *The. JL & Econ., 26*, 375–390.

DeZoort, F. T., Hermanson, D. R., & Archambeault, D. S., (2002). Audit structure effectiveness: A synthesis of the empirical audit structure literature. *Journal of Accounting Literature, 21*, 38–75.

Dibrell, C., Davis, P., & Craig, J., (2008). Fueling innovation through information technology in SMEs. *Journal of Small Business Management, 46*(2), 203–218.

DiMaggio, P. J., & Powell, W. W., (1983). The iron cage revisited: Institutional isomorphism and collective rationality in organizational fields. *American Sociological Review, 48*, 147–160.

Dimma, W. A., (1985). Competitive strategic planning. *Business Quarterly*, 22–26.

Dixit, A., (1980). The role of investment in entry-deterrence. *Economic Journal, 90*, 95–106.

Dixit, A., & Stiglitz, J., (1977). Monopolistic competition and optimum product diversity. *American Economic Review, 67*, 297–308.

Dong, L., (2008). Exploring the impact of top management support of enterprise systems implementation outcomes. *Business Process Management Journal, 14*, 204–218.

Drazin, R., & Rao, H., (2002). Harnessing managerial knowledge to implement product-line extensions: How do mutual fund families allocate portfolio managers to old and new funds? *Academy of Management Journal, 45*, 609–619.

Duménil, G., & Lévy, D., (2005). Costs and benefits of neoliberalism: A class analysis. In: Gerald, A. E., (ed.), *Financialization and the World Economy*. Cheltenham, Edward Elgar Publishing Limited, United Kingdom.

Dunning, J. H., (1988). The eclectic paradigm of international production: A restatement and some possible extensions. *Journal of International Business Studies, 19*(1), 1–31.

Dupuy, C., & Kechidi, M., (1996). Interpretability of rules and trust in organizational dynamics. *Sciences de la Société, 39*, 3–18.

Durand, T., (1997). Strategizing for innovation: Competence analysis in assessing strategic change. In: Heene, A., & Sanchez, R., (eds.), *Competence-Based Strategic Management* (pp. 127–150). John Wiley & Sons.

Durbin, J., (1954). *Errors in Variables* (Vol. 22, pp. 23–32). Review of the International Statistical Institute.

Dutton, J. E., & Duncan, R. B., (1987). The influence of the strategic planning process on strategic change. *Strategic Management Journal, 8*, 103–116.

Eisenhardt, K. M., (1989). Making fast strategic decisions in high-velocity environments. *Academy of Management Journal, 32*(3), 543–576.

Eisenhardt, K. M., & Martin, J. A., (2000). Dynamic capabilities: What are they? *Strategic Management Journal, 21*, 1105–1121.

Ekman, P., Raggio, R. D., & Thompson, S. M., (2016). Service network value co-creation: Defining the roles of the generic actor. *Industrial Marketing Management, 56*, 51–62.

Elsayed, K., (2011). Structure size and corporate performance: The missing role of structure leadership structure. *Journal of Management & Governance, 15*(3), 415–446.

Epstein, G. A., (2005). Introduction: Financialization and the world economy. In Gerald, A. E., (ed.), *Financialization and the World Economy*. Cheltenham, Edward Elgar Publishing Limited, United Kingdom.

Eraly, A., (1988). *The Structuring of the Firm: Rationality in Action*. Brussels: Éditions de l'Université de Bruxelles.

Evans, S. J., (1991). Strategic flexibility for high technology maneuvers: A conceptual framework. *Journal of Management Studies, 28*, 67–89.

Fahy, J., (2000). The resource-based view of the firm: Some stumbling-blocks on the road to understanding sustainable competitive advantage. *Journal of European Industrial Training, 24*, 94–104.

Falshaw, J. R., Glaister, K. W., & Tatoglu, E., (2006). Evidence on formal strategic planning and company performance. *Management Decision, 44*(1), 9–30.

Fama, E. F., (1970). Efficient capital markets: A review of theory and empirical work. *Journal of Finance, 25*(2), 383–417.

Fama, E. F., (1991). Efficient capital markets: II. *Journal of Finance, 46*(5), 1575–1617.

Fama, E. F., & French, K. R., (2007). Disagreement, tastes, Ans asset prices. *Journal of Financial Economics, 83*(3), 667–689.

Fama, E., & Jensen, M., (1983). Separation of ownership and control. *Journal of Law and Economics, 26*(2), 301–325.

Farmer, M., Archbold, S., & Alexandrou, G., (2013). CEO compensation and relative company productivity evaluation UK evidence. *Compensation & Benefits Review, 45*(2), 88–96.

Farooque, O., Van, Z. T., Dunstan, K., & Karim, A. W., (2007). Strategic planning in Bangladesh: Link between ownership and financial productivity. *Strategic Planning: An International Review, 15*(6), 1453–1468.

Favereau, O., (1994). Rules, organization and collective learning. *The Economic Analysis of Conventions* (pp. 113–137). A. Paris: PUF.

Ferguson, A., Francis, J. R., & Stokes, D. J., (2003). The effects of firm-wide and office-level industry expertise on audit pricing. *The Accounting Review, 78*(2), 429–448.

Filbeck, G., & Lee, S. K., (2006). Structure size and firm performance: The case of small firms. *Academy of Accounting and Financial Studies, 11*(1), 43.

Fornell, C., & Larcker, D. F., (1981). Evaluating structural equation models with unobservable variables and measurements. *Journal of Marketing Research, 18*, 33–50.

Fredrickson, J. W., (1984). The comprehensiveness of strategic decision processes: Extension, observations, future directions. *Academy of Management Journal, 27*, 445–466.

Fredrickson, J. W., (1986). The strategic decision process and organizational structure. *Academy Management Review, 11*(2), 280–297.

Fredrickson, J. W., & Mitchell, T. R., (1984). Strategic decision processes: Comprehensiveness and productivity in an industry with an unstable environment. *Academy of Management Journal, 27*(2), 399–423.

Freel, M. S., (2000). Do small innovating firms outperform non-innovators? *Small Business Economics, 14*, 195–210.

Friedberg, E., (1992). Les quatre dimensions de l'action organisée. *Revue Française de Sociologie, 33*, 531–557.

Friedberg, E., (1993). *Interview in Sciences Humaines*, p. 30.

Frieder, L., & Subrahmanyam, A., (2005). Brand perceptions and the market for common stock. *Journal of Financial and Qualitative Analysis, 40*(1), 57–85.

Friedlob, G. T., & Plewa, F. J., (1996). *Understanding Return on Investment*. Wiley.

Friedman, M., (1970). *The Social Responsibility of Business is to Increase its Profits* (pp. 122–126). New York Time Magazine.

Fudenberg, D., & Tirole, J., (1984). The fact-cat effect, the puppy-dog ploy, and the lean and hungry look. *American Economic Review Papers and Proceedings, 74*, 361–66.

Fulmer, R. M., & Rue, L. W., (1974). The practice and profitability of long-range planning. *Managerial Planning, 22*, 1–7.

Gabszewicz, J. J., & Thisse, J. F., (1979). "Price competition, quality and income disparities. *Journal of Economic Theory, 20*, 340–359.

Garnier, O., (2002). *The Place of Equities in Household Wealth Measurement and International Comparisons*. General Planning Commission.

Garriga, E., & Melé, D., (2004). Corporate social responsibility theories: Mapping the territory. *Journal of Business Ethics, 53*, 51–71.

Gauzente, C., (2000). Measuring business productivity in the absence of objective indicators: What validity? Analysis of the relevance of certain indicators. *Finance Control Strategy, 3*(2), 145–165.

Gavetti, G., Levinthal, D. A., & Rivkin, J. W., (2005). Strategy making in novel and complex worlds: The power of analogy. *Strategic Management Journal, 26*(8), 691–712.

Gendron, Y., & Bédard, J., (2006). On the constitution of audit structure effectiveness. *Accounting, Organizations and Society, 31*(3), 211–239.

Ghoshal, S., & Bartlett, C. A., (1994). Linking organizational context and managerial action: The dimensions of quality of management. *Strategic Management Journal, 15*, 91–112.

Giddens, A., (1984). *The Constitution of Society*. Cambridge: Polity Press.

Giddens, A., (1987). *The Constitution of Society*. Paris: PUF.

Giddens, A., (1994). *The Consequences of Modernity*. Paris: L'Harmattan.

Giddens, A., (1998). Meeting with Anthony Giddens. *Sciences Humaines, 84*, 38–41.

Gilgeous, V., (1997). *Operations and the Management of Change* (pp. 89–96). Pitman Publishing.

Giraud, F., Saulpic, O., Naulleau, G., Delmond, M. H., & Bescos, P. L., (2004). *Management Control and Productivity Steering*. Gualino editor, EJA Paris.

Glaister, K. W., & Falshaw, R. J., (1999). Strategic planning: Still going strong ?. *Long Range Planning, 32*(1), 107–116.

Godard, L., (2001). *Board Size: Determinants and Impact on Performance*. FARGO Notebook No. 1010702, 28.

Govin, J., & Slevin, D., (1989). Strategic management of small firms in hostile and benign environments. *Strategic Management Journal, 10*(1), 75–87.

Govindarajan, V., (1989). Implementing competitive strategies at the business unit level: Implications of matching managers to strategies. *Strategic Management Journal, 10*, 251–269.

Granovetter, M., (1985). "Economic action and social structure: The problem of embeddedness. *American Journal of Sociology, 91*(3), 481–510.

Granovetter, M., (1993). The nature of economic relationship. In Swedberg, R., (ed.), *Explorations in Economic Sociology* (pp. 3–41). New York: Russell Sage Foundation.

Granovetter, M., (1994). *Economic Institutions as Social Constructions: A Framework for Analysis* (pp. 79–94). The economic analysis of conventions. A. Paris: PUF.

Grant, R. M., (2003). Strategic planning in a turbulent environment: Evidence from the oil majors. *Strategic Management Journal, 24*(6), 491–517.

Grant, R., (1991). The resource-based theory of competitive advantage: Implications for strategy formulation. *California Management Review, 33*(3), 114–135.

Greenan, N., & Mairesse, J., (2006). Reorganizations, work changes and renewal of skills. *Economic Review, 57*(4).

Greenley, G. E., (1986). Does strategic planning improve company performance ?. *Long Range Planning, 19*(2), 101–109.

Griffin, D., & Tversky, A., (1992). The weighing of evidence and the determinants of confidence. *Cognitive Psychology, 24*, 411–435.

Groleau, C., (2000). Structuration theory applied to organizations: The case of technology studies. *Proceedings of the Conference Structuration et Management des organizations.* Paris: Université de Paris XII.

Guest, R., Hersey, P., & Blanchard, K. H., (1977). *Organizational Change Through Effective Leadership*. Prentice-Hall, Inc., New Jersey.

Gupta, A. K., & Govindarajan, V., (1984). Business unit strategy, managerial characteristics, and business unit effectiveness at strategy implementation. *Academy of Management Journal, 27*, 25–41.

Habermas, J., (1987). Theory of communicative action. *Rationality of Action and Rationalization of Society* (Vol. 1). Paris: Fayard.

Habib, R., (2014). Implication of good governance: A case study of bank of Khyber. *International Journal of Scientific and Engineering Research, 5*, 81–98.

Hall, R. H., (1980). Effectiveness theory and organizational effectiveness. *The Journal of Applied Behavioral Science, 16,* 536–545.

Hart, S. L., (1992). An integrative framework for strategy-making processes. *Academy of Management Review, 17*(2), 327–351.

Hatchuel, A., (1997). *Foundations of Knowledge and Legitimacy of Rules* (pp. 183–209). The limits of rationality. Tome 2: The figures of the collective. Paris: La Découverte.

Hausman, J. A., (1978). Specification tests in econometrics. *Econometrica: Journal of the Econometric Society,* 1251–1271.

Hermalin, B. E., & Weisbach, M. S., (2001). *Structures of Directors as an Endogenously Determined Institution: A Survey of the Economic Literature.* National Bureau of Economic Research.

Hermel, L., & Achard, P., (2007). *100 Questions to Understand and Act: Benchmarking.* AFNOR.

Hofer, C., & Schendel, D., (1978). *Strategy Formulation: Analytical Concepts*. West Publishing Company.

Hopkins, W. E., (1987). Developing the planning skills of engineers in management training. *Journal of Management in Engineering, 3,* 249–256.

Hopkins, W. E., & Hopkins, S. A., (1997). Strategic planning-financial performance relationships in banks: A causal examination. *Strategic Management Journal, 18*(8), 635–652.

Hossain, M., Prevost, A. K., & Rao, R. P., (2001). Strategic planning in New Zealand: The effect of the 1993 companies act on the relation between structure composition and firm performance. *Pacific-Basin Finance Journal, 9*(2), 119–145.

Hotelling, H. H., (1929). In: Stigler, G., &. Boulding, K., (eds.), *Stability in competition: Economic Journal* (Vol. 39, pp. 41–57). Readings in Price Theory–1952.

Hough, J. R., & White, M. A., (2003). Environmental dynamism and strategic decision-making rationality: An examination at the decision level. *Strategic Management Journal, 24*, 481–489.

Hult, G., Hurley, R., & Knight, G., (2004). Innovativeness: Its antecedents and impact on business productivity. *Industrial Marketing Management, 33*(5), 429–438.

Jacob, H., (1984). *Using Published Data: Errors and Remedies, Quantitative Applications in the Social Sciences* (Vol. 42, p. 33). Sage publications, London.

Jansen, J. J. P., Gorge, G., Van De, B. F. A. J., & Volberda, H. W., (2008). Senior team attributes and organizational ambidexterity: The moderating role of transformational leadership. *Journal of Management Studies, 45*, 982–1007.

Jansen, J., Van, D. B. F., & Volberda, H., (2006). Exploratory innovation, exploitative innovation, and productivity: Effects of organizational antecedents and environmental moderators. *Management Science, 52*(11), 1661–1674.

Jean-Yves, L., (2008). *Financial Communication: Building and Implementing a Financial Communication Strategy* (2nd edn.). Dunod (ed.) Paris.

Jensen, & Meckling, W. H., (1976). Theory of the firm: Managerial behavior, agency costs and ownership structure. *Journal of Financial Economics, 3*(4), 305–360.

Jensen, & Murphy, K. J., (1990). CEO incentives: It's not how much you pay, but how. *Harvard Business Review, 3*, 138–153.

Jensen, M. C., (1993). The modern industrial revolution, exit, and the failure of internal control systems. *The Journal of Finance, 48*(3), 831–880.

Jensen, M. C., & Meckling, W., (1976). Theory of the firm: Managerial behavior, agency cost, and capital structure. *Journal of Financial Economics, 3*, 305–360.

Jensen, M. C., & Murphy, K. J., (1990). Performance pay and top management incentives. *The Journal of Political Economy, 98*(2), 225–264.

John-Cramer, M. E., Cross, R. L., & Yan, A., (2003). Sources of fidelity in purposive organizational change: Lesson from a re-engineering case. *Journal of Management Studies, 40*, 1837–1870.

Joignant, A., (2000). Agent, structure, and cognition: Research questions from the sociology of Pierre Bourdieu and Anthony Giddens. *International Notebooks of Sociology*, 108.

Judge, W. Q., & Miller, A., (1991). Antecedents and outcomes of decision speed in different environmental contexts. *Acadeniy of Management Journal, 34*(2), 449–463.

Kahneman, D., & Tversky, A., (1979). Prospect theory: An analysis of decision under risk. *Econometrica, 47*(2), 263–291.

Kald, M., Nilsson, F., & Rapp, B., (2000). On strategy and value-based management: The importance of classifying the strategy of the business. *British Journal of Management, 11*, 197–212.

Kamoche, K., (1996). Strategic human resource management within a resource-capability view of the firm. *Journal of Management Studies, 33*, 213–233.

Kano, L., Tsang, E. W., & Yeung, H. W. C., (2020). Global value chains: A review of the multi-disciplinary literature. *Journal of International Business Studies*, 1–46.

Kechidi, M., (1998). Rationality and decision contexts: A return to H. Simon. *International Journal of Systemics, 12*(4, 5), 419–440.

Keung, C., Robin, A., & Tessoni, D., (2007). Sarbanes-Oxley: Are audit structures up to the task? *Managerial Auditing Journal, 22*(3), 255–267.

Khanchel, (2009). The role of the manager's discretionary power in improving optimization: The case of Tunisian companies. *The Journal of Management Sciences, 3*(237, 238), 95–103.

Klemperer, P. D., (1987). Entry deterrence in markets with consumer switching costs. *Economic Journal, 97*, 99–117.

Koeing, G., (1994). Organizational learning: Locating the places. *Revue Française de Gestion, 97*, 76–83.

Kren, L., & Kerr, J. L., (1997). The effects of outside directors and structure shareholdings on the relation between chief executive compensation and firm performance. *Accounting and Business Research, 27*(4), 297–309.

Krippner, G. R., (2005). The financialization of the American economy. *Socio-Economic Review, 3*(2), 173–208.

Krishnan, J., (2005). Audit structure quality and internal control: An empirical analysis. *The Accounting Review, 80*(2), 649–675.

Krueger, R. A., (1994). *Focus Group: A Practical Guide for Applied Research* (2nd edn., p. 31). Sage Publications, Thousand Oaks.

Kukalis, S., (1991). Determinants of strategic planning systems in large organizations: A contingency approach. *Journal of Management Studies, 28*(2), 143–160.

Lam, T. Y., & Lee, S. K., (2012). Family ownership, structure structures and firm performance: Evidence from Hong Kong. *Strategic planning, 12*(3), 353–366.

Langer, E., (1975). The illusion of control. *Journal of Personality and Psychology, 32*(2), 311–328.

Larcker, D. F., Richardson, S. A., & Tuna, I., (2007). Strategic planning, accounting outcomes, and organizational performance. *The Accounting Review, 82*(4), 963–1008.

Lee, G. K., (2008). Relevance of organizational capabilities and its dynamics: What to learn from the entrants' product portfolios about the determinants of entry timing. *Strategic Management Journal, 29,* 1257–1280.

Lemaire, J. P., & Petit, G., (2003). *Internationalization Strategy: International Business Development.* Paris, Dunod.

Lepoivre, F., (2005). *Benchmarking Concept and Methodology*. NEVAOCONSEIL V02/2005.

Leroux, F., (1998). Decoding changes in international capital markets in light of changing conventions. *Management International, 3*(1), 35–49.

Lichtenstein, S., Fischhoff, B., & Phillips, L. D., (1982). Calibration of probability: The state of the art to 1980. In: Kahneman F., Slovic, P., & Tversky, A., (eds.), *Judgment Under Uncertainty: Heuristics and Biases*. CUP.

Lintner, J., (1965). The valuation of risk assets and the selection of risky investments in stock portfolios and capital budgets. *Review of Economics and Statistics, 47*(1), 13–37.

Lipsey, M. W., (1990). *Design Sensitivity: Statistical Power for Experimental Research* (p. 11). Sage Publications.

Livet, P., (1994). *The Virtual Community: Action and Communication.* Combas: Editions de l' Eclat.

Livet, P., & Thevenot, L., (1994). *The Categories of Operational Action* (pp. 139–167). The economic analysis of conventions. A. Paris: PUF.

Lochridge, R. K., (1998). *Strategy in the 80s, in Perspectives on Strategy*. BCG, Wiley.

Lorino, P., (1992). *Strategic Management Control – Management by Activities*. Dunod, Paris.

Ma, H., (1999). Anatomy of competitive advantage: A select framework. *Management Decision, 37*(9), 709–718.

Mantera, S., (2008). Role expectations and middle manager strategic agency. *Journal of Management Studies, 45,* 294–316.

March, J., & Herbert, S., (1999). *Les Organizations*. Paris: Dunod.

Markowitz, H. M., (1991). Foundations of portfolio theory. *Journal of Finance, 46*(2), 469–477.

Markusen, J., (1981). Trade and gains from trade with imperfect competition. *Journal of international economic, 11*, 531–551.

Martin, S., (2001). *Advanced Industrial Economics* (2nd edn.). Oxford: Blackwell Publishing.
Megginson, W. L., Robert, C. N., & Matthias, V. R., (1994). The financial and operating productivity of newly privatized firms: An international empirical analysis. *Journal of Finance, 49*, 403–452.
Melcher, A. J., (1976). *Structure and Process of organizations: A System Approach*. Prentice-Hall, Inc.
Meyer, J. W., & Rowan, B., (1977). Institutionalized organizations: Formal structure as myth and ceremony. *American Journal of Sociology, 83*(2), 340–363.
Midler, C., (1994). *Rule Evolution and the Learning Process* (pp. 335–369). The economic analysis of conventions. A. Paris: PUF.
Miller, C. C., (2008). Decisional comprehensiveness and firm performance: Towards a more complete understanding, *Journal of Behavioral Decision Making, 21*, 598–620.
Miller, C. C., & Cardinal, L. B., (1994). Strategic planning and firm performance: A synthesis of more than two decades of research. *Academy of Management Journal, 37*(6), 1649–1665.
Miller, D. T., (1976). Ego involvement attributions for success and failure. *Journal of Personality and Social Psychology, 34*(5), 901–906.
Mintzberg, H., (1990). The design school: Reconsidering the basic premises of strategic management. *Strategic Management Journal, 11*, 171–195.
Mintzberg, H., (1991). Learning 1, planning 0: Reply to Igor Ansoff. *Strategic Management Journal, 12*, 463–466.
Mintzberg, H., (1994). *Rise and Fall of Strategy Optimization*. Dunod, Paris.
Mossin, J., (1966). Equilibrium in capital asset market. *Econometrica, 34*(4), 768–783.
Moureau, N., & Rivaud-Danset, D., (2004). *Uncertainty in Economic Theories*. Landmarks – The Discovery, Paris.
Mullenbach-Servayre, A., (2007). The contribution of stakeholder theory to the modeling of corporate social responsibility. *The Journal of Management Sciences, 1/2007* (223), 109–120.
Nayyar, D., (2006). Globalization, history and development: A tale of two centuries. *Cambridge Journal of Economics, 30*, 137–159.
Nickell, S. J., (1996). Competition and corporate performance. *Journal of Political Economy, 104*, 724–746.
Nickerson, R. S., (1998). Confirmation bias: A ubiquitous phenomenon in many guises. *Review of General Psychology, 2*(2), 175–220.
Nutt, P. C., (1986). Tactics of implementation. *Academy of Management Journal, 29*, 230–261.
Nutt, P. C., (1989). Selecting tactics to implement strategic plans. *Strategic Management Journal, 10*, 145–161.
O'Connell, V., & Cramer, N., (2010). The relationship between firm performance and structure characteristics in Ireland. *European Management Journal, 28*(5), 387–399.
Oldham, G. R., & Cummings, A., (1996). Employee creativity: Personal and contextual factors at work. *Academy of Management Journal, 39*, 607–634.
Orléan, A., (2004). Are financial markets rational? *Reflections and Perspectives, XLIII/2*, 35–41.
Orléan, A., (2005). *Reflections on the Fundamental Value Objectivity Assumption in Modern Financial Theory* (pp. 19–42). Beliefs, Collective Representations and Conventions in Finance, Paris, (eds.), Economica, Coll. Management Research.
Orléan, A., (2008). Is the notion of fundamental value essential to financial theory? *Perspectives on the Economy,* (3), 120–128.

Padioleau, J., (1986). *The Social Order*. Paris: L'Harmattan.

Pant, P. N., & Lachman, R., (1998). Value incongruity and strategic choice. *Journal of Management Studies, 35*, 195–212.

Papke-Shields, K. E., Malhotra, M. K., & Grover, V., (2002). Strategic manufacturing planning systems and their linkage to planning system success. *Decision Sciences, 33*, 1–30.

Papke-Shields, K. E., Malhotra, M. K., & Grover, V., (2006). Evolution in the strategic manufacturing planning process of organizations. *Journal of Operations Management, 24*, 421–439.

Paradeise, C., (1991). *Actor's Theories* (p. 248). French notebooks.

Pearce, J. II., Freeman, E., & Robinson, R., (1987). The tenuous link between formal strategic planning and financial performance. *Academy of Management Review, 12*(4), 658–675.

Pedhazur, E. J., & Pedhazur S. L. (1991). *Measurement, Design and Analysis: An Integrated Approach.* Psychology Press, pp. 840.

Peel, M. J., & Bridge, J., (1998). How planning and capital budgeting improve SME performance. *Long Range Planning, 31*(6), 848–856.

Peterson, D. K., & Pitz, G. F., (1988). Confidence, uncertainty, and the use of Information. *Journal of Experimental Psychology, 14,* 85–92.

Pfeffer, J., & Salancik, G. R., (1977). Organization design - The case for a coalitional model of organizations. *Organizational Dynamics, 6*(2), 15–29.

Pfeffer, J., & Salancik, G. R., (1978). *The External Control of Organizations: A Resource Dependence perspective.* New York: Harper and Row.

Pham, P. K., Suchard, J. A., & Zein, J., (2011). Strategic planning and alternative performance measures: Evidence from Australian firms. *Australian Journal of Management, 36*(3), 371–386.

Porter, M. E., (1996). What is strategy? *Harvard Business Review, 74*(10), 61–78.

Powell, T. C., (1992). Strategic planning as competitive advantage. *Strategic Management Journal, 13*, 551–558.

Powell, W. W., & DiMaggio, P. J., (1991). *The New Institutionalism in Organizational Analysis.* Chicago: University of Chicago Press.

Punjab, M., (2014). *Risk Management in Commercial Banking: An Analysis of the Process* (1st edn.). Patiala: Punjab University, School of Management Studies.

Quere, L., (1990). Acting in the public space: The intentionality of actions as a social phenomenon. Forms of action. In: P. Pharo, P., & Quéré, L., (eds.), *Semantics and Sociology* (pp. 85–112). Practical reasons. EHESS.

Quere, L., (1993). Do we really need the notion of convention? *Réseaux, 62*, 14–42.

Ramanujam, V., & Venkatraman, N., (1987). Planning system characteristics and planning effectiveness. *Strategic Management Journal, 8*, 453–468.

Rasmusen, E. B., Ramseyer, J. M., & Wiley, Jr. J. S., (1991). Naked exclusion. *American Economic Review, 81*, 1137–1145.

Reynaud, B., (2001). Following the rules in organizations. *Journal of Industrial Economics, 97*, 53–68.

Reynaud, J. D., (1997). *The Rules of the Game: Operational Action and Social Regulation.* Paris: Armand Colin.

Rhyne, L. C., (1986). The relationship of strategic planning to financial performance. *Strategic Management Journal, 7,* 423–436.

Robinson, R. B., & Pearce, J. A., (1983). The impact of formalized strategic planning on financial performance in small organizations. *Strategic Management Journal, 4*(3), 197–207.

Robinson, R. B., & Pearce, J. A., (1988). Optimized patterns of strategic behavior and their relationship to business-unit performance. *Strategic Management Journal, 9*(1), 43–60.

Rudd, J. M., Greenley, G. E., Beatson, A. T., & Lings, I. N., (2008). Strategic planning and productivity: Extending the debate. *Journal of Business Research, 61*, 99–108.

Salais, R., (1994). *Uncertainty and Work Interactions: From Products to Conventions* (pp. 371–403). The economic analysis of conventions. A. Paris: PUF.

Salop, S., (1979). Monopolistic competition in with outside goods. *Bell Journal of Economics, 10*, 141–156.

Sanchez, R., & Heene, A., (1997). Competence-based strategic management: Concepts and issues for theory, research, and practice. In Heene, A., & Sanchez, R., (eds.). *Competence-Based Strategic Management*. John Wiley & Sons.

Santos, M. F., & Eisenhardt, M. K., (2005). Organizational boundaries and theories of organization. *Organization Science, 16*(5), 491–508.

Saunders, M., Mann, R., & Smith, R., (2008). Implementing strategic initiatives: A framework of leading practices. *International Journal of Operations & Production Management, 28*, 1095–1123.

Schmalensee, R., (1988). Industrial economics: An overview. *The Economic Journal, 98*, 643–681.

Schwenk, C. R., & Shrader, C. B., (1993). Effects of formal strategic planning on financial performance in small firms: A meta-analysis. *Entrepreneurship: Theory and Practice, 17*(3), 53–64.

Segal, I. R., & Whinston, M. D., (2000). Naked exclusion: Comment. *American Economic Review, 90*, 296–309.

Segars, A. H., Grover, V., & Teng, J. T. C., (1998). Strategic information systems planning: Planning system dimensions, internal coalignment and implications for planning effectiveness. *Decision Sciences, 29*(2), 303–345.

Sewell, W. H. J., (1992). A theory of structure: Duality, agency, and transformation. *American Journal of Sociology, 98*(1), 1–29.

Shadish, R. W., Cook, D. T., & Campbell, D. T., (2002). *Experimental and Quasi-Experimental Designs for General Causal Inference*. Houghton Mifflin Company, Boston.

Shaked, A., & Sutton, J., (1982). Relaxing price competition through product differentiation. *Review of Economic Studies, 49*, 3–13.

Shaked, A., & Sutton, J., (1983). Natural oligopolies. *Econometrica, 51*, 1469–1484.

Shleifer, A., & Vishny, R. W., (1986). Large shareholders and corporate control. *The Journal of Political Economy*, 461–488.

Simons, R., (1994). How new top managers use control systems as levers of strategic renewal. *Strategic Management Journal, 15*, 169–189.

Smith, K. G., Collins, C. J., & Clark, K. D., (2005). Existing knowledge, knowledge creation capability, and the rate of new product introduction in high-technology firms. *Academy of Management Journal, 48*, 346–357.

Snider, B., Silveira, G. J. C., & Balakrishan, J., (2009). ERP implementation at SMEs: Analysis of five Canadian cases. *International Journal of Operations & Production Management, 29*, 4–29.

Soja, P., (2008). Examining the conditions of ERP implementations: Lessons learnt from adopters. *Business Process Management Journal, 14*, 106–123.

Spence, A. M., (1977). Entry, capacity investment, oligopolistic pricing. *Bell Journal of Economics 8*, 534–544.

Spence, M., (1976). Product selection, fixed costs and monopolistic competition. *Review of Economic Studies, 43,* 217–235.

Tapinos, E., Dyson, R. G., & Meadows, M., (2005). The impact of performance measurement in strategic planning. *International Journal of Productivity and Productivity Management, 54*(5, 6), 370–384.

Taylor, S. J., & Bogdan, R., (1984). *Introduction to Qualitative Research Methods: The Search for Meanings* (2nd edn., p. 5). John Wiley and sons.

Thomas, L. D., Autio, E., & Gann, D. M., (2014). Architectural leverage: Putting platforms in context. *Academy of Management Perspectives, 28*(2), 198–219.

Thompson, A. A. Jr., & Strickland, III. A. J., (1995). *Crafting & Implementing Strategy.* Irwin, Inc.

Thune, S. S., & House, R. J., (1970). Where long-range planning pays off: Findings of a survey of formal and informal planners. *Business Horizons, 13,* 81–87.

Tirole, J., (1988). *Theory of Industrial Organization.* Cambridge: MIT Press.

Tversky, A., & Kahneman, D., (1974). Judgment under uncertainty: Heuristics and biases. *Sciences, 185,* 1124–1130.

Tversky, A., & Kahneman, D., (1981). The framing of decisions and the psychology of choice. *Science, 211,* 457, 458.

Tversky, A., & Kahneman, D., (1986). Rational choice and the framing of decisions. *The Journal of Business, 59*(4), S251–S278.

Ughetto, P., (2001). Contradictory trends in the evolution of work and the renewal of analyses: An interpretation based on the representations of the employer. *IRES Review, 37,* 31–58.

Vafeas, N., & Waegelein, J., (2007). The association between audit structures, compensation incentives, and corporate audit fees. *Review of Quantitative Finance and Accounting, 28*(3), 241–255.

Veliyath, R., & Shortell, S. M., (1993). Strategic orientation, strategic planning system characteristics and productivity. *Journal of Management Studies, 30*(3), 359–381.

Villalonga, B., & Demsetz, H., (2001). Ownership structure and corporate productivity. *Journal of Corporate Finance, 7*(3), 209–233.

Vineeta, S., Vic, N., & Barry, L., (2009). Determinants of audit structure connexion frequency: Evidence from a voluntary governance system. *Accounting Horizons, 23*(3), 245–263.

Von, N. J., & Morgenstern, O., (1947). *Theory of Games and Economic Behavior* (1stedn.). Princeton University Press.

Wally, S., & Baum, J. R., (1994). Personal and structural determinants of the pace of strategic decision making. *Academy of Management Journal, 37*(4), 932–956.

Weimer, J., & Pape, J., (1999). A taxonomy of systems of strategic planning. *Strategic planning: An International Review, 7*(2), 152–166.

Whitehead, D. D., & Gup, B. E., (1985). Gup Bank and thrift probability: Does strategic planning really pay? *Economic Review Federal Reserve Bank of Atlanta, 70,* 14–25.

Wick, C. W., Pollock, R. V. H., Jefferson, A. M., & Flanagan, R. D., (2006). *The Six Disciplines of Break Through Learning* (p. 55). Pfeiffer.

Willekens, M., Bauwhede, H. V., & Gaeremynck, A., (2004). Voluntary audit structure formation and practices among Belgian listed companies. *International Journal of Auditing, 8*(3), 207–222.

Wintoki, M. B., Linck, J. S., & Netter, J. M., (2012). Endogeneity and the dynamics of internal strategic planning. *Journal of Financial Economics, 105*(3), 581–606.

Worchel, S., Wood, W., & Simpson, J. A., (1992). *Group Process and Productivity* (pp. 274–298). Sage Publication.

Wu, D. M., (1973). Alternative tests of independence between stochastic regressors and disturbances. *Econometrica: Journal of the Econometric Society*, 733–750.

Xu, X., & Wang, Y., (1999). Ownership structure and strategic planning in Chinese stock companies. *China Economic Review, 10*(1), 75–98.

Yatim, P., Kent, P., & Clarkson, P., (2006). Governance structures, ethnicity, and audit fees of Malaysian listed firms. *Managerial Auditing Journal, 21*(7), 757–782.

Yermack, D., (1996). Higher market valuation of companies with a small structure of directors. *Journal of Financial Economics, 40*(2), 185–211.

Index

E

F

K

L

M

N

O

P

W

Y